Proteomics for Biological Discovery

Timothy D. Veenstra

Laboratory of Proteomics
and Analytical Technologies
SAIC-Frederick, Inc.
Frederick, Maryland

John R. Yates

Department of Cell Biology
Scripps Research Institute
La Jolla, California

A John Wiley & Sons, Inc., Publication

Published by John Wiley & Sons, Inc., Hoboken, New Jersey
Published simultaneously in Canada

For general information on our other products and services or for technical support, please contact our Customer Care Department within the United States at (800) 762-2974, outside the United States at (317) 572-3993 or fax (317) 572-4002.

Wiley also publishes its books in a variety of electronic formats. Some content that appears in print may not be available in electronic formats. For more information about Wiley products, visit our web site at www.wiley.com.

Library of Congress Cataloging-in-Publication Data is available.

ISBN-13: 978-0-471-16005-2
ISBN-10: 0-471-16005-9   24314374

Printed in the United States of America

10 9 8 7 6 5 4 3 2 1

Contents

Contributors

CASSIO DA SILVA BAPTISTA
SAIC–Frederick, Inc.
National Cancer Institute at Frederick
Research Technology Program
Laboratory of Molecular Technology
Frederick, Maryland

KRISTY J. BROWN
CTL Bio Services
Rockville, Maryland

JULIO E. CELIS
Department of Proteomics in Cancer
Institute of Cancer Biology
Danish Cancer Society and Danish Centre for Translational Research in Breast
    Cancer
Copenhagen, Denmark

G. MARIUS CLORE
Laboratory of Chemical Physics
National Institute of Diabetes and Digestive and Kidney Diseases
National Institutes of Health
Bethesda, Maryland

THOMAS P. CONRADS
SAIC–Frederick, Inc.
National Cancer Institute at Frederick
Frederick, Maryland

AMY DAMBROWITZ
Department of Chemistry
University of Washington
Seattle, Washington

TRISHA N. DAVIS
Department of Biochemistry
University of Washington
Seattle, Washington

DON L. DEVOE
Department of Mechanical Engineering and Institute for System Research
University of Maryland
College Park, Maryland
and Calibrant Biosystems
Rockville, Maryland

NORMAN J. DOVICHI
Department of Chemistry
University of Washington
Seattle, Washington

CATHERINE FENSELAU
Department of Chemistry and Biochemistry
University of Maryland
College Park, Maryland

PAVEL GROMOV
Department of Proteomics in Cancer
Institute of Cancer Biology
Danish Cancer Society and Danish Cantre for Translational Research in Breast
    Cancer
Copenhagen, Denmark

IRINA GROMOVA
Department of Proteomics in Cancer
Institute of Cancer Biology
Danish Cancer Society and Danish Centre for Translational Research in Breast
    Cancer
Copenhagen, Denmark

BRIAN L. HOOD
SAIC–Frederick, Inc.
National Cancer Institute at Frederick
Frederick, Maryland

SHEN HU
Department of Chemistry
University of Washington
Seattle, Washington
Present address: UCLA School of Dentistry
Los Angeles, California

LEOPOLD L. ILAG
Department of Chemistry
University of Cambridge
Cambridge, United Kingdom

HALEEM J. ISSAQ
Laboratory of Proteomics and Analytical Technologies
SAIC–Frederick, Inc.
National Cancer Institute at Frederick
Frederick, Maryland

CHENG S. LEE
Department of Chemistry and Biochemistry
University of Maryland
College Park, Maryland 20742
and Calibrant Biosystems
Rockville, Maryland

YAN LI
Department of Chemistry and Biochemistry
University of Maryland
College Park, Maryland

DANIEL C. LIEBLER
Proteomics Laboratory
Mass Spectrometry Research Center
Vanderbilt University School of Medicine
Nashville, Tennessee

DANQIAN MAO
Department of Chemistry
University of Washington
Seattle, Washington

DAVID MICHELS
Department of Chemistry
University of Washington
Seattle, Washington

ERIC G. D. MULLER
Department of Biochemistry
University of Washington
Seattle, Washington

DAVID J. MUNROE
SAIC–Frederick, Inc.
National Cancer Institute at Frederick
Research Technology Program
Laboratory of Molecular Technology
Frederick, Maryland

DaRUE A. PRIETO
Laboratory of Proteomics and Analytical Technologies
SAIC–Frederick, Inc.
National Cancer Institute at Frederick
Frederick, Maryland

THIERRY RABILLOUD
DRDC/BECP
CEA-Grenoble
Grenoble, France

CAROL V. ROBINSON
Department of Chemistry
University of Cambridge
Cambridge, United Kingdom

DAVID L. TABB
Department of Genome Sciences
University of Washington
Seattle, Washington
Present address: Life Sciences Division
Oak Ridge National Laboratory
Oak Ridge, Tennessee

TIMOTHY D. VEENSTRA
Laboratory of Proteomics and Analytical Technologies
SAIC–Frederick, Inc.
National Cancer Institute at Frederick
Frederick, Maryland

JOHN R. YATES
Department of Cell Biology
The Scripps Research Institute
La Jolla, California

Foreword

The cell can arguably be viewed as the basic unit of life, and a key focus of biological research is therefore to understand how cells are put together. What are the design principles through which the molecular constituents of the cell are organized? How do they respond dynamically to a changing environment, and how do they associate to form tissues and organs within a multicellular animal? Equally important and puzzling, how do a mere 5000 genes or so provide sufficient information to build a viable, free-living cell with remarkably complex properties, and how do a paltry 30,000 genes specify a human being, containing cells as diverse in their functions as a lymphocyte, a neuron, or a myocyte?

In considering these challenges, it is worth recalling how far we have come over the last thirty years, both technically and conceptually. Thirty years ago we couldn't sequence DNA, molecular cloning was in its infancy, live imaging of cells was nonexistent, we had little understanding of the extent or functions of post-translational modifications of proteins, RNA splicing was not thought of, oligonucleotide-directed mutagenesis had not been conceived, the genetic manipulation of mammalian genomes was the stuff of science fiction, the term bioinformatics had not been coined, biological mass spectrometry and the yeast 2-hybrid system lay in the future, and solving a single protein structure was a Herculean effort. It was clear that an individual gene product, however, could have profound effects on many different aspects of cellular behavior.

This multieffect was especially evident for the proteins encoded by viral oncogenes that induce malignant transformation of cultured cells and tumors in vivo. The expression of a single oncoprotein such as v-Src, for example, causes changes in cell shape, adhesion, metabolism, growth, survival, and proliferation. This observation suggested that these distinct facets of cellular function must all be interconnected, be it directly or indirectly, and that it should be possible to define a logic that explains the inner working of the cell. For this enterprise, we need to know the complete coding potential of cellular genomes, and, more daunting, we need ways of globally investigating the expression, modifications, interactions, and subcellular locations of their products. Furthermore, we need databases, computational tools, and modeling approaches to collate and interpret this information, to investigate how cellular networks function to generate complex properties, and to provide new hypotheses regarding biological function that can be tested experimentally. This new area of science is not only explanatory in nature. By understanding the basic principles of cellular design, we can learn to reengineer cell signaling networks, and thus to endow cells with new properties. This synthetic approach to biology may be especially valuable in the treatment of diseases such as advanced cancer, in which the normal organization of cells and tissues becomes severely deranged, and in infection, in which there are complex interactions between the pathogen and the host. First, however, to quote an old recipe for rabbit stew, "catch your rabbit" (or in this case catch your proteins).

The extensive sequencing of cDNAs and genomic DNA has now given us a fairly comprehensive account of the protein coding potential of prokaryotic and eukaryotic genomes, although for most of the predicted proteins there remains a significant degree of uncertainty about their true identity, their splice variants, their functions, and their regulation. Nonetheless, we can argue that we have in hand an increasingly complete set of the protein building blocks through which cells are assembled. The primary amino acid sequence of a protein can potentially give us a large amount of information, in part because proteins are commonly constructed in a cassette-like fashion from multiple smaller domains with characteristic conserved sequences. These domains can have either an enzymatic activity (such as a protein kinase) or a binding function (such as a phosphotyrosine-binding SH2 domain) and have been used repeatedly in a wide range of proteins. It seems as though cells and organisms may have evolved primarily through the increasingly sophisticated use of a limited set of protein domains, joined in increasingly elaborate combinations, and have only occasionally resorted to the invention of entirely new biochemical functions. Thus, the presence of particular domains in a protein of unknown biological function can give us strong clues as to its physiological properties. In addition, as we better understand the abilities of signaling enzymes to modify their intracellular targets at specific sites, and learn the rules determining the binding of interaction domains to defined peptide motifs, we can search the proteome in silico for potential substrates and binding partners of a protein of interest. In silico analysis, however, cannot replace experimental analysis, and fortunately there has been a veritable revolution in our capacity to analyze protein expression, post-translational modifications, and interactions, that in aggregate has led to the burgeoning field of proteomics, a discipline that is proving essential for our understanding of cell biology.

Genome sequencing and associated techniques, such as microarray analysis of RNA expression, have as their underlying theme that cells and organisms cannot be fully understood by studying one gene or transcript at a time. This statement is especially true at the level of the proteome, which presents challenges with a heightened degree of difficulty related to its dynamic nature. Proteins are not equally stable; some have a half-life of a few minutes, while others persist for days. Indeed, a large family of proteins is dedicated to the selection of specific polypeptides for ubiquitination and degradation, often in response to changing cellular conditions. In addition, since a single gene can potentially encode many different products, each of which may have a distinct function, one must identify not simply an individual protein but the complement of related splice variants expressed in a particular cell or tissue. To complicate matters, proteins can undergo a number of modifications, such as phosphorylation, acetylation, methylation, hydroxylation, ubiquitination, and nitrosylation, among others. These modifications can alter a protein's enzymatic activity but also serve as switches to induce or antagonize modular protein–protein interactions and thus the assembly of regulatory complexes. This complexity is the tip of the proteomic iceberg, since a single protein can be modified simultaneously by several different groups, with each combination of modifications potentially generating a distinct biological function. This multi-modification phenomenon has been studied intensively in the context of proteins such as histones and p53, but there is every reason to suppose that it is the norm rather than the exception. Adding to the complexity, a single modification, such as ubiquitination, can come in several different flavors. Addition of a single ubiquitin to a lysine residue in a target protein creates a binding site for interaction domains, such as the ubiquitin interaction motif found in proteins involved in endocytosis and intracellular trafficking. The linking of further ubiquitins to the initial site of modification to form a polyubiquitin chain, however, can lead to recognition by the proteosome and degradation.

Fortunately, powerful new proteomic techniques have been introduced just at the moment when they are most needed to address these issues. The forerunner of this approach, still extraordinarily useful, is the yeast 2-hybrid technique, which allows the investigator to measure binary protein–protein interactions within the confines of the yeast cell. While initially used to search a library for binding partners of a single protein, it has more recently been employed for comprehensive screens involving entire proteomes or large subsets thereof. An interesting lesson from these efforts is that the use of orthogonal techniques can greatly increase the reliability of proteomic data. For example, combining a 2-hybrid screen involving all 28 yeast SH3 domains with data concerning their binding preferences for peptide motifs, identified by phage display analysis, has yielded a more reliable view of the interaction network controlled by SH3 domains in a yeast cell.

A parallel technique of exceptional power involves the use of mass spectrometry (MS) to analyze proteins, either in their intact state or, more commonly, following peptide digestion. Peptide fragmentation can give sufficient sequence information to unambiguously identify a protein by MS, by comparison with a database of potential products inferred from DNA sequence information. Through the use of isotopic labeling and selective modification with reagents such as isotope-coded affinity tags (ICATs), it is possible to use MS to compare protein expression and modifications in two related cell samples, and the use of an isotopically labeled reference peptide allows for quantitation of protein levels. In addition, through the affinity isolation of one protein, it is possible to identify its associated polypeptides, as demonstrated through analysis of the yeast interaction map (Ho, Gruhler, Heilbut et al. 2002. *Nature* 415: 180–183). While this latter approach has typically involved gel purification of the complex protein mixture prior to analysis, advances in peptide separation have enabled the use of gel-free techniques to analyze protein complexes, which will potentially enhance the speed and completeness with which sets of interacting proteins can be identified. This advancement will be a necessity as we approach the more complex proteomes of mammalian cells.

These advances have given us an unprecedented ability with which to explore the expression and modifications of cellular proteins and to establish a wiring diagram of the cell. To be truly useful, such proteomic data must be linked to functional analysis. In yeast this can readily be accomplished through the use of genetic deletion sets, in which each open reading frame has systematically been disrupted. Among other things, this tool has allowed a high-throughput screen of genetic interactions to complement the data regarding physical protein–protein interactions. In mammalian cells, gene targeting is much more laborious than in yeast, but short interfering (si) RNAs provide a powerful tool with which to down-regulate gene expression in cultured cells and in intact animals. Indeed, siRNAs can be used to analyze families of proteins and are being employed in genome-wide screens for proteins that control specific aspects of cellular function. Another source of functional information involves the use of genetically encoded fluorescent protein derivatives to track protein localization in live cells. This technique is suitable for automation and provides a further level of annotation regarding protein activity.

Finally, we need computational and mathematical tools to synthesize these data and build models that illuminate cellular organization and complexity. An important advantage should be our ability to compare data from different species, in an effort to identify common threads and significant differences between the proteomes and interaction maps of distinct organisms. We are entering a new era in biology, one in which we will finally have the tools and the data to understand how cells work, and what goes wrong in disease. Proteomics is at the forefront of this revolution.

Preface

Proteomics has come a long way. It was not that long ago that the mention of the term proteomics brought up images of a two-dimensional polyacrylamide gel electrophoresis (2D-PAGE) separation and a mass spectrometer. Tim Veenstra can remember, as a post-doctoral researcher in a molecular and cellular biology laboratory, his supervisor, Dr. Rajiv Kumar, discussing with him a company that could take cell extracts from control and 1,25-dihydroxyvitamin D_3 neural cells and separate them out on 2D-PAGE gels. Protein spots that were stained differentially when comparing the two gels could then be identified. "Being somewhat obtuse, I never realized until several years later that Dr. Kumar was proposing a classical proteomics study before the term even became popular," relates Tim. Times have quickly changed, as scientists from various areas of life sciences have come to embrace the capabilities of the ideas and technologies that have been developed in the field of proteomics.

While there are many reasons, the diversity of chapters presented in this book best describes why proteomics has become so popular. Obviously, no book on proteomics would be complete without a chapter describing proteomic analysis using 2D-PAGE. The remaining chapters, however, span a diverse group of disciplines. These disciplines include gel-free proteomics to measure protein abundances, characterization of intact noncovalent protein complexes, three-dimensional protein structure determination, protein localization by imaging techniques, and the emerging field of protein arrays. Proteomics would not exist as it does today without three other critical elements described in this book: separations, bioinformatics, and automation.

The increasing number of fields that proteomics touches will not plateau any time soon. We are just beginning to realize the potential to look at biological systems as a whole instead of individualized parts. Proteomics, along with genomics, transcriptomics, and metabolomics, will play a major role in this next scientific venture. The impact of proteomics in clinical research is just in its infancy. Clinical research is a vast area in which proteomics can have a huge impact both in the discovery of new diagnostics and therapeutics and also at the individual patient level by determining individually tailored courses of treatment. New ideas and technologies continue to be developed rapidly to meet these challenges.

The editors would like to thank the authors for the time and effort they put into their contributions. All authors were selected because they are recognized as leaders in their particular areas of proteomics and their special knowledge and experience are clearly reflected in each chapter. The editors are also grateful to the publisher, John Wiley & Sons, Inc., for patience in understanding the enormity of work required to put such a book together. We are confident that readers of this book will be enriched by insights provided by each of the authors.

Frederick, Maryland   Timothy D. Veenstra
La Jolla, California   John R. Yates

Part I

Foundations of Proteomics

1

Mass Spectrometry: The Foundation of Proteomics

Timothy D. Veenstra

Laboratory of Proteomics and Analytical Technologies, SAIC–Frederick, Inc., National Cancer Institute at Frederick, Frederick, Maryland

1.1 INTRODUCTION

Scientific direction can be driven by many factors. Obviously, science is still primarily hypothesis driven; however, the continuing technology developments have enabled a greater focus on discovery driven science. Hypothesis driven science formulates a question and then uses whatever technology is available to acquire the information necessary to answer that question. In contrast, discovery driven science collects the information first and then determines the questions (or answers) that can be formulated from the available data. While it may seem to function through a "shot-in-the-dark" mentality, present technological developments make discovery approaches quite logical. Never before in the history of science has there been the capacity to acquire the wealth of data on biological molecules as exists today. A great example of this data gathering capability is reflected within the human genome project. It was inconceivable two decades ago that sequencing of the entire human genome could be accomplished; yet here we are today with the capability of sequencing genomes of other organisms as a routine procedure. Fortunately, science was not content with being able to sequence genomes and soon after the capability to obtain global measurements on the relative abundances of gene transcripts was established. This capability has naturally progressed to the development of technologies to perform discovery driven studies on entire proteomes. This stage does not even represent the end of development, as significant progress is being made in metabolomics.

Proteomics for Biological Discovery, edited by Timothy D. Veenstra and John R. Yates.
Copyright © 2006 John Wiley & Sons, Inc.

The term proteomics has evolved over the past few years to almost replace what was once referred to as protein chemistry. The original, and still classical, connotation of proteomics, however, is the characterization of the complete set of proteins encoded by the genome of a given organism (Wilkins et al., 1996). In the early history of proteomics, proteins were fractionated by two-dimensional polyacrylamide gel electrophoresis (2D-PAGE) followed by visualization using protein stains such as Coomassie or silver stain (O'Farrell, 1975). To identify differences in the protein abundances of two distinct samples, each of their proteomes is fractionated and visualized on separate gels and those spots that reveal differences in their staining intensity are cored from the gel and identified, typically using mass spectrometry (MS). While it has been around for decades, the ability to use MS to characterize proteins has been the single largest force that has propelled proteomics. Many different facets of MS have led to its prominent position within the field of proteomics. The sensitivity of MS allows for the routine identification of proteins in the femtomole (fmol, 10^{-15} mol) to high attomole (amol, 10^{-18} mol) range (Moyer et al., 2003). The ability to identify proteins with confidence is aided by the mass measurement accuracy available using current MS technology. This accuracy is typically less than 50 parts per million (ppm) and is often routinely less than 5 ppm (Pasa-Tolic et al., 2004). The ability of tandem MS (MS/MS) to obtain partial sequence information in combination with on-line fractionation enables the confident identification of complex mixtures of peptides (Nesvizhskii and Aebersold, 2004). The throughput by which proteins can be identified by MS is unparalleled by any other biophysical technique—a critical parameter in the use of any technology to gather large datasets.

While used to characterize proteins, in reality it is peptides that MS is most adept at identifying. In a great majority of proteomics studies, the complex mixture of proteins is made even more complex by digesting the proteins into smaller peptides prior to MS analysis (Rappsilber and Mann, 2002). This digestion step is optimal for two main reasons. First, overall solubility of peptides in solution is much greater than that of intact proteins. Second, even though the mass measurement accuracy of MS is high, it is still not sufficient to confidently identify a protein *de novo* based solely on its molecular weight. Therefore, proteins are typically identified through peptides acting as surrogates for their parent protein of origin. One of the most common ways of identifying a protein is based on the mass spectrum of its peptide fragments that are produced by digestion using an enzyme such as trypsin. The resulting spectrum obtained from such a sample is referred to as a "peptide map" or a "peptide fingerprint" (Blackstock and Weir, 1999). To identify the protein, the collection of measured masses is compared to *in silico* peptide maps derived from a protein or genomic database (Figure 1.1). To identify a single protein within a simple mixture, peptide mapping works very well and it is quite easy to acquire the data necessary for obtaining the desired result. Peptide mapping of proteins within complex mixtures such as cell lysates is not possible since the peptide masses recorded in the mass spectrum will arise from a large number of different species and will not provide a conclusive identification. Fortunately the available instrumentation enables a greater depth of information to be obtained from peptide masses observed by MS. Instead of relying on the accurate mass of a specific peptide,

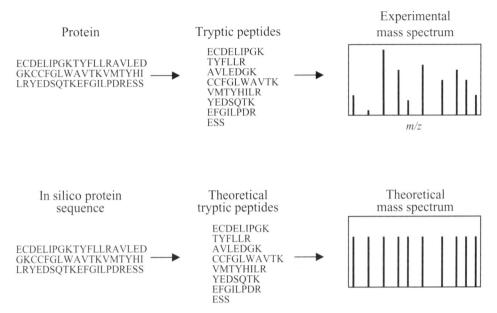

Protein	Tryptic peptides	Experimental mass spectrum

ECDELIPGKTYFLLRAVLED
GKCCFGLWAVTKVMTYHI
LRYEDSQTKEFGILPDRESS

ECDELIPGK
TYFLLR
AVLEDGK
CCFGLWAVTK
VMTYHILR
YEDSQTK
EFGILPDR
ESS

m/z

In silico protein sequence	Theoretical tryptic peptides	Theoretical mass spectrum

ECDELIPGKTYFLLRAVLED
GKCCFGLWAVTKVMTYHI
LRYEDSQTKEFGILPDRESS

ECDELIPGK
TYFLLR
AVLEDGK
CCFGLWAVTK
VMTYHILR
YEDSQTK
EFGILPDR
ESS

Figure 1.1. *Peptide mapping for protein identification. In peptide mapping, the protein of interest is proteolytically digested and the masses of the proteolytic peptides are measured using mass spectrometry (MS). To identify the correct protein, the sequences of all proteins within a specified database are digested in silico based on the specificity of the proteolytic enzyme used. The masses of the resulting peptides are calculated and theoretical mass spectra are constructed. The protein is identified based on the closest match between the experimental mass spectrum and the theoretical mass spectrum.*

individual peptide ions can be isolated and fragmented by collision induced dissociation (CID). After fragmentation of the peptide, the masses of the fragment ions are recorded and used to obtain partial or complete sequence information, as shown in Figure 1.2. This process is more commonly referred to as tandem MS or MS/MS (Martin et al., 1987). When peptides are subjected to MS/MS, they are not completely obliterated into their constituent amino acids, but instead an ensemble of fragments containing various lengths of the peptide is obtained. This information provides "sequence ladders" that enable partial primary sequence information of the peptide to be deduced. The raw data is then analyzed using software programs that can compare the experimental data to *in silico* MS/MS mass spectra calculated from the protein sequences in the database (Chamrad et al., 2004).

   Proteomics is conducted for many different purposes and at many different levels. Fortunately there are several different types of spectrometers available depending on the focus of the research being conducted. Obviously, if an investigation is focused on identifying simple protein mixtures, the instrument requirements would be different than if entire cell or tissue lysates were the sample of interest. In the following, a description of the various types of MS instrumental platforms available will be discussed with a focus on their application and mode of operation.

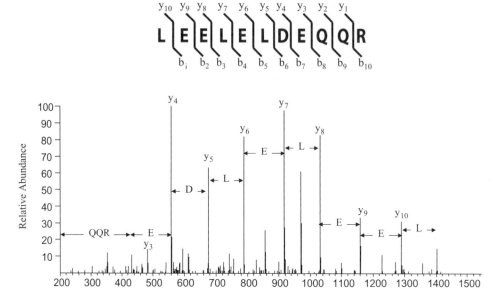

Figure 1.2. Tandem mass spectrometry (MS/MS) spectrum of a peptide observed from a tryptic digest of mitogen activated protein kinase kinase (MAPKK). Partial primary sequence information is determined by comparing the differences between major peaks in the spectrum with the calculated molecular masses of the amino acid monomers within the peptide.

1.2   IONIZATION METHODS

The mass spectrometer is made up of two major components: the ionization source and the mass analyzer. It is within the ionization source that the sample of interest is ionized and then desorbed into the gas phase. The mass analyzer acts to guide the gas-phase ions through the instrument to the detector. At the detector, the ions mass-to-charge (m/z) ratios are measured. While sometimes overlooked, many of the developments that have led to MS having a major impact on proteomics have been the invention of new ionization techniques.

The two most common methods to ionize biological molecules prior to their entrance into the analyzer region of the mass spectrometer are matrix-assisted laser desorption ionization (MALDI) (Karas et al., 1987) and electrospray ionization (ESI) (Fenn et al., 1989). While ESI and MALDI have enabled significantly larger proteins (i.e., greater than several hundred thousand daltons) to be analyzed, their greatest impact still remains in the analysis of peptides generated from proteolytic digests of larger species. One of the more significant advances enabled by ESI was the ability to interface separation methods such as liquid chromatography (LC) with MS. While separations are not discussed in this chapter, MS-based proteomics as it is practiced today would not be possible without the concurrent development of chromatographic and electrophoretic separation techniques.

1.2.1 Electrospray Ionization

ESI greatly enhanced the ability to characterize proteins and peptides by MS. Malcolm Dole, who conceived of using an electrospray process to produce intact high mass polymeric ions, provided the first description of ESI. He gained this insight from his knowledge of electrospraying automobile paint (Dole et al., 1968). These first experiments provided the basis of further studies by John Fenn (Fenn et al., 1989), who extended the use of ESI to measure biological molecules and was awarded the Nobel Prize in chemistry in 2002 for his discoveries.

The mechanism by which ESI works is relatively simple. ESI requires the sample of interest to be in solution so that it may flow into the ionization source region of the spectrometer (Figure 1.3A). Particulates or other insoluble entities in the

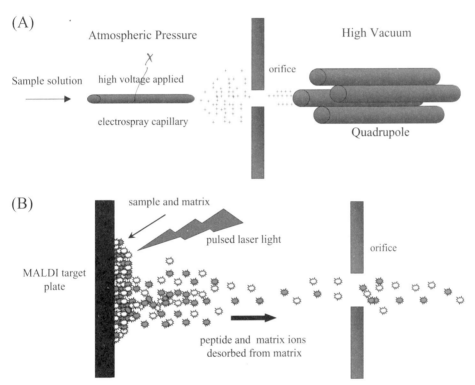

Figure 1.3. *(A) Electrospray ionization (ESI) of molecules for mass spectral characterization. The sample solution is passed through a stainless steel or other conductively coated needle. A high positive potential is applied to the capillary (cathode), causing positive ions to drift toward the tip with high voltage. The presence of a high electric field produces submicrometer-sized droplets upon the solution exiting the needle. The droplets travel toward the mass spectrometer orifice at atmospheric pressure and evaporate and eject charged analyte ions. The desolvated ions are drawn into the mass spectrometer by the relative low pressure maintained behind the orifice. (B) Principles of matrix-assisted laser desorption ionization (MALDI). The sample is cocrystallized with a large excess of matrix. Short pulses of laser light are focused onto the sample spot, causing the sample and matrix to volatilize. The matrix absorbs the laser energy, causing part of the illuminated substrate to vaporize. A rapidly expanding matrix plume carries some of the analyte into the vacuum with it and aids the sample ionization process. (See color insert.)*

sample will hamper ionization and cause the capillary through which the sample flows to become clogged. To ionize the sample, high voltage is applied to a stainless steel or other conductively coated needle through which the sample is flowing. The voltage results in charges being added to the sample, creating an ion that can be guided through the analyzer region of the instrument. The applied voltage can result in the sample becoming positively or negatively charged; however, positive ionization is used primarily in the analysis of proteins and peptides. As it exits the spray tip, the solution produces submicrometer-sized droplets containing both solute and analyte ions. The sample is then desorbed of solute prior to entering the analyzer region of the instrument. This desorption is achieved by evaporation of the solvent by passing the sample through a heated capillary or a curtain of drying gas, typically nitrogen. Since the desolvation of the ions occurs at atmospheric pressure and the mass analyzer region of the spectrometer is maintained at a lower pressure, the ions are drawn into the spectrometer based on this pressure differential.

What distinguished ESI from other ionization methods is its ability to produce multiply charged ions from large biological molecules. The number of charges that can be accepted by a particular molecule is dependent on many factors including its basicity and size. Depending on their size and the number of basic residues within, peptides typically exist as either singly, doubly, or triply charged ions. Since trypsin is the most commonly used protease in proteomics today, peptides are typically observed in both 1+ and 2+ charged states owing to the basic sites on the N terminus and the C-terminal lysine or arginine residues.

1.2.2 Matrix-Assisted Laser Desorption Ionization

Matrix-assisted laser desorption ionization (MALDI) is another "soft" ionization process that generates ions by irradiating a solid mixture with a pulsed laser beam. The solid mixture is comprised of the analyte of interest dissolved in an organic matrix compound. The laser pulse both indirectly ionizes and desorbs the analyte molecules from the solid mixture. A short-pulse (2–200 Hz) ultraviolet (UV) laser is typically used in MALDI; however, infrared irradiation has also been used (Tanaka et al., 1988; von Seggern et al., 2003). To prepare the solid mixture, an equal volume of the sample solution is combined with a saturated solution of matrix prepared in a solvent such as water, acetonitrile, acetone, or tetrahydrofuran. The matrix is a small, highly conjugated organic molecule (i.e., α-cyano-4-hydroxycinnamic acid (CHCA), 2,5-dihydroxybenzoic acid (DHB), and 3,5-dimethoxy-4-hydroxycinnamic acid (sinapinic acid)) that strongly absorbs energy in the (UV) region. A few microliters of the solid mixture is placed onto a MALDI target plate and allowed to dry. This drying procedure results in the incorporation of the peptides into a crystal lattice. The MALDI target plate is then inserted into the source region of the mass spectrometer followed by laser irradiation, as shown in Figure 1.3B. The MALDI source region of most spectrometers is maintained at a relatively high pressure, causing the ions to be drawn into the mass analyzer region of the instrument, which is maintained at a lower pressure. A recent development has been the design of MALDI sources that operate at atmospheric pressure (Moyer and Cotter, 2002). This ability to operate at atmospheric pressure enables MALDI sources to be interfaced to analyzers, such as ion traps and quadrupole time-of-flight analyzers. Such instruments have historically been interfaced with ESI sources.

Similar to ESI, MALDI can produce both positive and negative ions. Positive ions, which are typically the species of interest in peptide analysis, are formed by the acceptance of a proton as the analyte leaves the matrix. While yet to be absolutely determined, the prevailing theory is that analyte ionization occurs within the dense gas cloud that forms and expands supersonically into the vacuum region of the spectrometer. The analytes are protonated (or deprotonated) through collisions between analyte neutrals, excited matrix ions, and protons and cations. In MALDI, most analytes accept a single protein; therefore peptide and large biomolecular ions are singly charged. This singly charged character results in some molecules having large m/z values and therefore MALDI is typically interfaced with mass analyzers with large m/z ranges, such as time-of-flight (TOF) spectrometers.

1.2.3 Desorption Electrospray Ionization

While not yet applied to proteomic technology, a new method of desorption ionization has recently been described that allows the direct analysis of surfaces by MS. This ionization technique, called desorption electrospray ionization (DESI), was developed in the laboratory of R. Graham Cooks (Takats et al., 2004) and is illustrated in Figure 1.4. In this technique, electrosprayed droplets are directed toward a surface to be analyzed. The droplets produce gaseous ions of the material on the surface and these ions are sampled with a mass analyzer. The mass analyzer is equipped with an atmospheric interface connected via a flexible and extended ion transfer line. This ionization technique, while extremely new, has shown the

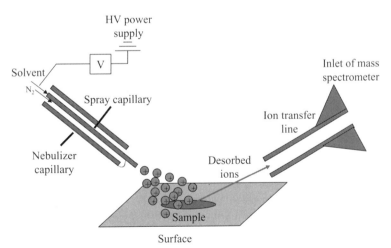

Figure 1.4. Schematic of desorption electrospray ionization (DESI) instrument. In DESI, electrosprayed droplets are directed to a surface. The impact of the charged droplets produces gaseous ions from the sample on the surface, which can be sampled using a commercially available mass analyzer equipped with an ion transfer line. (See color insert.)

capability of analyzing a range of compounds from nonpolar small molecules to polar peptides and proteins.

While the most fruitful uses of this new ionization technology are not clear, some of the demonstrated applications suggested a new exciting way to monitor things like drug distribution and surface analysis. In a novel experiment, 10 mg of loratadine, an over-the-counter antihistamine, was given to a patient and 40 minutes later DESI was able to detect the molecule directly from the skin surface and saliva of the individual. While the proteomic applications using this technology have not been clearly demonstrated, the potential exists for direct monitoring of proteins on the surface of cells in culture from tissue sections.

1.3   MASS ANALYZERS

1.3.1   Ion-Trap Mass Spectrometer

An ion-trap mass spectrometer functions just as its name implies: it traps ions. The ability to trap ions, however, does not explain the popularity of this mass analyzer for proteomic analysis. The popularity lies in the discovery and development of ways to manipulate the ions after they are trapped (Stafford et al., 1984). The first development was the mass-selective *instability* mode of operation. This mode allowed ions that were created and trapped over a given time period to be ejected sequentially into a conventional electron multiplier detector. Unlike the mass-selective *stability* mode of operation where only one m/z value could be stored, a wide range of m/z values could be stored. The second big development was showing that the addition of 1 mtorr of helium gas to the ion trap increased the mass resolution of the instrument. This increased resolution results from a reduction in the kinetic energy of the ions and causes the ion trajectories to contract to the center of the trap (Stafford et al., 1984). This phenomenon causes packets of ions of a given m/z to form, allowing them to be ejected faster and more efficiently than a diffuse cloud of ions.

The ion trap conducts repeated iterations of collecting, storing, and ejecting ions out of the trap. The true power of the ion-trap analyzer is its ability to isolate and fragment peptide ions (i.e., conduct MS/MS) from complex mixtures, such as found in many proteome analyses. To perform MS/MS analysis, specific ions are selected and the trapping voltages are adjusted to eject all other ions from the trap. The applied voltages are then increased to cause an increase in the energy of the remaining ions. These high-energy ions undergo collisions with He_2, causing them to fragment. These fragments are then trapped and scanned out according to their m/z values. Daughter ions resulting from the fragmentation of large ions can also be retained within the trap and subjected to further rounds of MS/MS (i.e., MS/MS/MS or MS^n), to obtain more structural information concerning the species of interest. While MS/MS/MS is not routinely used in the identification of peptides in a complex mixture, it has shown utility in the identification of phosphorylated peptides.

The ion-trap mass spectrometer enjoys a position of prominence as a true "workhorse" in global proteomic studies designed to characterize complex mixtures of proteins. When analyzing very complex mixtures, the ion trap operates in a data-dependent MS/MS mode in which each full MS scan is followed by a specific number (usually 3–5, but sometimes >10 when using a linear ion trap) of MS/MS scans, where the three most abundant peptide molecular ions are dynamically selected for fragmentation. Using such a mode of operation will often allow for the identification of over 1000 peptides in a single LC/MS/MS experiment.

1.3.2 Time-of-Flight Mass Spectrometer

Time-of-flight mass spectrometers (Olthoff et al., 1988) are an extremely popular choice of mass analyzer for proteomic research. The major attributes that make TOF-MS attractive are their high throughput, sensitivity, and resolution. TOF spectrometers measure the m/z ratios of ions based on the time it takes for the ions generated in the source to fly the length of the analyzer and strike the detector. The speed, and therefore the time, at which the ions fly down the analyzer tube is proportional to their m/z value. Larger ions have a slower speed compared to smaller ions and therefore take a longer time to reach the detector.

TOF analyzers have been used primarily to generate peptide fingerprints for identifying individual proteins. The simplicity of operation and robustness of MALDI-TOF analyzers have made them an excellent choice for such applications. With the development of MALDI-TOF/TOF instruments, these analyzers have been able to do true tandem MS through the inclusion of a collision cell separated by two TOF tubes. MALDI-TOF/TOF instruments are characterized by high throughput and high resolution and mass accuracy for both the MS and MS/MS modes. In this configuration, peptide ions generated in the source region are accelerated through the first TOF tube and are dissociated by introducing an inert gas (i.e., air or nitrogen) into a collision cell. Collisions between the gas and peptide ions cause fragmentation of the peptide. These fragment ions are then accelerated through a second TOF tube to the detector. This combination allows proteins to be identified through peptide fingerprinting and identification is confirmed through MS/MS of selected peptide species. In addition, proteins within complex mixtures can now be identified solely through MS/MS of specific peptide signals. Many standard TOF instruments do not contain a true collision cell to provide MS/MS sequence data. Instruments equipped with a reflectron, however, can measure fragmentation products through a process called "post-source decay" (PSD) (Kaufmann, 1995). In PSD, the reflectron voltage is adjusted during the analysis so that fragment ions generated during the ionization and acceleration of the peptide are focused and detected. Specifically, PSD produces immonium ions, which are useful indicators of the presence of a specific amino acid within a peptide (Kaufmann et al., 1996). While PSD analysis can be relatively slow and will not meet the high-throughput demands necessary for proteomics, it does provide useful complementary information to substantiate the identification of an intact peptide.

1.3.3 Triple Quadrupole Mass Spectrometer

The quadrupole mass spectrometer has been the most commonly used mass analyzer with ESI (Yost and Boyd, 1990). As its name implies, a quadrupole consists of four metal rods arranged in parallel, as shown in Figure 1.3A. Direct current and radiofrequency (rf) voltages are applied to these rods to guide and manipulate ions through the mass analyzer. Altering the voltage allows a specific m/z range of ions to pass through the quadrupole region of the analyzer and onto the detector. The two most common types of quadrupole mass spectrometers are single-stage and triple quadrupoles. Unfortunately single quadrupole analyzers have limited use in proteomics since they lack true MS/MS abilities, although in-source CID is possible. The triple quadrupole instrument, however, has true MS/MS capabilities since a collision cell is incorporated between two of the quadrupole regions.

To identify peptides using a triple quadrupole mass spectrometer requires switching the analyzer between two different scan modes. In the first "full-scan" mode, a broad m/z range of ions is allowed to pass through the first quadrupole. The ions that pass through are allowed to pass freely through the remaining two quadrupoles onto the detector. Essentially all of the ions produced in the source are measured. In the second scan mode, the first quadrupole is used as a mass filter and only a specific ion is allowed to pass through. The ion that is allowed through is then subjected to fragmentation within the second quadrupole by this region being filled with an inert gas. The resulting fragmentation ions then freely pass through the third quadrupole and are detected.

The versatility of the triple quadrupole analyzer is underscored by its ability to produce an ion, precursor ion, and neutral loss scanning. Triple quadrupoles have been used to identify proteins extracted from 2D-PAGE gels (Kuhn et al., 2004), in phosphopeptide characterization (Kocher et al., 2003), and in glycopeptide identification (Jiang et al., 2004). With a mass measurement accuracy of 0.5 amu, the fragment ions produced by MS/MS using triple quadrupole analyzers is sufficiently accurate to allow the identification of peptides by correlating the spectra with protein sequences obtained from biological databases.

1.3.4   Quadrupole Time-of-Flight Mass Spectrometer

The hybrid quadrupole time-of-flight mass spectrometer (QqTOF) is a versatile instrument that plays a key role in proteomic analysis (Chernushevich et al., 2001). The QqTOF mass spectrometer combines a mass-resolving quadrupole and collision cell with a TOF tube, as shown in Figure 1.5. This instrument combines the benefits of both types of mass analyzers: the ion selectivity and sensitivity of the quadrupole and the high mass resolution and mass accuracy of the TOF (Shevchenko et al., 2000). The high mass accuracy afforded with this configuration provides the potential for real *de novo* sequencing of peptides (Loboda et al., 2000). This instrument is also able to analyze samples using both ESI and MALDI methods, using sources that are readily interchangeable.

The usual QqTOF configuration is comprised of three quadrupoles, as in a triple quadrupole spectrometer, with the initial quadrupole acting as a rf-only quadrupole that serves to provide collisional damping. The following two quadrupoles perform as they would in a standard triple quadrupole mass analyzer; however, the third quadrupole is replaced by a reflecting TOF mass analyzer. For MS measurements, the mass filter is operated in a rf-only mode, permitting all of the ions to pass directly through onto the TOF tube. The resulting spectrum benefits from the high resolution and mass accuracy of the TOF tube. For MS/MS, the mass filter allows only the ion(s) of interest to pass to the collision cell, where it undergoes

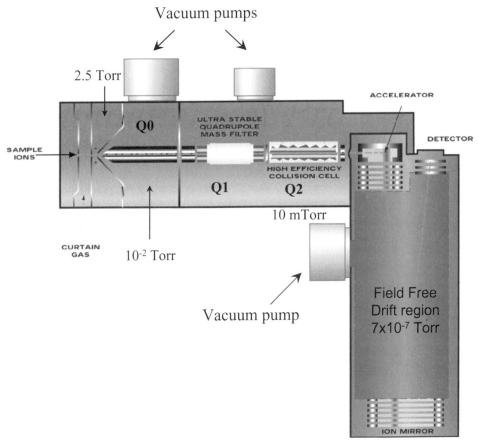

Vacuum pumps

2.5 Torr

Q0

ULTRA STABLE
QUADRUPOLE
MASS FILTER

ACCELERATOR

DETECTOR

SAMPLE
IONS

HIGH EFFICIENCY
COLLISION CELL

Q1

Q2

10 mTorr

CURTAIN
GAS

10^{-2} Torr

Vacuum pump

Field Free
Drift region
7×10^{-7} Torr

ION MIRROR

Figure 1.5. *Schematic of a quadrupole time-of-flight mass spectrometer (TOF-MS). In this instrument the third quadrupole of a triple quadrupole mass spectrometer has been replaced with a TOF tube. This combination gives this instrument the ion selection and tandem MS capabilities of a triple quadrupole MS with the high mass accuracy and resolution capabilities of a TOF. (See color insert.)*

fragmentation through collisions with a neutral gas. Due to their resulting high energies, the fragment ions are then collisionally cooled, followed by reacceleration and focusing prior to entering the TOF region. A pulsed electric field is applied at a frequency of several kilohertz (kHz) to push the ions into the accelerating column in an orthogonal direction to their original trajectory. The accelerated ions then enter the field-free drift space of the TOF analyzer, where they are separated based on their m/z and detected.

The utility of this mass analyzer in proteomics is represented by its ability to perform tandem MS for peptide identification as well as retain high resolution for separating isotopically the internal standard for peptides in quantitative studies. Some of the earliest studies in demonstrating the use of isotope-coded affinity tags for quantitative proteomics were done using a QqTOF mass analyzer equipped with a MALDI source. The resolution afforded from the QqTOF is critical to the quantitative analysis of other types of isotopically labeled proteomes such as the use of

[18]O labeling in which the mass difference between peptide pairs is only 4 Da. Doubly charged ions of this type will exhibit a m/z difference of only 2; a difference that is not very well resolved using a conventional ion-trap instrument.

1.3.5   Fourier Transform Ion Cyclotron Resonance Mass Spectrometry

Fourier transform ion cyclotron resonance (FTICR) mass spectrometry was developed over thirty years ago by Comisarow and Marshall (Comisarow and Marshall, 1974). It is only recently, however, that its many unrivaled attributes have been utilized to generate high-quality data within proteomics. A FTICR-MS functions somewhat like an ion-trap analyzer, with the trap being housed within a high-strength magnetic field, as shown in Figure 1.6 (Marshall et al., 1998). Ions within the trap resonate at their cyclotron frequency due to the presence of the magnetic field. A uniform electric field that oscillates at or near the cyclotron frequency of the trapped ions is applied to excite the ions into a larger orbit that can be measured as they pass by detector plates on opposite sides of the trap. The energy applied can also be adjusted to dissociate the ions or eject them from the trap. The detector measures the cyclotron frequencies of all of the ions in the trap and a Fourier transform is used to convert these frequencies into m/z values.

The ion trap of a FTICR-MS is placed in a magnetic field. Indeed, the characteristics of many FTICR systems are implied through the description of the magnetic field strength, much like the proton frequency is quoted when describing a

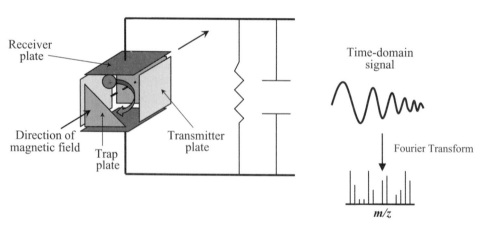

Figure 1.6. *Principles of Fourier transform ion cyclotron resonance (FTICR) mass spectrometry. In FTICR-MS the ion trap is placed in a strong magnetic field. The magnetic field causes ions captured within the trap to resonate at their cyclotron frequency. By applying the appropriate electric field energy, the ions are excited into a larger orbit that can be measured as they pass by detector plates on opposite sides of the trap. Energy can also be applied to dissociate the ions or to eject ions from the trap by accelerating them to a cyclotron radius larger than the radius of the trap. The detector measures the cyclotron frequency of all of the ions in the trap and uses a Fourier transform to convert these frequencies into m/z values. (See color insert.)*

NMR instrument (i.e., 500 MHz). Working at higher magnetic fields benefits many of the parameters related to FTICR performance. The two most critical parameters that are improved through the use of FTICR are resolution and mass accuracy. These analyzers have been proved experimentally to provide the highest resolution, mass accuracy, and sensitivity for peptide and protein measurements so far achieved (Page et al., 2004).

While FTICR has not been widely used in proteomics, that situation is rapidly changing. Its capabilities compared to other types of analyzers make it a potentially powerful tool in the characterization of global proteome mixtures. FTICR provides a wide dynamic range (i.e., $>10^3$) of measurements, enabling it to identify low abundance species in the presence of higher abundance components. The mass accuracy of FTICR is unequaled by other types of analyzers and is sufficiently high to enable multiple ions to be accumulated and fragmented simultaneously (Li et al., 2001). While previously FTICR was a very expensive and developmental technology, the advent of new commercially available instruments is making this technology available to more and more proteomic research centers.

1.4   CONCLUSION

Proteomics continues to follow many of the themes of the human genome project as investigators try to gain a greater understanding of the complexity and diversity of the human proteome. There are characterizations that proteomics is asked to make that require more than simple identification of proteins. Such characteristics as post-translational modifications play a critical role in cell function and will necessarily require identification if we hope to achieve a predictive capability concerning the effects of perturbations on a cell system. The complete characterization of all of the proteins expressed within a cell is becoming possible through a combination of developments in chromatography, mass analysis, and bioinformatics. Mass spectrometry has acted, and will undoubtedly continue to act, as the key analytical tool for the direct measurement of proteins in such a discovery driven science as proteomics. While we have tried to provide a description of some of the different types of mass spectrometers used in proteomic analysis, each type would require a book unto itself to describe each in depth. This chapter did not even attempt to include in-depth application of these different types of analyzers, as many such applications are more adequately described within other chapters in this book.

ACKNOWLEDGMENT

By acceptance of this article, the publisher or recipient acknowledges the right of the United States Government to retain a nonexclusive, royalty-free license and to any copyright covering the article. The content of this publication does not necessarily reflect the views or policies of the Department of Health and Human Services, nor does mention of trade names, commercial products, or organization imply endorsement by the U.S. Government. This project has been funded in whole or in part with federal funds from the National Cancer Institute, National Institutes of Health, under Contract No. NO1-CO-12400.

REFERENCES

Blackstock WP, Weir MP. 1999. Proteomics: quantitative and physical mapping of cellular proteins. *Trends Biotechnol* 17:121–127.

Chamrad DC, Korting G, Stuhler K, Meyer HE, Klose J, Bluggel M. 2004. Evaluation of algorithms for protein identification from sequence databases using mass spectrometry data. *Proteomics* 4:619–628.

Chernushevich IV, Loboda AV, Thomson BA. 2001. An introduction to quadrupole-time-of-flight mass spectrometry. *J Mass Spectrom* 36:849–865.

Comisarow MB, Marshall AG. 1974. Fourier transform ion cyclotron resonance spectroscopy. *Chem Phys Lett* 25:282–283.

Dole M, Ferguson LD, Hines RL, Mobley RC, Ferguson LD, Alice MB. 1968. Molecular beams of macroions. *J Chem Phys* 49:2240–2249.

Fenn JB, Mann M, Meng CK, Wong SF. 1989. Electrospray ionization for mass spectrometry of large biomolecules. *Science* 246:64–71.

Jiang H, Desaire H, Butnev VY, Bousfield GR. 2004. Glycoprotein profiling by electrospray mass spectrometry. *J Am Soc Mass Spectrom* 15:750–758.

Karas M, Bachmann D, Bahr U, Hillenkamp F. 1987. *Int J Mass Spectrom Ion Processes* 78:53–68.

Kaufmann R. 1995. Matrix-assisted laser desorption ionization (MALDI) mass spectrometry: a novel analytical tool in molecular biology and biotechnology. *J Biotechnol* 41:155–175.

Kaufmann R, Chaurand P, Kirsch D, Spengler B. 1996. Post-source decay and delayed extraction in matrix-assisted laser desorption/ionization-reflectron time-of-flight mass spectrometry. Are there trade-offs? *Rapid Commun Mass Spectrom* 10:1199–1208.

Kocher T, Allmaier G, Wilm M. 2003. Nanoelectrospray-based detection and sequencing of substoichiometric amounts of phosphopeptides in complex mixtures. *J Mass Spectrom* 38:131–137.

Kuhn E, Wu J, Karl J, Liao H, Zolg W, Guild B. 2004. Quantification of C-reactive protein in the serum of patients with rheumatoid arthritis using multiple reaction monitoring mass spectrometry and ^{13}C-labeled peptide standards. *Proteomics* 4:1175–1186.

Li L, Masselon CD, Anderson GA, Pasa-Tolic L, Lee SW, Shen Y, Zhao R, Lipton MS, Conrads TP, Tolic N, Smith RD. 2001. High-throughput peptide identification from protein digests using data-dependent multiplexed tandem FTICR mass spectrometry coupled with capillary liquid chromatography. *Anal Chem* 73:3312–3322.

Loboda AV, Krutchinsky AN, Bromirski M, Ens W, Standing KG. 2000. A tandem quadrupole/time-of-flight mass spectrometer with a matrix-assisted laser desorption/ionization source: design and performance. *Rapid Commun Mass Spectrom* 14:1047–1057.

Marshall AG, Hendrickson CL, Jackson GS. 1998. Fourier transform ion cyclotron resonance mass spectrometry: a primer. *Mass Spectrom Rev* 17:1–35.

Martin SA, Rosenthal RS, Biemann K. 1987. Fast atom bombardment mass spectrometry and tandem mass spectrometry of biologically active peptidoglycan monomers from *Neisseria gonorrhoeae*. *J Biol Chem* 262:7514–7522.

Moyer SC, Cotter RJ. 2002. Atmospheric pressure MALDI. *Anal Chem* 74:468A–476A.

Moyer SC, Budnik BA, Pittman JL, Costello CE, O'Connor PB. 2003. Attomole peptide analysis by high-pressure matrix-assisted laser desorption/ionization Fourier transform mass spectrometry. *Anal Chem* 75:6449–6454.

Nesvizhskii AI, Aebersold R. 2004. Analysis, statistical validation and dissemination of large-scale proteomics datasets generated by tandem MS. *Drug Discov Today* 9:173–181.

O'Farrell PH. 1975. High resolution two-dimensional electrophoresis of proteins. *J Biol Chem* 250:4007–4021.

Olthoff JK, Lys IA, Cotter RJ. 1988. A pulsed time-of-flight mass spectrometer for liquid secondary ion mass spectrometry. *Rapid Commun Mass Spectrom* 2:171–175.

Page JS, Masselon CD, Smith RD. 2004. FTICR mass spectrometry for qualitative and quantitative bioanalyses. *Curr Opin Biotechnol* 15:3–11.

Pasa-Tolic L, Masselon C, Barry RC, Shen Y, Smith RD. 2004. Proteomic analyses using an accurate mass and time tag strategy. *Biotechniques* 37:621–636.

Rappsilber J, Mann M. 2002. What does it mean to identify a protein in proteomics? *Trends Biochem Sci* 27:74–78.

Shevchenko A, Loboda A, Shevchenko A, Ens W, Standing KG. 2000. MALDI quadrupole time-of-flight mass spectrometry: a powerful tool for proteomic research. *Anal Chem* 72:2132–2141.

Stafford GC, Kelley PE, Syka JEP, Reynolds WE, Todd JFJ. 1984. Recent improvements in and analytical applications of advanced ion-trap technology. *Int J Mass Spectrom Ion Process* 60:85–98.

Takats Z, Wiseman JM, Gologan B, Cooks RG. 2004. Mass spectrometry sampling under ambient conditions with desorption electrospray ionization. *Science* 306:471–473.

Tanaka K, Waki H, Ido Y, Akita S, Yoshida Y, Yoshida T. 1988. Protein and polymer analyses up to *m/z* 100,000 by laser ionization time-of-flight mass spectrometry. *Rapid Commun Mass Spectrom* 2:151–153.

Von Seggern CE, Zarek PE, Cotter RJ. 2003. Fragmentation of sialylated carbohydrates using infrared atmospheric pressure MALDI ion trap mass spectrometry from cation-doped liquid matrixes. *Anal Chem* 75:6523–6530.

Wilkins MR, Sanchez JC, Gooley AA, Appel RD, Humphery-Smith I, Hochstrasser DF, Williams KL. 1996. Progress with proteome projects: why all proteins expressed by a genome should be identified and how to do it. *Biotechnol Genet Eng Rev* 13:19–50.

Yost RA, Boyd RK. 1990. Tandem mass spectrometry: quadrupole and hybrid instruments. *Methods Enzymol* 193:154–200.

2

Proteomic Analysis by Two-Dimensional Polyacrylamide Gel Electrophoresis

Pavel Gromov*, Irina Gromova, and Julio E. Celis*
Institute of Cancer Biology, Danish Cancer Society and Danish Centre for Translational Breast Cancer Research, Copenhagen, Denmark

2.1 INTRODUCTION

The sequencing of the human and other important genomes is only the beginning of the quest to understand the functionality of cells, tissues, and organs in both health and disease. Together with advances in bioinformatics, this development has paved the way for the revolution in biology and medicine that we are experiencing today. Science is rapidly moving from the study of single molecules to the analysis of complex biological systems, and the current explosion of emerging technologies within proteomics and functional genomics promises to elicit major advances in medicine in the near future. In particular, proteomic technologies are expected to play a key role in the study and treatment of diseases as they provide invaluable resources to define and characterize regulatory and functional networks. The ability of proteomic technology to define the precise molecular defect in diseased tissues and biological fluids will aid in the development of specific reagents to precisely pinpoint a particular disease or stage of a disease.

* To whom correspondence should be addressed.

Proteomics for Biological Discovery, edited by Timothy D. Veenstra and John R. Yates.
Copyright © 2006 John Wiley & Sons, Inc.

Proteomics can deal with problems that cannot be approached by genomic analysis, namely, relative abundance of the protein products, post-translational modifications, compartmentalization, turnover, as well as interactions and function. In addition, there are many biological samples that are not suitable for genomic or transcriptomic analysis. For example, biological fluids such as blood plasma, serum, urine, cerebral fluid, saliva, nipple fluid, and other secretions contain proteins of physiological and diagnostic significance that require direct analysis using proteomic technologies.

For the last three decades, high-resolution two-dimensional (2D) polyacrylamide gel electrophoresis (PAGE), often referred to as gel-based proteomics, has been the method of choice to analyze the protein composition of a given cell type and for monitoring changes in gene activity through the quantitative and qualitative analysis of the thousands of proteins that orchestrate various cellular functions (Cash and Bravo, 1984; Celis and Gromov, 1999; Ong and Pandey, 2001; Celis and Kroll, 2003; and references therein). This technique separates proteins in terms of both their isoelectric points (pI) and molecular weights (M_r) and is essentially a stepwise separation tool that combines isoelectric focusing and sodium dodecylsulfate (SDS)–PAGE.

Two-dimensional PAGE technology has been one of the driving forces behind the development of proteomics since it enables the separation of complex protein mixtures to the extent that thousands of individual components can be visualized on a single high-resolution gel. In addition, the separation and visualization capabilities of 2D-PAGE enable the protein expression profiles of paired samples (normal versus transformed cells, cells at different stages of growth or differentiation, etc.) to be readily compared. This capability allows proteins whose abundance levels are affected by some specified condition to be easily recognized. The differentially abundant proteins can then be identified using complementary proteomic technologies, including mass spectrometry (MS) and 2D-PAGE Western blotting. Furthermore, by carrying out studies in a systematic manner, one can store the information in comprehensive 2D-PAGE databases that record how genes are regulated in health and disease (http://proteomics.cancer.dk; http://expasy.hcuge.ch/ch2d/2d-index.html). As these databases achieve critical mass of data, they will become valuables sources of information for expediting the identification of signaling pathways and components that are affected in diseases (Celis et al., 1995; Celis et al., 1998; Gromov et al., 2002). In this chapter, we provide a brief description of the principles underlying high-resolution 2D-PAGE, as well as an appraisal of its current status and problems that challenge its central position in proteomics.

2.2   THE PROTEOME, 2D-PAGE, AND DYNAMIC RANGE OF PROTEIN EXPRESSION

Wilkins and colleagues originally defined the proteome as the complete set of prote*ins* coded by the gen*ome* (Wilkins et al., 1996), but the term has now been broadened to include the set of proteins expressed in both space and time under different physiological conditions. Currently, proteomics comprises a plethora of technologies to address the protein complement of complex biological samples. The protein universe of a cell, however, is extremely complex and highly dynamic, and any such analysis must take into consideration the dynamic range of expression, post-translation modifications, and protein–protein interactions, as well as functional aspects.

An ideal separation method would be able to resolve, in a single map, all of the proteins expressed by a single cell type. Current technologies, however, have not achieved this resolution yet, and today the only available technique that provides a global profile of a cell proteome is high-resolution 2D-PAGE (Klose, 1975; O'Farrell, 1975; O'Farrell et al., 1977). Unfortunately, 2D-PAGE suffers from various limitations imposed in part by the separation technology itself, but also by the lack of sensitive procedures to detect proteins expressed across a wide dynamic range. Approaches based on chromatography-MS (Link et al., 1999; Lin et al., 2003) allow high-throughput proteome characterization but are not ready yet for the study of complex samples such as tissue biopsies.

The impact of the proteomic technologies for large scale monitoring of protein expression depends very much on the number of proteins to be resolved. The actual number of messenger species in eukaryotic genomes far exceed the number of protein coding units (approximately 30,000; however, this estimate is constantly being revised) due to alternative RNA splicing, overlapping of transcription units, and *trans*-splicing RNA events (Labrador et al., 2001). Current estimates based on the various assemblies of expressed sequence tags (ESTs) suggest that the average number of mRNA species might be two to three times higher, yielding a putative number of 70,000–90,000 distinct primary translation products (Claverie, 2001 and references therein). In addition, many proteins have isoforms that arise from post-translational processing and modifications.

Data from a few laboratories, including our own, have shown that only a fraction of the whole genome is expressed at the protein level in a given cell type (Duncan and McConkey, 1982; Celis et al., 1991, 1998). With a few exceptions, single human cell types may express in the range of 5000–6000 different primary translation products plus their modified variants, which can be extensive in particular cases (Gooley and Packer, 1997). As much as 80–90% of these polypeptides may represent housekeeping proteins (components of metabolic pathways, cytoarchitecture elements, etc.) that are expressed by all cell types, albeit in variable amounts (Celis et al., 1995, 1998).

One of the main drawbacks of 2D-PAGE is that current detection procedures are not sensitive enough to reveal the whole proteome on a single gel. Abundant proteins such as actin may be present in approximately 10^8 molecules per cell while some low abundance proteins such as receptors or kinases are probably presented in 100–1000 copies per cell, yielding a dynamic range of protein concentration on the order of 10^5–10^6. Some biofluids such as a serum may possess an even higher dynamic range of protein concentration (10^8–10^9).

Gygi and coauthors calculated that 10^9 cells are required to provide 160 fmol of a protein that is present in 100 copies per cell (Gygi et al., 2000). For a 50 kilodalton (kDa) protein, this corresponds to 0.8 ng, an amount that is close to the lowest quantity required for identification using MS. Even though no losses may occur during sample solubilization and separation, the capacity of the current 2D-PAGE technology is not sufficient to permit such high loading on a single gel. Gygi and coauthors have reported that, despite the high sample load and extended electrophoretic separation, proteins expressed in yeast from genes with codon bias values of <0.1 (i.e., lower abundance proteins) were not detected by silver staining (Gygi et al., 1999). A number of pre-gel fractionation approaches have been developed to overcome this problem by increasing the chances of detecting low abundance proteins within a complex proteome (see Section 2.7).

By using the current 2D-PAGE technology, it is possible to obtain reproducible separations of nearly 4000–5000 [^{35}S]methionine labeled polypeptides from whole human cell extracts using broad pH gradients (Celis et al., 1991, 1995, 1997, 1999; see also procedures in http://proteomics.cancer.dk). Some authors have reported resolution of about 10,000 radiolabeled proteins in a single gel (Fey and Larsen, 2001), and even though this number is still short of the total number of polypeptides that may be present in a given cell type, it is sufficient to start proteomic applications to biological and clinical problems (Celis and Gromov, 2003; Celis et al., 2003).

2.3 OUTLINE OF THE 2D-PAGE PROCEDURE

Two-dimensional PAGE of proteins is essentially a stepwise combination of two electrophoretic techniques: isoelectric focusing (IEF) and SDS-PAGE. First, proteins are fractionated in a first-dimension pH gradient according to their electric charges. Being amphoteric, protein molecules migrate in an electric field along a continuous pH gradient until they reach their pI, that is, when their net charge is zero. Next, gel tubes or strips are then applied to the top of slab SDS-PAGE gels and the proteins are resolved on the basis of their M_r, yielding a two-dimensional protein separation.

The standard 2D-PAGE protocols comprise the following steps: (i) sample preparation, (ii) first dimension separation (IEF, nonequilibrium pH gradient electrophoresis (NEPHGE), or immobilized pH gradient (IPG)), (iii) second dimension separation (SDS-PAGE), (iv) protein visualization, (v) image analysis, and (vi) protein identification (Figure 2.1).

[handwritten annotation: Comasie + silver staining.   MS]

2.3.1 Sample Preparation

Sample solubilization is one of the most critical steps in gel-based proteomics (Herbert et al., 2001; Lilley et al., 2002; Simpson, 2003 and references therein). Since the majority of 2D-PAGE applications aim at the analysis of complex protein mixtures prepared directly from whole cells or tissue samples, it is crucial to disrupt all noncovalent bonds in protein complexes as well as aggregates. In addition, it is important to avoid or minimize polypeptide modifications during preparation, as the first-dimension separation is very sensitive to charge alterations. The electrophoretic separation also requires that the protein molecules remain fully solubilized throughout the entire procedure. To achieve this solubilization, the lysis cocktail usually contains a neutral chaotrope (urea, thiourea, or other urea derivatives), a nonionic detergent (Nonidet P-40, Triton X-100), and a reducing agent (β-mercaptoethanol, dithiothreitol (DTT)) (Simpson, 2003 and references therein). The lysis solution also contains carrier ampholytes to optimize the pH for protein solubilization and minimize protein aggregation.

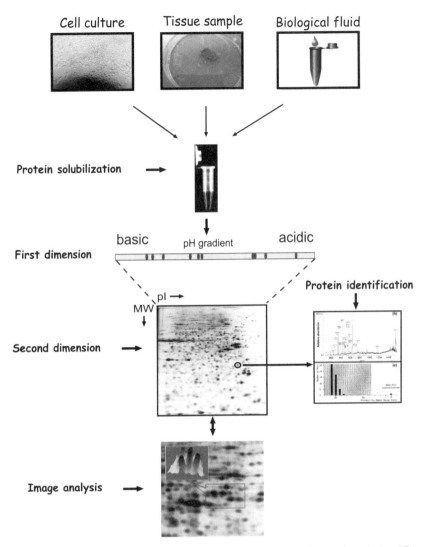

Figure 2.1. *Typical gel-based proteomic project scheme for generating and analyzing 2D protein images. (See color insert.)*

The originally developed lysis solution for 2D-PAGE (O'Farrell, 1975; O'Farrell et al., 1977) contains 9.5 M urea, 2% nonionic detergent (Nonidet P-40), 0.8–2.0% (w/v) ampholytes of various pH ranges, and 2% β-mercaptoethanol or DTT. This buffer has formed the basis for most of the subsequent developments of the 2D-PAGE solubilization procedure. This original formula gives satisfactory results for the majority of samples; however, some proteins are not properly dissolved by this solution, and special modifications are required. For example, highly hydrophobic proteins such as membrane proteins are difficult to solubilize and often streak during isoelectric focusing (Santoni et al., 1999, 2000). Recent developments such as the use of various chaotropes (thiourea) (Musante et al., 1998), zwitterionic detergents (Perdew et al., 1983; Adessi et al., 1997; Chevallet et al., 1998; Rabilloud et al., 1990, 1994, 1997), and the selective application of organic solvents (Deshusses et al., 2003) have improved the solubilization of hydrophobic proteins, although there is still much work to do. The efficiency of many nonionic and zwitterionic detergents as membrane protein solubilizers in 2D electrophoresis has recently been evaluated by Luche and coauthors (Luche et al., 2003), who emphasize that it is important to combine various sample treatments to resolve soluble as well as hydrophobic proteins (Santoni et al., 1999).

SDS, an anionic detergent, is the most effective surfactant to solubilize and extract proteins while simultaneously inactivating proteolytic enzymes. Unfortunately, the use of such charge-shifting reagents in 2D-PAGE is limited as the first dimension (IEF) in the separation of proteins based on their charge. The effect of SDS, however, can be diminished by using an excess of nonionic detergent (final ratio should be at least $8:1$), and by reducing the SDS final concentration to 0.25% (Ames and Nikado, 1976; Harder et al., 1999). The latter can be achieved by diluting the SDS-protein sample with a nonionic surfactant (NP-40, Triton X-100), an operation that concomitantly decreases the final concentration of the protein and therefore limits the amount that can be loaded to the first-dimension gel (Molloy, 2000). Accordingly, this recipe is restricted mainly to analytical applications (Harder et al., 1999).

Dithiothreitol, the most commonly used reducing agent for 2D-PAGE, is charged and is eliminated out of the pH gradient during IEF, resulting in decreased solubility of some proteins. It has been reported that tributyl phosphine (TBP) may be more effective for protein solubilization than traditional reducing agents such as β-mercaptoethanol or DTT (Rabilloud et al., 1997; Herbert et al., 1998), although it is rather unstable in aqueous solutions (Simpson, 2003).

Nonprotein contaminants, such as small ionic molecules, nucleic acids, and lipids, may interfere with protein separation in the first dimension as well as with subsequent protein staining. These contaminants can be eliminated by including additional steps such as pre-gel organic solvent protein precipitation (i.e., to remove lipids), dialysis (i.e., desalting of sample), and nuclease treatment (i.e., to remove DNA and RNA). It should be emphasized that sample solubilization depends very much on the type of sample being analyzed, and therefore the composition of the lysis solution should carefully be optimized in each case to provide useful separations. In general, sample preparation is easier for cultured cells than for tissue samples, as the latter are much more complex and adequate procedures have not yet been developed in a systematic fashion.

2.3.2 First Dimension: Available Systems

Today, 2D-PAGE can be carried out using traditional conventional IEF gels or IPG strips. In the first case, the pH gradient is generated and maintained by special amphoteric compounds, carrier ampholytes that migrate and stack according to their pI when an electric field is applied. At the beginning of electrophoresis, IEF gels have a uniform pH, corresponding to the average pI of the different carrier ampholytes used, and the gradient is made by prerunning the gels. The amphoteric nature of protein molecules causes them to migrate in a continuous pH gradient until their net charge is zero. Proteins can also be separated by using NEPHGE, but neither the pH gradient nor the proteins reach the equilibration state in this case.

Immobilized pH gradient strips, on the other hand, are an integral part of polyacrylamide matrix. This integration is achieved by copolymerization of a set of nonamphoteric buffering species with various pK values within the fibers of a gel (Bjellqvist et al., 1982). Commercial precast IPG strips of various ranges of pH (3–10, 4–6, 5–7, 8–9, and others) and different lengths (6–18 cm) are currently available commercially. When the electric field is applied, only the protein molecules and any unbound ions migrate in the electric field. Both carrier ampholytes and IPGs are widely used in many laboratories; however, the latter have the advantage of avoiding cationic drift and the near-isoelectric protein precipitation that results from the low ionic strength of amphoteric buffers, which can induce protein smearing. In addition, IPGs allow higher sample loads of up to 10 mg of crude protein sample without loss of resolution (Bjellqvist et al., 1993).

Conventional IEF and NEPHGE To date, a large collection of recipes for running various types of narrow and broad pH gradients have been published (Celis et al., 1997). Here, we briefly describe the protocols that have been used for many years in our laboratory to perform proteomic studies of cultured cells and tissues as well as for establishing 2D-PAGE protein databases (Celis et al., 1998; http://proteomics.cancer.dk).

To run IEF or NEPHGE gels using carrier ampholytes, gels are polymerized in thin tubes (inner diameter of about 0.2 cm) of 10–14 cm in length in the presence of various combinations of carrier ampholytes encompassing the required pH range. Isoelectric focusing gels contain 4% polyacrylamide, 9.5 M urea, 2% NP-40, 4% carrier ampholytes of pH range 5–7, and 1.3% carrier ampholytes of pH range 3–10. Following polymerization, the tubes are inserted into the upper chamber (cathode, 20 mM NaOH). The bottom chamber (anode) is filled with 10 mM H_3PO_4. Isoelectric focusing gels are usually prerun at room temperature for 15 min at 200 V, 30 min at 300 V, and 60 min at 400 V. After prerunning, the protein samples are loaded onto the gel tubes, and IEF is carried out for 19 h at 400 V.

Nonequilibrium pH gradient electrophoresis (NEPHGE) gels contain 4% polyacrylamide, 9.5 M urea, 2% NP-40, 2.3% carrier ampholytes of pH range 7–9, and 0.3% carrier ampholytes of pH range 9–11. These gels are not prerun, and the proteins migrate from the anode (10 mM H_3PO_4) to the cathode (20 mM NaOH) for 4.5 h at room temperature. Gels are then removed from the tubes with the aid of a syringe and placed in an appropriate volume of equilibration solution (0.06 M Tris-HCl [pH 6.8], 2% SDS, 100 mM DTT, and 10% glycerol) for 10 min. Gels can be applied directly to the second dimension or kept at −20 °C until use.

Immobilized pH Gradients (IPGs) The 2D-PAGE technology based on IPGs has been well described in reviews by Görg (1999) and Görg and Weiss (2000). Immobilized pH gradient slab gels of a chosen pH gradient can be cast with the aid of a standard two-vessel mixer using a density gradient to stabilize the Immobiline pH gradient. After polymerization, the IPG gels are washed extensively with deionized water and dried. Prior to running the first-dimensional separation, the desired number of dried IPG gel strips are cut from the slab gel and rehydrated to their original thickness in a rehydration solution containing 8 M urea, 0.5%

CHAPS, 15 mM DTT, and 0.2% Pharmalite (pH 3–10). Commercially available precast, dried IPG gel strips cover several pH ranges and provide quite reproducible 2D protein patterns. The rehydrated strips are placed on the cooling block of the electrofocusing chamber, and the samples are loaded into sample applicators placed on the surface of the IPG strip near the anode. Several variants of in-gel sample application have been described (Sanchez et al., 1997) that have improved the resolution and simplified the procedure. An important development has been the introduction of gel rehydration for loading protein samples into IPG strips. In this case, the dehydrated IPG strips are swollen directly in the rehydration solution that contains the protein sample (Rabilloud et al., 1994; Sanchez et al., 1997). The running conditions depend very much on the pH gradient and the length of the IPG strips used. After IEF, IPG strips are equilibrated twice in a solution containing 6 M urea, 0.05 M Tris-HCl, (pH 8.8), 2% SDS, and 30% glycerol for 15 min. During the second equilibration, 260 mM iodoacetamide is added to the equilibration buffer to remove excess DTT. The IPG strips that are not used immediately for a second dimension can be stored in a plastic bag at −80 °C for several months.

2.3.3 Second Dimension: SDS-PAGE

The second dimension separates proteins on the basis of molecular weight in a SDS acrylamide gel essentially as described by Laemmli (1970).

Application of the First-Dimension IEF/NEPHGE Tube Gels The tube gel is thawed, if frozen in equilibration buffer, laid on a piece of Parafilm, and placed carefully on the top of the vertical 15% slab gel (Celis et al., 1997). The tube gel is thereafter covered with melted 0.5–1% agarose solution containing 0.06 M Tris-HCl (pH 6.8), 2% SDS, 100 mM DTT, and 0.002% bromophenol blue. After cooling, the slab gel is placed in the electrophoresis apparatus.

Application of the First-Dimension IPG Strip Gels After equilibration (see above), IPG strips are placed on a piece of filter paper to remove excess equilibration buffer. The blotted IPG strips are placed gel side down onto the surface of the horizontal gradient SDS gel adjacent to the cathode (Görg and Weiss, 1997). Immobilized pH gradient strips can also be applied to the top of the vertical SDS gel in the same way as IEF and NEPHGE tube gels.

In our hands, broad range IPGs (3–10, Figure 2.2A) and carrier ampholytes (3.5–10, Figure 2.2B) resolve similar numbers of proteins (about 2500). This resolution is illustrated in Figure 2.2 with the separation of whole protein extracts from human keratinocytes labeled with [^{35}S]methionine (Celis et al., 1995; Celis and Gromov, 1999). Even though both formats yield similar results, confirmation of the identity of the proteins exhibiting similar mobility requires the use of standard protein identification techniques (Nawrocki et al., 1998). Immobilized pH gradient strips with very narrow ranges (i.e., less than 0.5 pH units, also referred as ultra zoom gels) can also be prepared and can resolve polypeptides having only slight differences in their electric charge (Wildgruber et al., 2000; Zuo and Speicher, 2002).

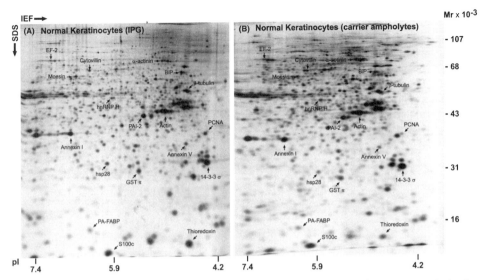

Figure 2.2. *Whole proteins expressed in noncultured normal human keratinocyte separated using (A) IPG (3–10) and (B) carrier ampholytes (3.5–10). A few proteins are indicated for reference (from Bjellqvist et al., 1994).*

Recently, major progress has been made to solve problems related to reproducibility and poor protein resolution in the basic pH range (Görg, 1999; Hoving et al., 2002). For example, Görg has improved considerably the separation of very basic proteins using wide-range immobilized pH 4–12 gradients that yield highly reproducible protein patterns with focused spots up to pH 12 (Görg, 1999). Basic proteins having a pI up to 11 can also be separated using carrier ampholytes with the NEPHGE format, as shown in Figure 2.3 (Celis et al., 1997; http://proteomics. cancer.dk).

2.4　PROTEIN VISUALIZATION

Protein spots resolved on 2D-PAGE gels can be visualized using a variety of methods, including organic dye staining, silver nitrate, isotope labeling, immunoblotting, overlay procedures, and several others.

2.4.1　Organic Dye Staining

Organic dyes such as Coomassie Blue R (GBR) and G types (GBG) are often used to visualize proteins separated by 1D- or 2D-PAGE. Coomassie Blue is easy to use, shows linearity of detection over a 10–50-fold range of concentration, and is highly compatible with downstream protein identification methods such as MS (Patton, 2002; Simpson, 2003 and references therein). The low sensitivity of the Coomassie Blue dyes (8–10 ng/spot and 40–100 ng/spot for conventional and for colloidal stains, respectively), however, limits their applicability to the visualization

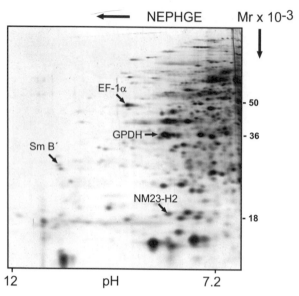

Figure 2.3. *Two-dimensional patterns of whole protein extracts from a breast tumor. Proteins were separated by 2D-PAGE–NEPHGE and visualized by silver staining. The identity of a few proteins is indicated for reference.*

of only the most abundant proteins. Several improvements, however, have been described to decrease the background and to increase the sensitivity (Neuhoff et al., 1988; 1990).

2.4.2 Silver Stain-Based Methods

Switzer and coauthors introduced silver staining in 1979, a technique that today provides a very sensitive tool for protein visualization with a detection level down to the 0.3–10 ng level (Switzer et al., 1979). Silver staining protocols can be divided into two general categories: (i) silver amine or alkaline methods and (ii) silver nitrate or acidic methods (Merril, 1990). The silver amine or alkaline methods usually have lower background and, as a result, are more sensitive but require extended procedures. Acidic protocols, on the other hand, are faster but slightly less sensitive (Mortz et al., 2001; Sorensen et al., 2002).

   Until recently, most of the silver staining protocols used glutaraldehyde-based sensitizers in the fixing and sensitization step, which introduced crosslinking of lysine residues within protein chains affecting MS analysis by hampering trypsin digestion. These reagents also reduced protein extraction from the gel (Rabilloud, 1990). To overcome this drawback, several modifications of the silver nitrate staining have been developed for visualizing proteins that can subsequently be digested, can be recovered from the gel, and are compatible with MS analysis (Shevchenko et al., 1996a; Yan et al., 2000). A representative 2D-PAGE gel (NEPHGE) of proteins from breast carcinomas stained with such a procedure is shown in Figure 2.3.

In general, silver staining is a rather complex multistep procedure and gel-to-gel reproducibility may be problematic (Quadroni and James, 1999). It is necessary to remember that not all proteins are equivalently stained by this technique. For example, several classes of highly negative charged proteins, including proteogly-cans and mucins that contain high levels of sulfated sugar residues as well as some very acidic proteins, are poorly detected by this staining method (Goldberg and Warner, 1997). Also, the linear dynamic range of silver stain is restricted to approx-imately a tenfold range, which makes this method unsuitable for quantitative analysis.

2.4.3 Radioactive Labeling Methods

Silver staining can detect only moderate and high abundance proteins, while metabolic labeling with amino acids labeled with specific isotopes may reveal less abundant components. Radiolabeling is generally carried out by incorporating radiolabeled amino acids into proteins expressed within cultured cells and fresh tissue samples (Celis et al., 1997). Radioactive label-based detection may be applied in parallel with fluorescent dye or silver staining to simultaneously assess total protein expression levels along with protein synthesis rate, as well as to locate low abundance proteins on a preparative loaded stained gel (Westbrook et al., 2001).

Cells or tissue samples can be labeled with [^{35}S]methionine, [^{35}S]cysteine, or a [^{14}C]amino acid mixture under conditions that allow the detection of approximately 3000–4000 proteins within a standard size of 2D gel autoradiogram (Figure 2.2). The lowest level of detection for [^{35}S]methionine-labeled proteins corresponds to polypeptides that are present in about 40,000 or more copies per cell. A brief overview of autoradiography of 2D gels can be found in the article by Link (1999). More accurate quantitative data, however, can be obtained by using a storage phosphor imager that provides higher detection sensitivity and a larger dynamic range.

Limitations of the radiolabeling approach include (i) lack of labeling of some proteins due to low turnover rates and (ii) problems associated with safety regula-tions and disposal. In addition, it has been shown that incubation of cells with [^{35}S]methionine under even standard doses and experimental conditions may lead to DNA damage, increased p53 levels, cell morphology alterations, and aberrant cell cycle arrest and/or apoptosis (Yeargin and Hass, 1995; Hu and Heikka, 2000; Hu et al., 2001).

2.4.4 Fluorescence-Based Staining Methods

Highly sensitive fluorescence-based protein detection techniques use dyes from the SYPRO family. Fluorophors such as SYPRO Orange and SYPRO Red exhibit low variations in fluorescent intensity between different proteins and have a wide linear range of detection between intensity and protein volume in the spot. This wide linear range provides accurate quantitation of both high and low abundance pro-teins (Steinberg et al., 1996; Patton, 2002). Both SYPRO dyes are commercially available and have been used to analyze whole protein lysates from bacterial and mammalian cells. Their sensitivity is comparable to that of silver nitrate staining, but not as high as radiolabeling (Steinberg et al., 1996).

SYPRO Ruby is a highly sensitive, ruthenium-based metal chelate fluorescent stain that can accurately quantitate protein expression levels and is compatible with standard fluorescent visualization systems and downstream identification techniques such as MS (Lopez et al., 2000; Patton, 2000, 2001, 2002). SYPRO Ruby detects 20% more proteins as compared to silver nitrate, is more reproducible, and has a linear dynamic range that covers three orders of magnitude (Lopez et al., 2000). Some additional advantages over silver staining include short staining time and low background, and the gels do not need to be fixed prior to staining. However, the dye is rather expensive and requires a fluorescence scanner for low level detection (1 ng protein/spot).

Fluorescence technologies also offer the possibility for multicolor labeling and detection. A pair of two succinimidyl ester derivatives of the fluorescent cyanin dyes, Cy3 and Cy5, which exhibit different excitation and emission spectra, are being used in a new modification of the 2D-PAGE technique, termed differential gel electrophoresis (DIGE). This staining technique allows quantitative differences in protein abundances between two samples to be measured on a single gel (Unlu et al., 1997; Gharbi et al., 2002). Briefly, two protein samples are labeled separately in vitro with the fluorophors, mixed, and run within the same 2D-PAGE gel. The protein pattern is revealed using a laser scanner, equipped with appropriate excitation/emission filters, that generates two separate images, one for each of the Cy3- and Cy5-labeled proteomes, that are then analyzed by a computer-assisted overlay method (Gharbi et al., 2002). A further development of this technology that incorporates an internal pooled standard labeled with a third fluorophor, Cy2, has also been developed (Alban et al., 2003).

2.4.5 Detection of Specific Groups of Proteins

Even though everyday research requires detection techniques that can be applied routinely to a large number of resolved proteins, there are also several methods suitable for detecting specific groups of proteins that share common functional features.

Ligand Blot Overlay   Protein blot overlay assays (also known as "Far-Western," "Western–Western," "ligand," or "affinity" blotting) are very powerful techniques for detecting and analyzing proteins or protein motifs involved in cellular targeting processes. These methods are based on the fact that proteins or protein fragments resolved by electrophoresis and transferred to an immobilizing matrix such as nitrocellulose or a nylon membrane can be probed with putative binding partners followed by subsequent detection of the complexes. Many probes can be used including metal ions, nucleotides, nucleic acids, hormones, proteins, antibodies, viruses, and cells.

Overlay procedures have been applied successfully to a wide range of protein–protein interactions, such as kinase anchoring (Hausken et al., 1998) and ligand–protein interactions using Ca^{2+} (Hoffmann et al., 1998), GTP (Gromov and Celis, 1997), and nucleic acids (Dejgaard and Celis, 1997). Subfamilies of small GTP-binding proteins expressed in various human cells and tissues as revealed by the [32P]orthophosphate-labeled GTP overlay are portrayed in Figure 2.4. Drawbacks

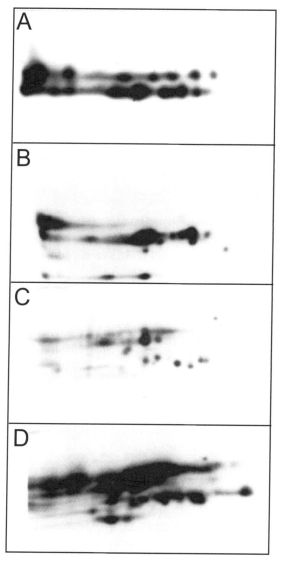

Figure 2.4. *Small GTP-binding proteins expressed in several human cells and tissues. (A) Differentiated human keratinocytes, (B) SV40-transformed (K14) human keratinocytes, (C) fetal lung, and (D) fetal brain as revealed by the [α³²P]GTP overlay technique. Proteins were separated by 2D-PAGE (IEF), electroblotted onto a nitrocellulose membrane, and overlayed with 1 μCi of [α³²P]GTP per milliliter. Only fractions of the 2D gel blot ³²P-autoradiographs are shown.*

of the method include (i) weak binding of some proteins to the immobilizing matrix; (ii) loss of reactivity due to denaturing effects; and (iii) the fact that some protein complexes are only formed in specific microenvironments that are difficult to reproduce in vitro. Despite these limitations, however, protein overlay can readily be used for confirming suspected interactions as well as for identifying new binding partners.

2D Gel Profiling of Post-translationally Modified Proteins Using in Vivo Isotope Labeling Two-dimensional detection of chemically co- and post-translationally altered protein variants can be made by in vivo labeling of the proteins with the corresponding isotope-labeled metabolite followed by its identification by MS. Once a radiolabeled ligand is covalently bound to the protein, it can be detected on a gel or in a blot by autoradiography or phosphor imaging. Two-dimensional gel profiles of phosphorylated and prenylated (farnesylated and geranylgeranylated) proteins expressed in transformed human amnion cells (AMA) are shown in Figures 2.5 and 2.6, respectively. Radiolabeled protein spots can be subjected to matrix-assisted laser desorption ionization time-of-flight (MALDI-TOF) mass fingerprinting or partial tandem MS peptide sequencing both for protein identification as well as to verify the type of modification.

Several new fluorescence-based methods have recently been introduced for detecting post-translational modified proteins. For example, a green fluorophor glycoprotein-specific stain, referred to as Pro-Q Emerald 300 and 488 dye, can detect as little as 300 pg of glycoprotein per spot depending on the degree of glycosylation (Steinberg et al., 2001; Hart et al., 2003). The Pro-Q Diamond phosphoprotein dye technology is suitable for the fluorescent detection of phosphoserine-, phosphothreonine-, and phosphotyrosine-containing proteins directly in 1D and 2D gels (Martin et al., 2003; Steinberg et al., 2003). The fluorescence signal intensity correlates with the number of phosphorylated residues on the protein and can detect as little as 1–2 ng of beta-casein, a pentaphosphorylated protein, and 8 ng of pepsin, a monophosphorylated protein. The technology is suitable for determining protein kinase and phosphatase substrate preference.

The multiplex proteomics approach is based on the serial staining of proteins using different technologies to detect various functional subgroups. This discovery platform generates several images from a single 2D gel and allows parallel determination of the total protein expression levels (silver stain, SYPRO Ruby), and subproteomic or altered post-translational modification patterns (Pro-Q Diamond, Pro-Q Emerald, in vivo metabolic radiolabeling, phosphoamino acid-selective antibodies) within a single 2D gel (Steinberg et al., 2003). The main advantage of multiplex detection is that it provides a broad platform of information and enhances matching of 2D multiple images as these are generated from the same gel.

2.5 IMAGE ANALYSIS

Protein profiles can be scanned and analyzed using appropriate software to search for changes in the levels of preexisting proteins and induction of new products as well as coregulated polypeptides. Moreover, using the phosphor-imaging technology, it is possible to enhance the sensitivity and linearity of detection. Various software packages for the analysis and processing of 2D gel protein images have been developed including PDQUEST (BioRad), MELANIE (Swiss Institute of Bioinformatics), ProXPRESS (PerkinElmer), PHORETIX (Nonlinear Dynam-

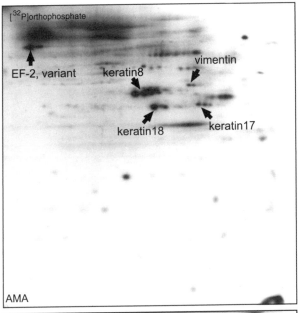

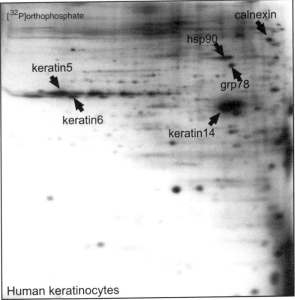

Figure 2.5. *Two-dimensional (IEF) autoradiographs of whole protein extracts from AMA cells (top) and human keratinocytes (bottom) labeled with [^{32}P]orthophosphate. Several phosphoproteins are indicated as references.*

ics), Z3 (Compugen), and Protein Mine (Schimagix). These software packages provide various levels of automation and speed. To our knowledge, none of the currently available software packages on the market, however, provide full automatic analysis of 2D gel images. In addition, there are still serious problems

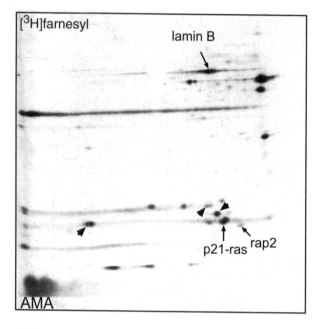

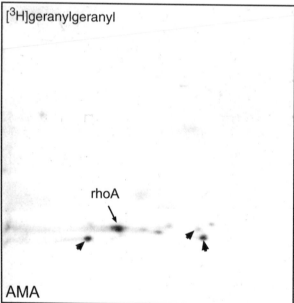

Figure 2.6. *Two-dimensional (IEF) fluorographs of whole protein extracts from AMA cells labeled with [³H]farnesyl-PP (top) and [³H]geranylgeranyl-PP (bottom). Several prenylated proteins are indicated with arrows. Proteins labeled with both isoprenoids are indicated with solid arrowheads.*

associated with the quantitative analysis of 2D-PAGE gels caused by a number of factors that include lack of reproducibility of the 2D-PAGE technology and problems associated with spot recognition and matching, as well as normalization of the background and spot intensities.

2.6 PROTEIN IDENTIFICATION

2.6.1 Mass Spectrometry

The technique of choice for protein identification after 2D-PAGE fractionation is MS, as it requires picomole levels of protein. There are several formats of this technology, namely, MALDI-TOF, quadrupole (Q)-TOF, electrospray (ESI)-TOF, and several others that have been developed during the last years with the goal of achieving sensitive and high-throughput analysis (Roepstorff, 2000; Wilm, 2000; Mann et al., 2001 and references therein). Two ionization techniques that allow production of gas-phase molecular ions of proteins or peptides either from the liquid phase (i.e., ESI) (Fenn et al., 1989) or from the solid phase (i.e., MALDI) (Karas and Hillenkamp, 1988) were developed in the late 1980s.

Both MALDI-TOF/MS and ESI/MS technologies are the methods of choice when a systematic comparison of multiple biological samples is required. These methods are highly complementary, but MALDI-TOF/MS is more straightforward and sensitive. It almost exclusively produces singly charged ions, can tolerate relatively high buffer and salt concentrations in the analyte mixture, and is usually the first method of choice in any protein study (Shevchenko et al., 1996b; Jensen et al., 1998). MALDI-TOF/MS is often used for the analysis of gel-separated proteins or for proteins that have been transferred to nitrocellulose membranes and probed with antibodies or specific overlay assays (Klarskov and Naylor, 2002). The later is very promising as most proteins are often found as multiple forms on 2D gels due to post-translational processing and chemical modifications (phosphorylation, glycosylation, methylation, acetylation, myristoylation, palmitoylation, sulfation, ubiquination, etc.) (Wilkins et al., 1999; Fryksdale et al., 2002; Mann et al., 2002). Modified forms can also be characterized by combining MS approaches with a number of biochemical procedures (Pandey and Mann, 2000 and references therein).

The process of identifying a 2D-PAGE separated protein is shown in Figure 2.7. MALDI-TOF/MS and ESI/MS are highly dependent on gene and protein bioinformatics as the precise mass of any given protein, peptide, or ion fragment is a unique feature that can be matched with the theoretically calculated mass obtained from sequence information available in protein, EST, or genomic databases. In gel-based proteomics, the protein of interest is excised from the gel and digested with a sequence-specific protease that produces a set of peptides that serves as a unique fingerprint (Patterson and Aebersold, 1995; Pappin, 1997). The experimentally obtained peptide masses are then compared to the calculated peptide masses resulting from the theoretical digestion of every full-length protein sequence entry in protein, cDNA, or genomic databases. Peptides can also be sequenced using tandem MS (MS/MS), a fact that allows the identification of unknown proteins.

Recently, Gygi and colleagues described a novel approach based on the isotope-coded affinity tag (ICAT) labeling of proteins that highly enhances an accurate quantification and identification of individual proteins in complex mixtures (Gygi et al., 1999). Smolka and coauthors (2002) have shown that proteins labeled with the isotopically heavy and normal forms of the reagent, respectively, precisely comigrate during 2D-PAGE gels. In addition, they have also shown that the ratio of protein abundance can accurately be determined from the relative MS signal

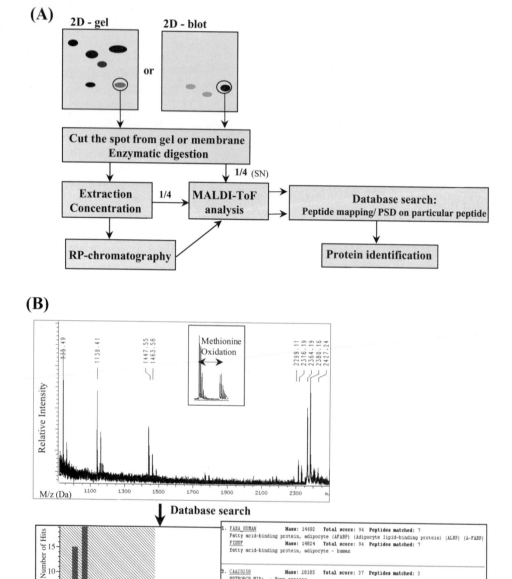

Figure 2.7. (A) Strategy for identifying proteins using MALDI-TOF/MS. (B) Identification of proteins excised from 2D gels followed by in situ digestion and MALDI-TOF analysis. The upper panel shows the MALDI-TOF mass spectrum. The pick list provides the input data for protein identification obtained by searching the MCDB database with the MASCOT software. The lower panels present the identification probability plot from the database search. Probability scores greater than 57 are considered significant (p < 0.05). (See color insert.)

intensities of the heavy and normal forms of the labeled cysteine-containing peptides.

Modern mass spectrometers can routinely analyze peptides at the femtomole level (Wilm et al., 1996; Stensballe and Jensen, 2001) with a mass accuracy of 0.01–0.1% (Takach et al., 1997; Gobom et al., 2002). Lately, a number of developments have been described that minimize even more the amount of starting material needed and that have pushed the sensitivity to the low attomole range. These innovations make use of nanoliter bed volume reversed-phase columns of different configurations, alone or in combination with prestructured sample supports (Schuerenberg et al., 2000; Johnson et al., 2001).

2.6.2 Western Immunoblotting

Western immunodetection is an invaluable tool for the identification of proteins, in particular, low abundance polypeptides. For example, we have shown that 2D-PAGE immunoblotting in combination with enhanced chemoluminescence (ECL) can detect as little as 100–500 molecules per cell in unfractionated extracts of cultured cells if highly specific antibodies are available (Celis and Gromov, 2000). The procedure also reveals post-translational modification(s) (Figure 2.8) and can effectively complement MS and metabolic labeling with specific radioactive precursors to determine the nature of the modification (see above). Several high-sensitivity substrates, such as the SuperSignal West substrates (Pierce Biotechnology, Inc.), Lumigen PS-3, and TMA-6 (Lumigen, Inc.), have been developed to enhance the ECL sensitivity and are particularly useful for detecting very low abundance proteins or for cases in which limited amounts of antibody are available.

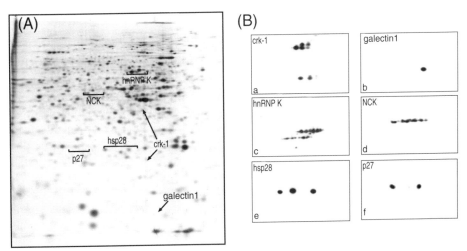

Figure 2.8. Two-dimensional ECL immunoblots of proteins from a bladder transitional cell carcinoma labeled with [^{35}S]-methionine (A) and reacted with antibodies (B) against (a) crk-1, (b) galectin1, (c) hnRNP K, (d) NCK, (e) hsp28, and (f) p27.

It should be emphasized that antibodies are invaluable to validate expression data obtained by proteomics and cDNA array approaches (Celis and Gromov, 2003; Celis et al., 2003). In general, the strategy is based on immunohistochemistry and is critically important when gel-based proteomics is used for the analysis of complex tissue samples composed of different cell types.

2.7   PRE-GEL FRACTIONATION

Pre-gel fractionation procedures may enhance the detection of low abundance proteins that are often missed in unfractionated 2D gel separations. A complex protein sample can be fractionated to yield several fractions of lower complexity, each of which is analyzed separately by 2D-PAGE (Corthals et al., 1997; Molloy et al., 1998; Ramsby et al., 1994).

Several procedures that utilize isoelectric prefractionation on immobilized pH gradients have been described (Herbert and Righetti, 2000; Righetti et al., 2001; Zuo and Speicher, 2000, 2002). Extraction procedures using Triton X-114 and alkaline media in combination with sequential extraction by different zwitterionic detergents have also been used to enhance the separation and visualization of soluble and hydrophobic proteins (Santoni et al., 1999). Subcellular fractionation is another promising approach by which sample complexity can be reduced. A systematic identification of the protein components of cell organelles is currently underway and has been reviewed elsewhere (Jung et al., 2000).

Even though prefractionation prior to 2D-PAGE appears to be a highly promising strategy, the application of this approach to a large number of samples as required for large scale projects in translational cancer research may be limited (Celis and Gromov, 2003; Celis et al., 2003). Also, fractionation of clinical relevant tissue samples may be restricted by the size of the biopsy, and additional fractionation steps may introduce intersample variability and adversely affect quantitative measurements.

2.8   TWO-DIMENSIONAL POLYACRYLAMIDE GEL ELECTROPHORESIS PROTEIN DATABASES

By carrying out 2D-PAGE studies in a systematic fashion, it is possible to establish proteomic databases that link protein and DNA mapping and sequence information and offer an effective route to drug discovery by pinpointing signaling pathways and components that are deregulated in particular diseases. Such a systematic analysis requires, in addition to a reproducible 2D gel system, computer-assisted technology to scan the gels, make synthetic images, assign numbers to individual spots, and match spots as well as enter and retrieve information (Panek and Vohradsky, 1999; Pleissner et al., 1999; Smilansky, 2001; Dowsey et al., 2003). A number of 2D protein databases are available via the World Wide Web that contain information on a variety of cells, tissues, and body fluids (http://proteomics.cancer.dk; SWISS-2DPAGE; SIENA-2DPAGE, among others). A full list of 2D protein databases available online can be found on the WORLD-2DPAGE Index http://us.expasy.org/ch2d/2d-index.html.

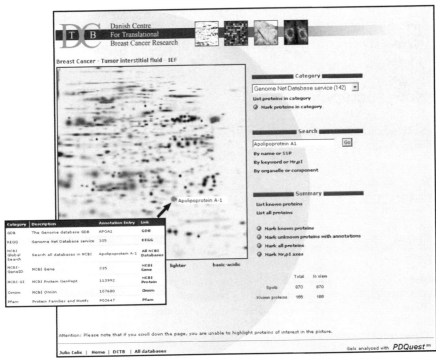

Figure 2.9. *Master synthetic image of breast tumor interstitial fluid proteins separated by IEF 2D-PAGE as depicted on the Internet (http://proteomics.cancer.dk). By clicking on any spot, it is possible to open a file that contains protein information available in the database as well as links to other related Web sites. Only part of the file for Apolipoprotein A1 is shown.*

Our laboratory has established several comprehensive 2D gel protein databases that focus on skin biology (human keratinocyte databases), bladder cancer (TCCs and SCCs) and breast cancer (tumor interstitial fluid and adipose tissue) (http://proteomics.cancer.dk; Figure 2.9). With the integrated approach offered by these databases, it will be possible to reveal and identify phenotype-specific proteins and to study regulatory properties and function of groups of proteins that are co-coordinately expressed in both health and disease.

2.9 CONCLUSION

Despite the development of many promising parallel technologies in proteomics, 2D-PAGE will most likely remain in extensive use for the foreseeable future. Various aspects of the 2D-PAGE technology including resolution, sample preparation, detection, and image analysis for supporting gel comparisons and databasing, however, will continue to be improved. Further development should also be expected in the design of new computerized scanning devices and densitometers. As the number of 2D gel images available on the Internet expands, algorithms and software for scanning will be created that will make it possible to compare directly on the Internet 2D gel patterns generated across different experiments and sites.

ACKNOWLEDGMENTS

This work was supported by a grant from the Danish Cancer Society and from the Danish Medical Research Council.

REFERENCES

Adessi C, Miege C, Albrieux C, Rabilloud T. 1997. Two-dimensional electrophoresis of membrane proteins: a current challenge for immobilized pH gradients. *Electrophoresis* 18:127–135.

Alban A, David SO, Bjorkesten L, Andersson C, Sloge E, Lewis S, Currie I. 2003. A novel experimental design for comparative two-dimensional gel analysis: two-dimensional difference gel electrophoresis incorporating a pooled internal standard. *Proteomics* 3:36–44.

Ames GF, Nikaido K. 1976. Two-dimensional gel electrophoresis of membrane proteins. *Biochemistry* 15:616–623.

Bjellqvist B, Ek K, Righetti PG, Gianazza E, Görg A, Westermeier R, Postel W. 1982. Isoelectric focusing in immobilized pH gradients: principle, methodology and some applications. *J Biochem Biophys Methods* 6:317–339.

Bjellqvist B, Sanchez JC, Pasquali C, Ravier F, Paquet N, Frutiger S, Hughes GJ, Hochstrasser D. 1993. Micropreparative two-dimensional electrophoresis allowing the separation of samples containing milligram amounts of proteins. *Electrophoresis* 14:1375–1378.

Bjellqvist B, Basse B, Olsen E, Celis JE. 1994. Reference points for comparisons of two-dimensional maps of proteins from different human cell types defined in a pH scale where isoelectric points correlate with polypeptide compositions. *Electrophoresis* 15:529–539.

Cash P, Kroll JS. 2003. Protein characterization by two-dimensional gel electrophoresis. *Methods Mol Med* 71:101–118.

Celis JE, Bravo R (Eds). 1984. *Two-Dimensional Gel Electrophoresis of Proteins: Methods and Applications.* Academic Press, New York.

Celis JE, Gromov PS. 1999. 2D protein electrophoresis: Can it be perfected? *Curr Opin Biotechnol* 10:16–21.

Celis JE, Gromov P. 2000. High-resolution two-dimensional gel electrophoresis and protein identification using Western blotting and ECL detection. *EXS* 88:55–67.

Celis JE, Gromov P. 2003. Proteomics in translational cancer research: toward an integrated approach. *Cancer Cell* 3:9–15.

Celis JE, Rasmussen HH, Leffers H, Madsen P, Honore B, Gesser B, Dejgaard K, Vanderkerckhove J. 1991. Human cellular protein patterns and their link to genome DNA sequence data: usefulness of two-dimensional gel electrophoresis and microsequencing. *FASEB J* 5:2200–2208.

Celis JE, Rasmussen HH, Gromov P, Olsen E, Madsen P, Leffers H, Honore B, Dejgaard K, Vorum H, Kristensen DB, Ostergaard M, Haunso A, Jensen NA, Celis A, Basse B,

Lauridsen JB, Ratz GP, Anderson AH, Walbum E, Kjaergaard I, Andersen I, Puype M, Van Damme J, Vanderkerckhove J. 1995. The human keratinocyte two-dimensional gel protein database (update 1995): mapping components of signal transduction pathways. *Electrophoresis* 16:2177–2240.

Celis JE, Ratz G, Basse B, Lauridsen JB, Celis A, Jensen NA, Gromov P. 1997. High-resolution two-dimensional gel electrophoresis of proteins: isoelectric focusing (IEF) and nonequilibrium pH gradient electrophoresis (NEPHGE). In: Celis JE, Carter N, Hunter T, Shotton D, Simons K, Small JV (Eds), *Cell Biology. A Laboratory Handbook*, Vol. 4, Academic Press, New York, pp 375–385.

Celis JE, Østergaard M, Jensen NA, Gromova I, Rasmussen HH, Gromov P. 1998. Human and mouse proteomic databases: novel resources in the protein universe. *FEBS Lett* 430:64–72.

Celis JE, Gromov P, Gromova I, Moreira JM, Cabezon T, Ambartsumian N, Grigorian M, Lukanidin E, Thor Straten P, Guldberg P, Bartkova J, Bartek J, Lukas J, Lukas C, Lykkesfeldt A, Jaattela M, Roepstorff P, Bolund L, Orntoft T, Brunner N, Overgaard J, Sandelin K, Blichert-Toft M, Mouridsen H, Rank FE. 2003. Integrating proteomic and functional genomic technologies in discovery-driven translational breast cancer research. *Mol Cell Proteomics* 2:369–377.

Chevallet M, Santoni V, Poinas A, Rouquié D, Fuchs A, Kieffer S, Rossignol M, Lunardi J, Garin J, Rabilloud T. 1998. New zwitterionic detergents improve the analysis of membrane proteins by two-dimensional electrophoresis. *Electrophoresis* 19:1901–1909.

Claverie J-M. 2001. What if there are only 30,000 human genes? *Science* 291:1255–1257.

Corthals GL, Molloy MP, Herbert BR, Williams KL, Gooley AA. 1997. Prefractionation of protein samples prior to two-dimensional electrophoresis. *Electrophoresis* 18:317–323.

Dejgaard K, Celis JE. 1997. Two-dimensional Northwestern blotting. In: Celis JE, Carter N, Hunter T, Shotton D, Simons K, Small JV (Eds), *Cell Biology. A Laboratory Handbook*, Vol. 4, Academic Press, New York, pp 482–486.

Deshusses JM, Burgess JA, Scherl A, Wenger Y, Walter N, Converset V, Paesano S, Corthals GL, Hochstrasser DF, Sanchez JC. 2003. Exploitation of specific properties of trifluoroethanol for extraction and separation of membrane proteins. *Proteomics* 3:1418–1424.

Dowsey AW, Dunn MJ, Yang GZ. 2003. The role of bioinformatics in two-dimensional gel electrophoresis. *Proteomics* 3:1567–1596.

Duncan R, McConkey EH. 1982. How many proteins are there in a typical mammalian cell? *Clin Chem* 28:749–755.

Fenn JB, Mann M, Meng CK, Wong SF, Whitehouse CM. 1989. Electrospray ionization for MS of large biomolecules. *Science* 246:64–71.

Fey SJ, Larsen PM. 2001. 2D or not 2D. *Curr Opin Chem Biol* 5:26–33.

Fryksdale BG, Jedrzejewski PT, Wong DL, Gaertner AL, Miller BS. 2002. Impact of deglycosylation methods on two-dimensional gel electrophoresis and matrix assisted laser desorption/ionization-time of flight-MS for proteomic analysis. *Electrophoresis* 23:2184–2193.

Gharbi S, Gaffney P, Yang A, Zvelebil MJ, Cramer R, Waterfield MD, Timms JF. 2002. Evaluation of two-dimensional differential gel electrophoresis for proteomic expression analysis of a model breast cancer cell system. *Mol Cell Proteomics* 1:91–98.

Gobom J, Mueller M, Egelhofer V, Theiss D, Lehrach H, Nordhoff E. 2002. A calibration method that simplifies and improves accurate determination of peptide molecular masses by MALDI-TOF MS. *Anal Chem* 74:3915–3923.

Goldberg HA, Warner KJ. 1997. The staining of acidic proteins on polyacrylamide gels: enhanced sensitivity and stability of "Stains-all" staining in combination with silver nitrate. *Anal Biochem* 251:227–233.

Gooley AA, Packer NH. 1997. The importance of protein co- and post-translational modifications in proteome projects. In: Wilkins MR, Williams KL, Appel RD, Hochstrasser DF (Eds), *Proteome Research: New Frontiers in Functional Genomics* (*Principles and Practice*), Springer Verlag, New York.

Görg A. 1999. IPG-Dalt of very alkaline proteins. *Methods Mol Biol* 112:197–209.

Görg A, Weiss W. 1997. High-resolution two-dimensional electrophoresis of proteins using immobilised pH gradient. In: Celis JE, Carter N, Hunter T, Shotton D, Simons K, Small JV (Eds), *Cell Biology. A Laboratory Handbook*, Vol. 4, Academic Press, New York. pp 386–397.

Gromov PS, Celis JE. 1994. Several small GTP-binding proteins are strongly down-regulated in simian virus 40 (SV40) transformed human keratinocytes and may be required for the maintenance of the normal phenotype. *Electrophoresis* 15:474–481.

Gromov P, Celis JE. 1997. Blot overlay assay for the identification of GTP-binding proteins. In: Celis JE, Carter N, Hunter T, Shotton D, Simons K, Small JV (Eds), *Cell Biology. A Laboratory Handbook*, Vol. 4, Academic Press, New York. pp 454–457.

Gromov P, Ostergaard M, Gromova I, Celis JE. 2002. Human proteomic databases: a powerful resource for functional genomics in health and disease. *Prog Biophys Mol Biol* 80:3–22.

Gygi SP, Rist B, Gerber SA, Turecek F, Gelb MH, Aebersold R. 1999. Quantitative analysis of complex protein mixtures using isotope-coded affinity tags. *Nat Biotechnol* 17:994–999.

Gygi SP, Corthals GL, Zhang Y, Rochon Y, Aebersold R. 2000. Evaluation of two-dimensional gel electrophoresis-based proteome analysis technology. *Proc Natl Acad Sci USA* 97:9390–9395.

Harder A, Wildgruber R, Nawrocki A, Fey SJ, Larsen PM, Görg A. 1999. Comparison of yeast cell protein solubilization procedures for two-dimensional electrophoresis. *Electrophoresis* 20:826–829.

Hart C, Schulenberg B, Steinberg TH, Leung WY, Patton WF. 2003. Detection of glycoproteins in polyacrylamide gels and on electroblots using Pro-Q Emerald 488 dye, a fluorescent periodate Schiff-base stain. *Electrophoresis* 24:588–598.

Hausken ZE, Coghlan VM, Scott JD. 1998. Overlay, ligand blotting and band-shift techniques to study kinase anchoring. In: Clegg RA (Ed), *Protein Targeting Protocols. Methods in Molecular Biology*, Vol. 88, Humana Press, Totowa, NJ, pp 47–64.

Herbert B, Righetti PG. 2000. A turning point in proteome analysis: sample prefractionation via multicompartment electrolyzers with isoelectric membranes. *Electrophoresis* 21:3639–3648.

Herbert BR, Molloy MP, Gooley AA, Walsh BJ, Bryson WG, Williams KL. 1998. Improved protein solubility in two-dimensional electrophoresis using tributyl phosphine as a reducing agent. *Electrophoresis* 19:845–851.

Herbert BR, Harry JL, Packer NH, Gooley AA, Pedersen SK, Williams KL. 2001. What place for polyacrylamide in proteomics? *Trends Biotechnol* 19:S3–S9.

Hoffmann HJ, Gromov P, Celis JE. 1998. Calcium overlay assay. In: Celis JE, Carter N, Hunter T, Shotton D, Simons K, Small JV (Eds), *Cell Biology. A Laboratory Handbook*, Vol. 4, Academic Press, New York, pp 450–453.

Hoving S, Gerrits B, Voshol H, Muller D, Roberts RC, van Oostrum J. 2002. Preparative two-dimensional gel electrophoresis at alkaline pH using narrow range immobilized pH gradients. *Proteomics* 2:127–134.

Hu VW, Heikka DS. 2000. Radiolabeling revisited: metabolic labeling with (35)S-methionine inhibits cell cycle progression, proliferation, and survival. *FASEB J* 14:448–454.

Hu VW, Heikka DS, Dieffenbach PB, Ha L. 2001. Metabolic radiolabeling: experimental tool or Trojan horse? (35)S-methionine induces DNA fragmentation and p53-dependent ROS production. *FASEB J* 15:1562–1568.

Jensen ON, Larsen MR, Roepstorff P. 1998. Mass spectrometric identification and micro-characterization of proteins from electrophoretic gels: strategies and applications. *Proteins Suppl* 2:74–89.

Johnson T, Bergquist J, Ekman R, Nordhoff E, Schurenberg M, Kloppel KD, Muller M, Lehrach H, Gobom J. 2001. A CE-MALDI interface based on the use of prestructured sample supports. *Anal Chem* 73:1670–1675.

Jung E, Heller M, Sanchez JC, Hochstrasser DF. 2000. Proteomics meets cell biology: the establishment of subcellular proteomes. *Electrophoresis* 21:3369–3377.

Karas M, Hillenkamp F. 1988. Laser desorption ionization of proteins with molecular masses exceeding 10,000 daltons. *Anal Chem* 60:2299–2301.

Klarskov K, Naylor S. 2002. India ink staining after sodium dodecylsulfate polyacrylamide gel electrophoresis and in conjunction with Western blots for peptide mapping by matrix-assisted laser desorption/ionization time-of-flight MS. *Rapid Commun Mass Spectrom* 16:35–42.

Klose J. 1975. Protein mapping by combined isoelectric focusing and electrophoresis of mouse tissues. A novel approach to testing for induced point mutations in mammals. *Humangenetik* 26:231–243.

Labrador M, Mongelard F, Plata-Rengifo P, Baxter EM, Corces VG, Gerasimova TI. 2001. Protein encoding by both DNA strands. *Nature* 409:1000.

Laemmli UK. 1970. Cleavage of structural proteins during the assembly of the head of bacteriophage T4. *Nature (London)* 227:680–685.

Lilley KS, Razzaq A, Dupree P. 2002. Two-dimensional gel electrophoresis: recent advances in sample preparation, detection and quantitation. *Curr Opin Chem Biol* 6:46–50.

Lin D, Tabb DL, Yates JR III. 2003. Large-scale protein identification using MS. *Biochim Biophys Acta* 1646:1–10.

Link AJ, Eng J, Schieltz DM, Carmack E, Mize GJ, Morris DR, Garvik BM, Yates JR III. 1999. Direct analysis of protein complexes using MS. *Nat Biotechnol* 17:676–682.

Link AJ. 1999. Autoradiography of 2-D gels. *Methods Mol Biol* 112:289–290.

Lopez MF, Berggren K, Chernokalskaya E, Lazarev A, Robinson M, Patton W. 2000. A comparison of silver stain and SYPRO Ruby protein gel stain with respect to protein detection in two-dimensional gels and identification by peptide mass profiling. *Electrophoresis* 21:3673–3683.

Luche S, Santoni V, Rabilloud T. 2003. Evaluation of nonionic and zwitterionic detergents as membrane protein solubilizers in two-dimensional electrophoresis. *Proteomics* 3:249–253.

Mann M, Hendrickson RC, Pandey A. 2001. Analysis of proteins and proteomes by MS. *Annu Rev Biochem* 70:437–473.

Mann M, Ong SE, Gronborg M, Steen H, Jensen ON, Pandey A. 2002. Analysis of protein phosphorylation using MS: deciphering the phosphoproteome. *Trends Biotechnol* 20:261–268.

Martin K, Steinberg TH, Cooley LA, Gee KR, Beechem JM, Patton WF. 2003. Quantitative analysis of protein phosphorylation status and protein kinase activity on microarrays using a novel fluorescent phosphorylation sensor dye. *Proteomics* 3:1244–1255.

Merril CR. 1990. Silver staining of proteins and DNA. *Nature* 343:779–780.

Molloy MP. 2000. Two-dimensional electrophoresis of membrane proteins using immobilized pH gradients. *Anal Biochem* 280:1–10.

Molloy MP, Herbert BR, Walsh BJ, Tyler MI, Traini M, Sanchez JC, Hochstrasser DF, Williams KL, Gooley AA. 1998. Extraction of membrane proteins by differential solubilization for separation using two-dimensional gel electrophoresis. *Electrophoresis* 19:837–844.

Mortz E., Krogh TN, Vorum H, Gorg A. 2001. Improved silver staining protocols for high sensitivity protein identification using matrix-assisted laser desorption/ionization-time of flight analysis. *Proteomics* 1:1359–1363.

Musante L, Candiano G, Ghiggeri GM. 1998. Resolution of fibronectin and other characterized proteins by 2-D-PAGE with thiourea. *J Chromatogr B* 705:351–356.

Nawrocki A, Larsen MR, Podtelejnikov AV, Jensen ON, Mann M, Roepstorff P, Görg A, Fey SJ, Larsen PM. 1998. Correlation of acidic and basic carrier ampholyte and immobilized pH gradient two-dimensional gel electrophoresis patterns based on mass spectrometric protein identification. *Electrophoresis* 19:1024–1035.

Neuhoff V, Arold N, Taube D, Ehrhardt W. 1988. Improved staining of proteins in polyacrylamide gels including isoelectric focusing gels with clear background at nanogram sensitivity using Coomassie Brilliant Blue G-250 and R-250. *Electrophoresis* 9:255–262.

Neuhoff V, Stamm R, Pardowitz I, Arold N, Ehrhardt W, Taube D. 1990. Essential problems in quantification of proteins following colloidal staining with Coomassie Brilliant Blue dyes in polyacrylamide gels, and their solution. *Electrophoresis* 11:101–117.

Ong SE, Pandey A. 2001. An evaluation of the use of two-dimensional gel electrophoresis in proteomics. *Biomol Eng* 18:195–205.

O'Farrell PH. 1975. High resolution two-dimensional electrophoresis of proteins. *J Biol Chem* 250:4007–4021.

O'Farrell PZ, Goodman HM, O'Farrell PH. 1977. High resolution two-dimensional electrophoresis of basic as well as acidic proteins. *Cell* 12:1133–1141.

Pandey A, Mann M. 2000. Proteomics to study genes and genomes. *Nature* 405:837–846.

Panek J, Vohradsky J. 1999. Point pattern matching in the analysis of two dimensional gel electrophoregrams. *Electrophoresis* 20:3483–3491.

Pappin DJ. 1997. Peptide mass fingerprinting using MALDI-TOF MS. *Methods Mol Biol* 64:165–173.

Patterson SD, Aebersold R. 1995. Mass spectrometric approaches for the identification of gel-separated proteins. *Electrophoresis* 16:1791–1814.

Patton WF. 2000. A thousand points of light: the application of fluorescence technologies to two-dimensional gel electrophoresis and proteomics. *Electrophoresis* 21:1123–1144.

Patton WF. 2001. Detecting proteins in polyacrilamide gels and on electroblot membranes. In: Pennington SR, Dunn MJ (Eds), *Proteomics: From Protein Sequence to Function*, BIOS Scientific, Oxford, UK, Springer-Verlag, New York, pp 65–86.

Patton WF. 2002. Detection technologies in proteome analysis. *J Chromatogr B Analyt Technol Biomed Life Sci* 771:3–31.

Perdew GH, Schaup HW, Selivonchik DP. 1983. The use of a zwitterionic detergent in two-dimensional gel electrophoresis of trout liver microsomes. *Anal Biochem* 135:453–455.

Pleissner KP, Hoffmann F, Kriegel K, Wenk C, Wegner S, Sahlström A, Oswald H, Alt H, Fleck E. 1999. New algorithmic approaches to protein spot detection and pattern matching in two-dimensional electrophoresis gel databases. *Electrophoresis* 20:755–765.

Quadroni M, James P. 1999. Proteomics and automation. *Electrophoresis* 20:664–677.

Rabilloud T. 1990. Mechanisms of protein silver staining in polyacrylamide gels: a 10-year synthesis. *Electrophoresis* 10:785–794.

Rabilloud T, Gianazza E, Cattò N, Righetti PG. 1990. Amidosulfobetaines, a family of detergents with improved solubilization properties: application for isoelectric focusing under denaturing conditions. *Anal Biochem* 185:94–102.

Rabilloud T, Valette C, Lawrence JJ. 1994. Sample applications by in-gel rehydration improves the resolution of two-dimensional electrophoresis with immobilized pH gradients in the first dimension. *Electrophoresis* 15:1552–1558.

Rabilloud T, Adessi C, Giraudel A, Lunardi J. 1997. Improvement of solubilisation of proteins in two-dimensional electrophoresis with immobilized pH gradients. *Electrophoresis* 18:307–316.

Ramsby ML, Makowski GS, Khairallah EA. 1994. Differential detergent fractionation of isolated hepatocytes: biochemical, immunochemical and two-dimensional gel electrophoresis characterization of cytoskeletal and noncytoskeletal compartments. *Electrophoresis* 15:265–277.

Righetti PG, Castagna A, Herbert B. 2001. Prefractionation techniques in proteome analysis. *Anal Chem* 73:320A–326A.

Roepstorff P. 2000. MALDI-TOF MS in protein chemistry. *EXS* 88:81–97.

Sanchez JC, Rouge V, Pisteur M, Ravier F, Tonella L, Moosmayer M, Wilkins MR, Hochstrasser DF. 1997. Improved and simplified in-gel sample application using reswelling of dry immobilized pH gradients. *Electrophoresis* 18:324–327.

Santoni V, Rabilloud T, Doumas P, Rouquié D, Mansion M, Kieffer S, Garin J, Rossignol M. 1999. Towards the recovery of hydrophobic proteins on two-dimensional electrophoresis gels. *Electrophoresis* 20:705–711.

Santoni V, Molloy M, Rabilloud T. 2000. Membrane proteins and proteomics: un amour impossible? *Electrophoresis* 21:1054–1070.

Schuerenberg M, Luebbert C, Eickhoff H, Kalkum M, Lehrach H, Nordhoff E. 2000. Prestructured MALDI-MS sample supports. *Anal Chem* 72:3436–3442.

Shevchenko A, Wilm M, Vorm O, Mann M. 1996a. Mass spectrometric sequencing of proteins silver-stained polyacrylamide gels. *Anal Chem* 68:850–858.

Shevchenko A, Wilm M, Vorm O, Jensen ON, Podtelejnikov AV, Neubauer G, Shevchenko A, Mortensen P, Mann M. 1996b. A strategy for identifying gel-separated proteins in sequence databases by MS alone. *Biochem Soc Trans* 24:893–896.

Simpson RJ. 2003. *Proteins and Proteomics. A Laboratory Manual.* Cold Spring Harbor Laboratory Press, New York.

Smilansky Z. 2001. Automatic registration for images of two-dimensional protein gels. *Electrophoresis* 22:1616–1626.

Smolka M, Zhou H, Aebersold R. 2002. Quantitative protein profiling using two-dimensional gel electrophoresis, isotope-coded affinity tag labeling, and MS. *Mol Cell Proteomics* 1:19–29.

Sorensen BK, Hojrup P, Ostergard E, Jorgensen CS, Enghild J, Ryder LR, Houen G. 2002. Silver staining of proteins on electroblotting membranes and intensification of silver staining of proteins separated by polyacrylamide gel electrophoresis. *Anal Biochem* 304:33–41.

Steinberg TH, Jones LJ, Haugland RP, Singer VL. 1996. SYPRO Orange and SYPRO Red protein gel stains: one-step fluorescent staining of denaturing gels for detection of nanogram levels of protein. *Anal Biochem* 239:223–237.

Steinberg TH, Pretty On Top K, Berggren KN, Kemper C, Jones L, Diwu Z, Haugland RP, Patton WF. 2001. Rapid and simple single nanogram detection of glycoproteins in polyacrylamide gels and on electroblots. *Proteomics* 1:841–855.

Steinberg TH, Agnew BJ, Gee KR, Leung WY, Goodman T, Schulenberg B, Hendrickson J, Beechem JM, Haugland RP, Patton WF. 2003. Global quantitative phosphoprotein analysis using multiplexed proteomics technology. *Proteomics* 3:1128–1144.

Stensballe A, Jensen ON. 2001. Simplified sample preparation method for protein identification by matrix-assisted laser desorption/ionization MS: in-gel digestion on the probe surface. *Proteomics* 1:955–966.

Switzer RC 3rd, Merril CR, Shifrin S. 1979. A highly sensitive silver stain for detecting proteins and peptides in polyacrylamide gels. *Anal Biochem* 98:231–227.

Takach EJ, Hines WM, Patterson DH, Juhasz P, Falick AM, Vestal ML, Martin SA. 1997. Accurate mass measurements using MALDI-TOF with delayed extraction. *J Protein Chem* 16:363–369.

Unlu M, Morgan ME, Minden JS. 1997. Difference gel electrophoresis: a single gel method for detecting changes in protein extracts. *Electrophoresis* 18:2071–2077.

Westbrook JA, Yan JX, Wait R, Dunn MJ. 2001. A combined radiolabelling and silver staining technique for improved visualization, localization, and identification of proteins separated by two-dimensional gel electrophoresis. *Proteomics* 1:370–376.

Wildgruber R, Harder A, Obermaier C, Boguth G, Weiss W, Fey SJ, Larsen PM, Gorg A. 2000. Towards higher resolution: two-dimensional electrophoresis of *Saccharomyces cerevisiae* proteins using overlapping narrow immobilized pH gradients. *Electrophoresis* 21:2610–2616.

Wilkins MR, Sanchez JC, Gooley AA, Appel RD, Humphery-Smith I, Hochstrasser DF, Williams KL. 1996. Progress with proteome projects: why all proteins expressed by a genome should be identified and how to do it. *Biotechnol Genet Eng Rev* 13:19–50.

Wilkins MR, Gasteiger E, Gooley AA, Herbert BR, Molloy MP, Binz PA, Ou K, Sanchez JC, Bairoch A, Williams KL, Hochstrasser DF. 1999. High-throughput mass spectrometric discovery of protein post-translational modifications. *J Mol Biol* 289:645–657.

Wilm M. 2000. Mass spectrometric analysis of proteins. *Adv Protein Chem* 54:1–30.

Wilm M, Shevchenko A, Houthaeve T, Breit S, Schweigerer L, Fotsis T, Mann M. 1996. Femtomole sequencing of proteins from polyacrylamide gels by nano-electrospray MS. *Nature* 379:466–469.

Yan JX, Wait R, Berkelman T, Harry RA, Westbrook JA, Wheeler CH, Dunn MJ. 2000. A modified silver staining protocol for visualization of proteins compatible with matrix-assisted laser desorption/ionization and electrospray ionization-MS. *Electrophoresis* 17:3666–3672.

Yeargin J, Haas M. 1995. Elevated levels of wild-type p53 induced by radiolabeling of cells leads to apoptosis or sustained growth arrest. *Curr Biol* 5:423–431.

Zuo X, Speicher DW. 2000. A method for global analysis of complex proteomes using sample prefractionation by solution isoelectrofocusing prior to two-dimensional electrophoresis. *Anal Biochem* 284:266–278.

Zuo X, Speicher DW. 2002. Comprehensive analysis of complex proteomes using microscale solution isoelectrofocusing prior to narrow pH range two-dimensional electrophoresis. *Proteomics* 2:58–68.

3

Isotope Labeling in Quantitative Proteomics

Kristy J. Brown
CTL Bio Services, Rockville, Maryland

Catherine Fenselau*
University of Maryland, College Park, Maryland

3.1   INTRODUCTION

The Human Genome Project (http://www.ornl.gov/hgmis) began in 1990 with the goal of sequencing the entire human genome. The enormity of this project astonished most scientists; however, completion of the first draft came sooner than expected, in February 2001, when the public and private draft sequences were simultaneously published in *Nature* and *Science* (Lander et al., 2001; Venter et al., 2001). There are approximately 24,000 genes in the human genome, which is much smaller than the original estimates. That number is only roughly double the number of genes in a worm or fly (Lander et al., 2001). But, encoded in the human genome is approximately half a million proteins. This large number of proteins, or genome products, is what accounts for the complexity of humans (Brower, 2001). The fruits of the Human Genome Project provided researchers with a "blueprint" that has revolutionized biological research in academic, government, and industrial fields. Unfortunately, the genome cannot provide answers for all questions regarding human development, health, and disease. Since the complexity of human beings cannot simply be assigned to their genome, the proteome must also be considered (Fields, 2001).

* To whom correspondence should be addressed.

Proteomics for Biological Discovery, edited by Timothy D. Veenstra and John R. Yates.
Copyright © 2006 John Wiley & Sons, Inc.

The proteome is the complete protein complement to the genome (Wilkins et al., 1996). The genome cannot predict the protein content of a particular cell or the presence of post-translational modifications, so the proteins must be studied directly (Haynes and Yates, 2000). Technologies have been developed to quantitate mRNA levels to obtain quantitative data regarding transcription, but the mRNA abundance levels do not always correlate with the protein concentration in a cell (Anderson and Seilhamer, 1997; Gygi et al., 1999b). The lack of correlation between mRNA and protein data can be attributed to varying protein half-lives, post-translational modifications, or cotranslational degradation (Turner and Varshavsky, 2000).

Even prior to the completion of the Human Genome Project, proteomics had emerged as a dynamic field of research. The complexity of the proteome is many orders of magnitude greater than the genome. Regrettably, no amplification technologies for proteins exist, as are available for genomic studies. Also, unlike the genome, the proteome is dynamic and constantly changing. All of these characteristics make complete characterization of a cellular proteome infinitely more challenging than that of a cellular genome.

The two main areas of proteomic interest can be subclassified as functional and comparative proteomics. More specifically, functional proteomic studies focus on protein structures and complexes using a variety of techniques, while comparative proteomic studies quantitatively compare protein levels from two or more samples. Proteomic studies are expanding the catalog of human proteins as well as providing information about function and concentration. Enabling the rapid development of the field of proteomics is the rapidly evolving instrumentation of mass spectrometry. Mass spectrometry combined with bioinformatics has provided the basis for identification and, in many cases, quantitation of proteins.

Comparative proteomic studies aim to identify, as well as quantitate, proteins in two or more samples. To date, the most commonly used method for comparative proteomic studies is two-dimensional gel electrophoresis (2DE) combined with mass spectrometric identification. This method is discussed at length in another chapter of the book. More recently, gel-free "shotgun" methods have been proposed and practiced in proteomics research (Opiteck et al., 1997; Link et al., 1999; Spahr et al., 2000; Washburn et al., 2001).

A shotgun method involves initial digestion of all the proteins in the mixture, followed by chromatographic or electrophoretic separation of peptides, and detection by mass spectrometry. In such an experiment, partial sequence information is obtained from the tandem mass spectra of the peptides and is used to search databases (Eng et al., 1994; Mann and Wilm, 1994) to obtain the identity of the precursor protein. Stable isotope labeling methods have provided the shotgun method with the capability for relative quantitation. The combination of mass spectrometry and stable isotope labeling is a logical union for carrying out comparative proteomic studies (Sechi and Chait, 1998). Although mass spectrometry is not usually considered an absolute quantitative method in proteomic studies, relative quantitation can be achieved if the samples of interest are analyzed simultaneously. Clearly, the samples must be differentiated in some way, and the most practical distinction for mass spectrometry is a change in mass. This idea has lead to the development of many stable isotope labeling techniques for comparative proteomic studies. Stable isotope labeling experiments require the incorporation of a stable isotope such as ^{2}H, ^{13}C, ^{15}N, or ^{18}O. In these experiments, one protein or peptide pool remains unlabeled, while a second is labeled with a stable isotope(s). The mass-differentially labeled samples can then be analyzed by mass spectrometry. Quantitative information is derived by comparing peak heights or areas that are recorded simultaneously from the labeled and unlabeled samples (Aebersold and Mann, 2003; Goshe and Smith, 2003). Isotope labels can be introduced for comparative quantitation by in vitro labeling and by in situ or metabolic labeling. Absolute quantitation of proteins can be accomplished by the introduction of isotope labeled standards (Gerber et al., 2003). Currently, most experimentalists try to work with ratios close to 1:1; however, studies are beginning to appear that demonstrate more extreme isotope ratios (Cahill et al., 2003).

3.2 IN VITRO LABELING

An in vitro stable isotope labeling experiment incorporates the isotope label after sample collection, that is, following tissue biopsy or cell harvest, and after the release of proteins from the tissue or cells. Strategies for in vitro labeling are discussed below.

3.2.1 Isotope-Coded Affinity Tags

The stable isotope labeling strategy known as isotope-coded affinity tags or ICAT was developed in the Aebersold lab and first published in 1999 (Gygi et al., 1999a). Isotope labels are incorporated via a thiol-specific reactive group connected to a linker containing various numbers of stable isotopes. Also attached to the linker is a biotin group for the isolation of the modified peptides via immobilized avidin.

In an ICAT experiment, two pools of proteins are denatured and reduced, and the cysteine residues are derivatized with either the "heavy" or "light" ICAT reagent. The labeled pools are then combined, cleaned up to remove excess reagent, and digested (usually with trypsin). The cysteine-containing peptides, carrying "heavy" and "light" isotope tags, are then captured on an avidin column via the reactive biotin moiety. Finally, these peptides are eluted and analyzed by mass spectrometry. The pairs of labeled and unlabeled peptides, which differ by 6–9 Da according to what reagent is used, are readily observable by high and low resolution mass spectrometry. An example of a labeled and an unlabeled peptide pair stemming from an ICAT experiment is shown in Figure 3.1 (Hansen et al., 2003).

The ICAT strategy has been well characterized. Only the cysteine-containing peptides are labeled in this method, which is both a plus and a minus. On the plus side, the sample complexity is greatly reduced. On the minus side, only cysteine-containing peptides are analyzed, so there is usually only a single peptide available to identify and quantitate each protein. One criticism that has been made of the method is that the sample is usually contaminated by peptides bound nonspecifically to the avidin column or beads. The first incarnation of ICAT, in which eight deuterium atoms were incorporated in the heavy linker, suffered from several drawbacks, which have encouraged the development of variations. The original ICAT reagents modified cysteine residues with a 442.2 Da tag, which often frag-

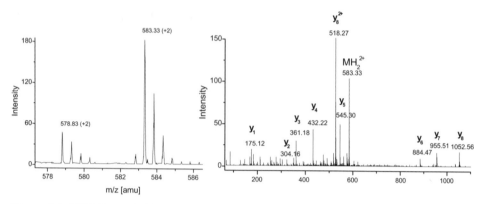

Figure 3.1. *ESI-MS and MS/MS spectra of ICAT light and heavy labeled calgranulin A peptide LLETEC*PQYIR. Reprinted with permission from Hansen et al. (2003).*

mented off during MS/MS analysis (Smolka et al., 2001). Also, the heavy and light derivatized peptides did not coelute when using reverse phase chromatography. This differential elution complicated the analysis since simultaneous ionization facilitates quantitation using mass spectrometry. To address these issues, Aebersold and co-workers developed a photo cleavable ICAT reagent bound to a solid support, thus eliminating the biotin tag (Zhou et al., 2002). Since the tag is cleavable, the mass of the alkylation derivative is reduced to 170 Da. Also, the solid support simplifies the affinity selection and wash steps, reducing the complexity of the experiment. A version of the ICAT reagent that uses acid-labile isotope-coded extractants has also been developed (Qiu et al., 2002). Finally, to ensure chromatographic coelution of peptide pairs, reagents utilizing ^{13}C isotopes in place of ^2H have been adopted (see Figure 3.1). Multiple variations in the mass of the tag are envisioned to permit analysis of three or more time points or cell varieties. A variety of laboratories have reported successful use of the ICAT strategy (Yu et al., 2002; Hansen et al., 2003; Kubota et al., 2003) and Tao and Aebersold (2003) have reviewed the various ICAT reagents and their application.

3.2.2 ^{18}O Labeling

Another stable isotope labeling method developed for application in shotgun comparative proteomics utilizes the ^{18}O isotope (Yao et al., 2001). All shotgun strategies require digestion of the proteins to peptides at some point in their method. The ^{18}O method combines the digestion and labeling steps. Two pools of proteins are digested in parallel, one in $H_2{}^{16}O$ and the other in $H_2{}^{18}O$ (greater than 95% ^{18}O). In the latter, two atoms of ^{18}O are incorporated into the carboxyl terminus of each peptide, resulting in a peptide pool differing by 4 Da. The two pools are mixed and analyzed by mass spectrometry. Since the isotope labels are localized to the C terminus, the y-ions in a tandem mass spectrum appear as labeled and unlabeled doublets, while the b-ions do not contain a label. This doublet signature for y-ions greatly simplifies MS/MS data interpretation. The mass difference between the heavy and light peptides is only 4 Da, so it was originally assumed high resolution mass spectrometry was necessary to obtain quantitative data, but recently it has been shown that useful data can be obtained on a low resolution instrument such as an ion trap (Heller et al., 2003).

A proteolytic enzyme catalyzes isotope labeling using this method, so it is specific and does not result in any side reactions such as side chain labeling. The first ^{18}O atom is incorporated during protein cleavage. The enzyme then rebinds the peptide product and incorporation of a second ^{18}O atom occurs when the enzyme–substrate intermediate is hydrolyzed. In one view, the enzyme is working twice as hard, so additional enzyme and digestion time are required for this method to ensure complete double ^{18}O incorporation. An arsenal of enzymes are available for use, including trypsin, chymotrypsin, Lys-C, and Glu-C (Yao et al., 2001; Reynolds et al., 2002; Yao et al., 2003). An example in which a glycoprotein was treated with both Glu-C and PNGase F in >95% $H_2{}^{18}O$ is shown in Figure 3.2. As can be seen in the spectrum, the deglycosylated glycopeptide now contains three atoms of ^{18}O, while the other peptides have incorporated only two (Reynolds et al., 2002).

Adding to the versatility of the method, both ^{18}O labels may be incorporated at the peptide level (Yao et al., 2003). This decoupling of the digestion and labeling steps enables the user to digest the protein pools using standard methods without having to lyophilize the proteins. Subsequently, the sample is lyophilized at the peptide level, then resolubilized in $H_2{}^{18}O$ with additional enzyme added to facilitate

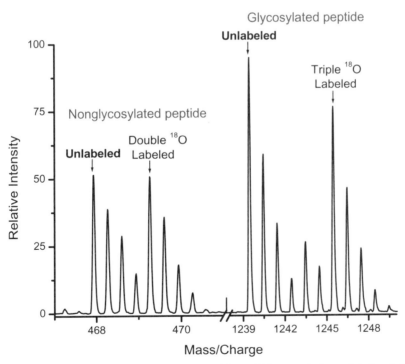

Figure 3.2. Electrospray mass spectrum of two peptides, which were deglycosylated and digested in $H_2{}^{18}O$ in parallel with the $H_2{}^{16}O$ counterparts. A triply charged peptide that did not contain a glycosylation site and a singly charged peptide that had been glycosylated are detected with the respective 4 Da and 6 Da mass increments. Reprinted with permission from Reynolds et al. (2002).

the isotope labeling. This technique is an improvement, since peptides are easier to resuspend following lyophilization than proteins. Proteins, which precipitate in low salt buffer, can be digested with Lys-C at higher urea concentrations, and then the peptide products can be labeled with trypsin (Yao et al., 2001).

An inverse ^{18}O labeling strategy has also been proposed (Wang et al., 2001). In this experiment the control and experimental protein pools are labeled in $H_2^{16}O$ and $H_2^{18}O$, respectively; then the same experiment is carried out inversely (i.e., with the control protein pool now in $H_2^{18}O$). This strategy is designed to gain confidence in detecting extreme concentration changes.

The proteolytic ^{18}O labeling method has been criticized in the literature for being limited by the amount of urea that can be used, for producing complex samples, and for incomplete double ^{18}O incorporation. The urea concentration must be at a level that does not denature the enzyme, a requirement in the digestion steps of all shotgun methods. The labels are stable, and the labeled and unlabeled peptides coelute (Reynolds et al., 2002), so the complex sample can be fractionated via affinity selection, such as in the ICAT strategy, or any combination of chromatographic strategies. Another alternative that facilitates all labeling strategies is to prefractionate at the protein level (Shefcheck et al., 2003). This fractionation results in a less complex protein mixture and consequently a simplified peptide pool. By reducing complexity at the protein level, not the peptide level, no protein information such as post-translational modifications is lost. Also, multiple peptides from each protein are present for quantitation, allowing for standard deviations to be calculated for the average quantitative measurement of the protein as a whole. Finally, for complete double ^{18}O incorporation, sufficient enzyme concentration and digestion times are necessary. In order to ensure this, the use of immobilized enzymes has been applied. Immobilized enzymes enable the use of very high enzyme concentrations, while limiting autolysis products. By using immobilized enzymes, the time for complete double ^{18}O incorporation is less than that required for a simple digestion in which solubilized, free proteolytic enzyme is added to the sample. Applications of this method have been reported from a variety of laboratories (Shevchenko, 2001; Stewart et al., 2001; Wang et al., 2001; Back et al., 2002; Bonenfant et al., 2003; Heller et al., 2003).

3.2.3   Amino Group Acylation

Another in vitro stable isotope labeling method, developed by Regnier and co-workers, among others, uses acylation chemistry with deuterated acids to modify peptide primary amines (Geng et al., 2000; Chakraborty and Regnier, 2002; Liu and Regnier, 2002). Like ^{18}O labeling, this method labels every peptide, and consequently Regnier and co-workers have employed the term Global Internal Standard Technology or GIST to describe this technology (Chakraborty and Regnier, 2002). This laboratory has employed $[^2H_4]$succinic anhydride and N-acetoxy-$[^2H_3]$succinimide as acylation reagents to label peptides isotopically.

In a GIST experiment, two protein pools are fractionated, as necessary, and digested into peptides. The peptide pools are acylated in parallel with deuterated and unlabeled reagents. The peptide pools are mixed and analyzed together by mass spectrometry. Unfortunately, stable isotope labeling strategies that utilize ^2H result in reagents that do not behave identically in reverse phase chromatography, as demonstrated by the original ICAT reagents. The substitution of ^{13}C-modified reagents is a logical alternative, although this substitution increases the cost of the experiment substantially.

Acylation modifies primary amines, which include each peptide N terminus as well as lysine residue side chains. The modification of lysine side chains reduces the basicity and consequently can affect the ionization efficiency of the peptide. Also, N-terminally blocked peptides are not labeled. To address this issue, Liu and Regnier (2002) have combined a GIST experiment with ^{18}O labeling. The body of work stemming from the Regnier laboratory also details extensive chromatographic and affinity capture applications to simplify samples prior to mass spectrometric analysis. The importance of fractionation in proteomics should not be underestimated.

James and co-workers have also employed acylation chemistry using [^{2}H$_4$]nicotinyl-*N*-hydroxysuccinimide to provide both a labeled derivative and one that favors formation of N-terminal ions (Munchbach et al., 2000). To ensure that peptides are only labeled at the N terminus, the lysine residues on the peptides are blocked prior to labeling. In this case, the N-terminal fragment ions appear as doublets. These authors also employed partial ^{18}O labeling to facilitate de novo sequencing via the easily interpreted y-ions generated from digestion in 50% H$_2$^{18}O. Like the ^{18}O strategy, these acylation methods provide global labeling, thus providing a larger window on post-translational modifications, and they may readily be combined with a variety of affinity capture or chromatographic fractionation techniques.

3.2.4   Other Chemical Modifications

All of the reactive functional groups in different amino acid residues are potential sites for isotope labeling reactions in vitro. For example, chemistry specific for tryptophan has also been evaluated for differential isotope labeling (Kuyama et al., 2003). Tryptophan residues in proteins are derivatized with 2-nitrobenzenesulfenyl chloride carrying six ^{1}H or six ^{2}H. Peptides are obtained by tryptic or other proteolytic digestion and those carrying the hydrophobic derivative are enriched using Sephadex LH-20.

Guanidination and similar modification of lysine residues have been reported that incorporate isotope labels (Brancia et al., 2001; Peters et al., 2001; Cagney and Emili, 2002). These approaches harness the strategy of a GIST experiment with the improvement that lysine residues are labeled in a way that enhances ionization of the peptide. As yet these methods have not been put to wide use by proteomics laboratories.

Esterification chemistry has been proposed for peptides by Goodlett et al. (2001). In this method, carboxyl groups of aspartic and glutamic acid residues as well as the C terminus are modified using methanolic HCl. While this method is inexpensive, it is limited by not knowing the number of labels incorporated for a given sample and the reaction conditions are harsh, which can lead to unexpected modifications.

3.2.5   Post-translational Modifications

Isotope labeling strategies are under development to help locate post-translational modifications and to quantitate their abundances. Phosphorylation and glycosylation sites have received the most attention thus far. The incorporation of ^{18}O from

$H_2^{18}O$ in the transformation of asparagine to aspartic acid at the site of N-glycosylation when treated with PNGase F is well known (Gonzalez et al., 1992) and has been incorporated into the ^{18}O labeling strategy (Reynolds et al., 2002) and into a lectin-based strategy (Kaji et al., 2003) to identify N-linked glycopeptides and provide relative quantitation. A more complex method has also been suggested in which succinylation of immobilized glycopeptides provides a 4 Da mass difference for comparative quantitation (Zhang et al., 2003).

A number of methods have been reported to analyze phosphorylation sites, with the objectives of stabilizing or replacing the labile phosphate group and introducing isotope labels (McLachlin and Chait, 2001). These labeling methods are often combined with isolation of phosphopeptides by antibody or metal ion affinity (Bonenfant et al., 2003). Weckwerth et al. (2000) proposed chemical dehydrophosphorylation of phosphoserine and phosphothreonine residues, followed by thioethylation with undeuterated and $^2H_5$-deuterated reagents. Subsequently, strategies have been proposed in which the phosphate group is replaced by a thiol derivative that carries an isotope label linked to a biotin moiety (Goshe et al., 2001; Oda et al., 2001). This modification allows affinity enrichment and also relative quantitation. The latter group has proposed the acronym PhIAT for phosphopeptide isotope-coded affinity tag. Zhou et al. (2001) have immobilized phosphopeptides to solid supports via the phosphate group and incorporated isotope tags by derivatizing carboxylate groups in the peptides. Synthetic standard phosphopeptides have been used to quantitate cell cycle dependent phosphorylation at one site in the human separase protein (Gerber et al., 2003). Efforts continue to improve the β-elimination and subsequent chemical steps, to reduce side reactions and increase yield and sensitivity for analysis of this very important protein modification.

3.3 IN SITU OR METABOLIC LABELING

The second major category for introducing stable isotopes is metabolic or in situ labeling. This strategy requires that the isotope label be incorporated during sample growth, that is, during cell culture, as has been done for many years with radioisotope labeled phosphate. Clearly, this method is advantageous because labeled and unlabeled cell samples can be mixed immediately following harvest, and thereby errors due to sample handling are minimized. These methods are less readily applicable to biopsy or other clinical material. Because proteins are labeled, rather than peptides, this approach can be combined with 1D and 2D gel separations, as well as gel-free (i.e., shotgun) analysis.

3.3.1 ^{15}N Incorporation from Ammonium Salts

The first reports of in situ stable isotope labeling appeared from two laboratories in 1999, which used ^{15}N-enriched media to grow *Saccharomyces cerevisiae* (Oda et al., 1999) and *Escherichia coli* (Pasa-Tolic et al., 1999). Incorporation of isotopic labels could be analyzed at either the protein or peptide level. The ^{15}N atoms are metabolically incorporated into the sample during sample growth or culture, eventually replacing all natural isotopic abundance (i.e., ^{14}N) nitrogen atoms. The corresponding mass shift is unpredictable, since it is dependent on the number of nitrogen atoms present in each peptide or protein. This unpredictable mass shift complicates sample analysis and requires high resolution mass spectrometry, such as Fourier transform ion cyclotron resonance (FTICR), for the analysis (Conrads et al., 2001). The value of reverse labeling has also been evaluated for this labeling method (Wang et al., 2002). More recently, this inexpensive method of heavy isotope incorporation has been exercised in a high-throughput analyses of *S. cerevisiae* (Washburn et al., 2002, 2003).

3.3.2 Incorporation of Isotope Labeled Amino Acids

Incorporation of isotopically labeled amino acids during culture of bacteria and yeast was demonstrated shortly after the successful incorporation of ^{15}N (Chen et al., 2000; Veenstra et al., 2000; Jiang and English, 2002). In one particularly interesting study, turnover rates were defined for yeast proteins labeled in situ (Pratt et al., 2002). Although isotope labeled amino acids are expensive, their use has allowed metabolic labeling to be extended to cultured mammalian cells. The Mann laboratory has applied the strategy to mouse fibroblast cells and proposed the term SILAC for stable isotope labeling by amino acids in cell culture (Ong et al., 2002, 2003a). Commercially available amino acids that have been used include Leu-D_{10}, Leu-D_3, Met-D_3, Ser-D_3, Tyr-D_2, Lys-$^{13}C_6$, and Arg-$^{13}C_6$.

In these types of experiments, two sets of cells are grown in parallel, one in a cultured medium that contains natural isotopic abundance amino acids, and the other in a medium with one or more isotopically substituted amino acids. Ong et al. (2002) employed [$^2H_3$]leucine during culture of mouse fibroblast cells in amino acid deficient medium to which appropriate amino acids had been added. At least five growth passages were required to ensure complete incorporation. Mann and co-workers observed no differences in cell morphology as a result of using isotopically labeled medium, an important consideration when using this technology to specifically measure the effect of a perturbation on the cell's protein expression profile. This fibroblast cell line is particularly robust and does not require the presence of growth hormones and other serum components to grow. A spectrum of a peptide pair obtained from MCF-7 breast cancer cells grown in medium containing labeled ($^{13}C_6$) and unlabeled arginine is shown in Figure 3.3. Equal amounts of cells were mixed, and cytosolic proteins were isolated and fractionated by 2D gel electrophoresis.

For analysis, the labeled cells can be combined with the unlabeled cells. Since the mixing step occurs so early, errors due to differential sample processing are minimized. The mixed pools can be fractionated using any method, including 2DE, without alteration of the protein ratios. Here, as earlier, when ^2H isotopes are used, chromatographic separation is a problem for accurate quantitation of peptides. Mann and co-workers recognized this problem and have subsequently demonstrated that the essential amino acid [$^{13}C_6$]arginine can be metabolically incorporated (Ong et al., 2003a). In this study some labeled proline was found to be incorporated, since this amino acid is a catabolite of arginine. When trypsin is used as the enzyme in a [$^{13}C_6$]arginine experiment, the label is localized to the C terminus of arginine-containing peptides, while lysine-terminated peptides are unlabeled. Alternatively, if both ^{13}C labeled lysine and arginine amino acids are employed during culture, the experiment would have benefits similar to the ^{18}O

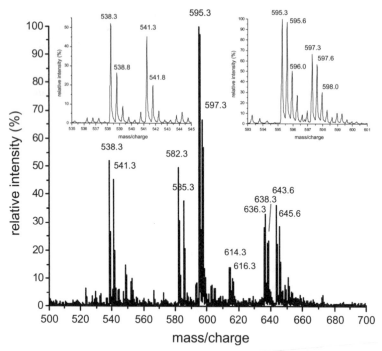

Figure 3.3. *Partial electrospray mass spectrum of tryptic peptides from a protein identified as human heat shock protein 27, isolated by 2D gel electrophoresis from the purified cytosol from human MCF-7 breast cancer cells grown with unlabeled amino acids and grown in media containing Lys-$^{13}C_6$ and Arg-$^{13}C_6$. The two insets show expanded peak regions for a doubly (left) and a triply (right) charged peptide in the mixture. This figure was provided by Drs. Marion Gehrmann and Yetrib Hathout, University of Maryland.*

labeling method (Ong et al., 2003b). Following tryptic digestion, arginine- and lysine-terminating peptides, or all peptides except the original C terminus, are labeled at the C terminus, and therefore y-ions are readily observable in tandem mass spectra as doublets. In situ or metabolic labeling offers great promise for future experiments and is rapidly being modified and applied in other laboratories. Stegmann and colleagues have begun to explore the range through which isotope ratios can be varied, that is, the extent of labeling that must be achieved in the labeled cell line to obtain accurate relative quantitation (Vogt et al., 2003).

3.4 CONCLUSION

There are a number of criteria for selecting an isotope labeling strategy. The relative importance of these will vary, according to the analytical objective. In addition to cost and ease/speed of introduction, the effect on chromatographic separation, the number of peptides available from each protein, and the mass difference between the labeled and unlabeled peptides may be important. In many cases, it will be preferable to introduce labels that increment peptide masses homogeneously. Thought should be given to possible changes in hydrophobicity and desorption sensitivity. Most procedures involve chemical reactions, and these should proceed to completion without producing side products. It is desirable that the process is applicable to small amounts of sample proteins in mixtures with a wide dynamic range, and applicable in a high-throughput mode. Similarly, strategies are preferable that permit parallel samples to be combined as early in the process as possible. This last point is the major argument for incorporating labels metabolically during cell culture. However, in vitro labeling will be most directly applicable to clinical, animal, and plant specimens.

It should be noted that the precision of the measurement of the ratio and the dynamic range available for measuring the ratio (i.e., relative quantitation) will largely be determined by the mass spectrometer, and not by the isotopes or labeling strategy used. The slower scanning speeds available with laser desorption, compared to HPLC-electrospray, provide better ion statistics and thus more accurate isotope ratios. Isotope profiles may be better envisioned when higher resolution is used, for example, in Fourier transform mass spectrometry. However, relative quantitation is often accurate when lower resolution analyzers are used.

Stable isotope labeling provides a compelling strategy for comparative proteomic studies. All reagents, or their precursors, for the methods discussed are commercially available. In these strategies, mass spectrometry provides stable isotope ratios, as it has been doing since 1919 (Aston, 1919).

REFERENCES

Aebersold R, Mann M. 2003. Mass spectrometry-based proteomics. *Nature* 422:198–207.

Anderson L, Seilhamer J. 1997. A comparison of selected mRNA and protein abundances in human liver. *Electrophoresis* 18:533–537.

Aston FW. 1919. A positive-ray spectrograph. *Philos Mag* 38:707–715.

Back JW, Notenboom V, de Koning LJ, Muijsers AO, Sixma TK, de Koster CG, de Jong L. 2002. Identification of cross-linked peptides for protein interaction studies using mass spectrometry and ^{18}O labeling. *Anal Chem* 74:4417–4422.

Bonenfant D, Schmelzle T, Jacinto E, Crespo JL, Mini T, Hall MN, Jenoe P. 2003. Quantitation of changes in protein phosphorylation: a simple method based on stable isotope labeling and mass spectrometry. *Proc Natl Acad Sci USA* 100:880–885.

Brancia FL, Butt A, Beynon RJ, Hubbard SJ, Gaskell SJ, Oliver SG. 2001. A combination of chemical derivatization and improved bioinformatic tools optimizes protein identification for proteomics. *Electrophoresis* 22:552–559.

Brower V. 2001. The promise of proteomics: leading the way to 21st century medicine. Presented at The Promise of Proteomics, New York Academy of Sciences, New York.

Cagney G, Emili A. 2002. De novo peptide sequencing and quantitative profiling of complex protein mixtures using mass-coded abundance tagging. *Nat Biotechnol* 20:163–170.

Cahill MA, Wozny W, Schwall G, Schroer K, Holzer K, Poznanovic S, Hunzinger C, Vogt JA, Stegmann W, Mathies H, Schrattenholz A. 2003. Analysis of relative isotopologue abundances for quantitative profiling of complex protein mixtures labeled with the acrylamide/D3-acrylamide alkylation tag system. *Rapid Commun Mass Spectrom* 17:1283–1290.

Chakraborty A, Regnier FE. 2002. Global internal standard technology for comparative proteomics. *J Chromatogr A* 949:173–184.

Chen X, Smith LM, Bradbury EM. 2000. Site-specific mass tagging with stable isotopes in proteins for accurate and efficient protein identification. *Anal Chem* 72:1134–1143.

Conrads TP, Alving K, Veenstra TD, Belov ME, Anderson GA, Anderson DJ, Lipton MS, Pasa-Tolic L, Udseth HR, Chrisler WB, Thrall BD, Smith RD. 2001. Quantitative analysis of bacterial and mammalian proteomes using a combination of cysteine affinity tags and ^{15}N-metabolic labeling. *Anal Chem* 73:2132–2139.

Eng JK, McCormack AL, Yates JR III. 1994. An approach to correlate tandem mass spectral data of peptides with amino acid sequences in a protein database. *J Am Soc Mass Spectrom* 5:976–989.

Fields S. 2001. Proteomics in genomeland. *Science* 291:1221–1224.

Geng M, Ji J, Regnier FE. 2000. Signature-peptide approach to detecting proteins in complex mixtures. *J Chromatogr A* 870:295–313.

Gerber SA, Rush J, Stemman O, Kirschner MW, Gygi SP. 2003. Absolute quantification of proteins and phosphoproteins from cell lysates by tandem MS. *Proc Natl Acad Sci USA* 100:6940–6945.

Gonzalez J, Takao T, Hori H, Besada V, Rodriguez R, Padron G, Shimonishi Y. 1992. A method for determination of N-glycosylation sites in glycoproteins by collision-induced dissociation analysis in fast atom bombardment mass spectrometry: identification of the positions of carbohydrate-linked asparagine in recombinant alpha-amylase by treatment with peptide-*N*-glycosidase F in ^{18}O-labeled water. *Anal Biochem* 205:151–158.

Goodlett DR, Keller A, Watts JD, Newitt R, Yi EC, Purvine S, Eng JK, von Haller P, Aebersold R, Kolker E. 2001. Differential stable isotope labeling of peptides for quantitation and de novo sequence derivation. *Rapid Commun Mass Spectrom* 15:1214–1221.

Goshe MB, Smith RD. 2003. Stable isotope-coded proteomic mass spectrometry. *Curr Opin Biotechnol* 14:101–109.

Goshe MB, Conrads TP, Panisko EA, Angell NH, Veenstra TD, Smith RD. 2001. Phosphoprotein isotope-coded affinity tag approach for isolating and quantitating phosphopeptides in proteome-wide analyses. *Anal Chem* 73:2578–2586.

Gygi SP, Rist B, Gerber SA, Turecek F, Gelb MH, Aebersold R. 1999a. Quantitative analysis of complex protein mixtures using isotope-coded affinity tags. *Nat Biotechnol* 17:994–999.

Gygi SP, Rochon Y, Franza BR, Aebersold R. 1999b. Correlation between protein and mRNA abundance in yeast. *Mol Cell Biol* 19:1720–1730.

Hansen KC, Schmitt-Ulms G, Chalkley RJ, Hirsch J, Baldwin MA, Burlingame AL. 2003. Mass spectrometric analysis of protein mixtures at low levels using cleavable 13-C-isotope-coded affinity tag and multidimensional chromatography. *Mol Cell Proteomics* 2:299–314.

Haynes PA, Yates JR. 2000. Proteome profiling—pitfalls and progress. *Yeast* 17:81–87.

Heller M, Mattou H, Menzel C, Yao X. 2003. Trypsin catalyzed ^{16}O-to-^{18}O exchange for comparative proteomics: tandem mass spectrometry comparison using MALDI-TOF, ESI-QTOF and ESI-ion trap mass spectrometers. *J Am Soc Mass Spectrom* 14:704–718.

Jiang H, English AM. 2002. Quantitative analysis of the yeast proteome by incorporation of isotopically labeled leucine. *J Proteome Res* 1:345–350.

Kaji H, Saito H, Yamauchi Y, Shinkawa T, Taoka M, Hirabayashi J, Kasai K, Takahashi N, Isobe T. 2003. Lectin affinity capture, isotope-coded tagging and mass spectrometry to identify N-linked glycoproteins. *Nat Biotechnol* 21:667–672.

Kubota K, Wakabayashi K, Matsuoka T. 2003. Proteome analysis of secreted proteins during osteoclast differentiation using two different methods: two-dimensional electrophoresis and isotope-coded affinity tags analysis with two-dimensional chromatography. *Proteomics* 3:616–626.

Kuyama H, Watanabe M, Toda C, Ando E, Tanaka K, Nishimura O. 2003. An approach to quantitative proteome analysis by labeling tryptophan residues. *Rapid Commun Mass Spectrom* 17:1642–1650.

Lander ES, Linton LM, Birren B, et al. 2001. Initial sequencing and analysis of the human genome. *Nature* 409:860–921.

Link AJ, Eng J, Schieltz DM, Carmack E, Mize GJ, Morris DR, Garvik BM, Yates JR. 1999. Direct analysis of protein complexes using mass spectrometry. *Nat Biotechnol* 17:676–682.

Liu P, Regnier FE. 2002. An isotope coding strategy for proteomics involving both amine and carboxyl group labeling. *J Proteome Res* 1:443–450.

Mann M, Wilm M. 1994. Error-tolerant identification of peptides in sequence databases by peptide sequence tags. *Anal Chem* 66:4390–4399.

McLachlin DT, Chait BT. 2001. Analysis of phosphorylated proteins and peptides by mass spectrometry. *Curr Opin Chem Biol* 5:591–602.

Munchbach M, Quadroni M, Miotto G, James P. 2000. Quantitation and facilitated de novo sequencing of proteins by isotopic N-terminal labeling of peptides with a fragmentation-directing moiety. *Anal Chem* 72:4047–4057.

Oda Y, Huang K, Cross FR, Cowburn D, Chait BT. 1999. Accurate quantitation of protein expression and site-specific phosphorylation. *Proc Natl Acad Sci USA* 96:6591–6596.

Oda Y, Nagasu T, Chait BT. 2001. Enrichment analysis of phosphorylated proteins as a tool for probing the phosphoproteome. *Nat Biotechnol* 19:379–382.

Ong SE, Blagoev B, Kratchmarova I, Kristensen DB, Steen H, Pandey A, Mann M. 2002. Stable isotope labeling by amino acids in cell culture, SILAC, as a simple and accurate approach to expression proteomics. *Mol Cell Proteomics* 1:376–386.

Ong S, Kratchmarova I, Mann M. 2003a. Properties of ^{13}C-substituted arginine in stable isotope labeling by amino acids in cell culture (SILAC). *J Proteome Res* 2:173–181.

Ong SE, Foster LJ, Mann M. 2003b. Mass spectrometric-based approaches in quantitative proteomics. *Methods* 29:124–130.

Opiteck GJ, Lewis KC, Jorgenson JW, Anderegg RJ. 1997. Comprehensive on-line LC/LC/MS of proteins. *Anal Chem* 69:1518–1524.

Pasa-Tolic L, Jensen PK, Anderson GA, Lipton MS, Peden KK, Martinovic S, Tolic N, Bruce JE, Smith RD. 1999. High throughput proteome-wide precision measurements of protein expression using mass spectrometry. *J Am Chem Soc* 121:7949–7950.

Peters EC, Horn DM, Tully DC, Brock A. 2001. A novel multifunctional labeling reagent for enhanced protein characterization with mass spectrometry. *Rapid Commun Mass Spectrom* 15:2387–2392.

Pratt JM, Petty J, Riba-Garcia I, Robertson DH, Gaskell SJ, Oliver SG, Beynon RJ. 2002. Dynamics of protein turnover, a missing dimension in proteomics. *Mol Cell Proteomics* 1:579–591.

Qiu Y, Sousa EA, Hewick RM, Wang JH. 2002. Acid-labile isotope-coded extractants: a class of reagents for quantitative mass spectrometric analysis of complex protein mixtures. *Anal Chem* 74:4969–4979.

Reynolds KJ, Yao X, Fenselau C. 2002. Proteolytic [18]O labeling for comparative proteomics: evaluation of endoprotease Glu-C as the catalytic agent. *J Proteome Res* 1:27–33.

Sechi S, Chait BT. 1998. Modification of cysteine residues by alkylation. A tool in peptide mapping and protein identification. *Anal Chem* 70:5150–5158.

Shefcheck K, Yao X, Fenselau C. 2003. Fractionation of cytosolic proteins on an immobilized heparin column. *Anal Chem* 75:1691–1698.

Shevchenko A. 2001. Evaluation of the efficiency of in-gel digestion of proteins by peptide isotopic labeling and MALDI mass spectrometry. *Anal Biochem* 296:279–283.

Smolka MB, Zhou H, Purkayastha S, Aebersold R. 2001. Optimization of the isotope-coded affinity tag-labeling procedure for quantitative proteome analysis. *Anal Biochem* 297:25–31.

Spahr CS, Susin SA, Bures EJ, Robinson JH, Davis MT, McGinley MD, Kroemer G, Patterson SD. 2000. Simplification of complex peptide mixtures for proteomic analysis: reversible biotinylation of cysteinyl peptides. *Electrophoresis* 21:1635–1650.

Stewart II, Thomson T, Figeys D. 2001. ^{18}O labeling: a tool for proteomics. *Rapid Commun Mass Spectrom* 15:2456–2465.

Tao WA, Aebersold R. 2003. Advances in quantitative proteomics via stable isotope tagging and mass spectrometry. *Curr Opin Biotechnol* 14:110–118.

Turner GC, Varshavsky A. 2000. Detecting and measuring cotranslational protein degradation in vivo. *Science* 289:2117–2120.

Veenstra TD, Martinovic S, Anderson GA, Pasa-Tolic L, Smith RD. 2000. Proteome analysis using selective incorporation of isotopically labeled amino acids. *J Am Soc Mass Spectrom* 11:78–82.

Venter JC, Adams MD, Myers EW, et al. 2001. The sequence of the human genome. *Science* 291:1304–1351.

Vogt JA, Schroer K, Holzer K, Hunzinger C, Klemm M, Biefang-Arndt K, Schillo S, Cahill MA, Schrattenholz A, Matthies H, Stegmann W. 2003. Protein abundance quantification in embryonic stem cells using incomplete metabolic labeling with ^{15}N amino acids, matrix-assisted laser desorption/ionization time-of-flight mass spectrometry, and analysis of relative isotopologue abundances of peptides. *Rapid Commun Mass Spectrom* 17:1273–1282.

Wang YK, Ma Z, Quinn DF, Fu EW. 2001. Inverse ^{18}O labeling mass spectrometry for the rapid identification of marker/target proteins. *Anal Chem* 73:3742–3750.

Wang YK, Ma Z, Quinn DF, Fu EW. 2002. Inverse ^{15}N-metabolic labeling/mass spectrometry for comparative proteomics and rapid identification of protein markers/targets. *Rapid Commun Mass Spectrom* 16:1389–1397.

Washburn MP, Wolters D, Yates JR. 2001. Large-scale analysis of the yeast proteome by multidimensional protein identification technology. *Nat Biotechnol* 19:242–247.

Washburn MP, Ulaszek R, Deciu C, Schieltz DM, Yates JR. 2002. Analysis of quantitative proteomic data generated via multidimensional protein identification technology. *Anal Chem* 74:1650–1657.

Washburn MP, Koller A, Oshiro G, Ulaszek RR, Plouffe D, Deciu C, Winzeler E, Yates JR. 2003. Protein pathway and complex clustering of correlated mRNA and protein expression analyses in *Saccharomyces cerevisiae*. *Proc Natl Acad Sci USA* 100:3107–3112.

Weckwerth W, Willmitzer L, Fiehn O. 2000. Comparative quantification and identification of phosphoproteins using stable isotope labeling and liquid chromatography/mass spectrometry. *Rapid Commun Mass Spectrom* 14:1677–1681.

Wilkins MR, Pasquali C, Appel RD, Ou K, Golaz O, Sanchez JC, Yan JX, Gooley AA, Hughes G, Humphery-Smith I, Williams KL, Hochstrasser DF. 1996. From proteins to proteomes: large scale protein identification by two-dimensional electrophoresis and amino acid analysis. *Biotechnology* 14:61–65.

Yao X, Freas A, Ramirez J, Demirev PA, Fenselau C. 2001. Proteolytic ^{18}O labeling for comparative proteomics: model studies with two serotypes of adenovirus. *Anal Chem* 73:2836–2842.

Yao X, Afonso C, Fenselau C. 2003. Dissection of proteolytic ^{18}O labeling: endoprotease-catalyzed ^{16}O to ^{18}O exchange of truncated peptide substrates. *J Proteome Res* 2:147–152.

Yu LR, Johnson MD, Conrads TP, Smith RD, Morrison RS, Veenstra TD. 2002. Proteome analysis of camptothecin-treated cortical neurons using isotope-coded affinity tags. *Electrophoresis* 23:1591–1598.

Zhang H, Li XJ, Martin DB, Aebersold R. 2003. Identification and quantitation of N-linked glycoproteins using hydrazide chemistry, stable isotope labeling and mass spectrometry. *Nat Biotechnol* 21:660–666.

Zhou H, Watts JD, Aebersold R. 2001. A systematic approach to the analysis of protein phosphorylation. *Nat Biotechnol* 19:375–378.

Zhou H, Ranish JA, Watts JD, Aebersold R. 2002. Quantitative proteome analysis by solid-phase isotope tagging and mass spectrometry. *Nat Biotechnol* 20:512–515.

4

Mass Spectrometric Characterization of Post-translational Modifications

Thomas P. Conrads, Brian L. Hood, and Timothy D. Veenstra*

SAIC-Frederick, Inc., National Cancer Institute at Frederick, Frederick, Maryland

4.1 INTRODUCTION

Proteomics represents a vast field of study that now employs technologies designed to maximize the information content that can be derived from high-throughput analyses of complex protein samples. A few of the major foci are illustrated in Figure 4.1. The high-throughput identification of proteins is probably the best developed and validated of these techniques; however, this level of characterization represents only a small fraction of knowledge required to fully describe a proteome. To gain a complete understanding of the proteome requires identification, quantitation, and localization of the proteins within the cell. Beyond these characteristics, the protein interactions within the cell need to be discerned. Great strides have been made toward measurement of the relative abundances of proteins from different systems, as exemplified in the many two-dimensional polyacrylamide gel electrophoresis (2D-PAGE)-based proteomic studies that have been published, as well as others that use stable isotopes combined with mass spectrometry (MS) (Conrads et al., 2003). In addition, a detailed description of the cellular localization of a large percentage of proteins within yeast utilizing epitope-tagged protein has recently been reported (Kumar et al., 2002). The next frontier that is likely to be

* To whom correspondence should be addressed.

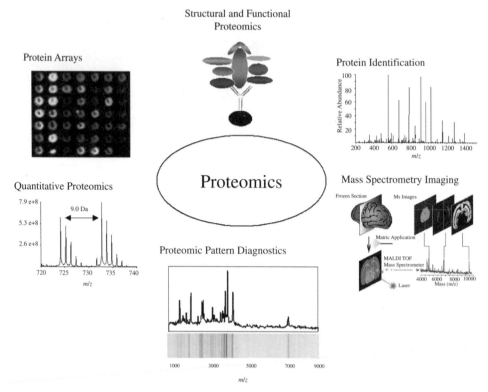

Figure 4.1. *A schematic of some of the various solutions used to gain a greater understanding of the proteome. Of these methods, protein identification is the best developed technology; however, exciting strides are being made in the field of diagnostic medicine through the use of proteomic pattern methods and mass spectrometry imaging. Extraordinary progress has been made in structural and functional proteomics that promise greater understanding of protein interactions and three-dimensional structure. While protein arrays are still not in widespread use, this developing technology has a very tangible future in proteome characterization.(See color insert.)*

conquered is the characterization of post-translational modifications (PTMs) on a global level. While there are more than 200 known, the major PTMs that occur within the eukaryotic cells (and currently of the most interest in proteomics) are phosphorylation and glycosylation (Griffin et al., 2001).

The importance of phosphorylation and glycosylation of proteins to the overall function of the cell cannot be underestimated. A keyword search of PubMed using "phosphorylation" and "glycosylation" results in the approximately 7500 and 1300 papers, respectively, just during the first three-quarters of the year 2003. Phosphorylation is the key signal through which proteins propagate signals within a cell (Cohen, 2002). It has been estimated that at least one-third of all of the proteins in a eukaryotic cell are phosphorylated at any given time. While it is unlikely that every potential phosphorylation site is central to protein function, their sheer number reflects their overall importance in cell homeostasis. Glycosylation influences the physiochemical properties of a protein; in such ways as altering its resistance to proteolysis, it solubility, stability, overall structure, and immunogenicity (Haltiwanger and Lowe, 2004). Glycosylation is also commonly observed within

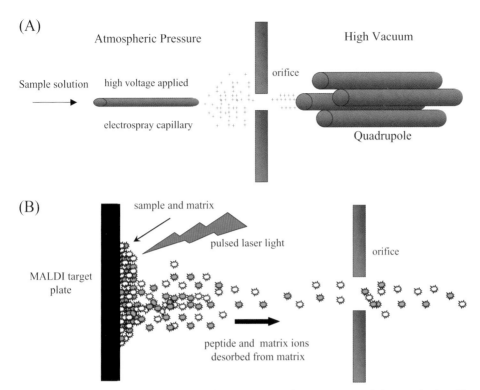

Figure 1.3. (A) Electrospray ionization (ESI) of molecules for mass spectral characterization. The sample solution is passed through a stainless steel or other conductively coated needle. A high positive potential is applied to the capillary (cathode), causing positive ions to drift toward the tip with high voltage. The presence of a high electric field produces submicrometer-sized droplets upon the solution exiting the needle. The droplets travel toward the mass spectrometer orifice at atmospheric pressure and evaporate and eject charged analyte ions. The desolvated ions are drawn into the mass spectrometer by the relative low pressure maintained behind the orifice. (B) Principles of matrix-assisted laser desorption ionization (MALDI). The sample is cocrystallized with a large excess of matrix. Short pulses of laser light are focused onto the sample spot, causing the sample and matrix to volatilize. The matrix absorbs the laser energy, causing part of the illuminated substrate to vaporize. A rapidly expanding matrix plume carries some of the analyte into the vacuum with it and aids the sample ionization process.

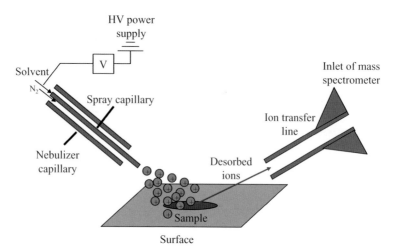

Figure 1.4. *Schematic of desorption electrospray ionization (DESI) instrument. In DESI, electro-sprayed droplets are directed to a surface. The impact of the charged droplets produces gaseous ions from the sample on the surface, which can be sampled using a commercially available mass analyzer equipped with an ion transfer line.*

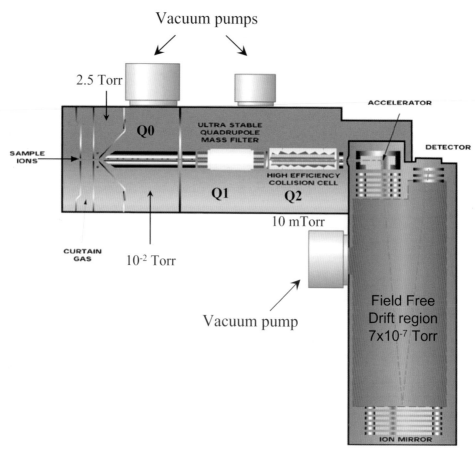

Figure 1.5. *Schematic of a quadrupole time-of-flight mass spectrometer (TOF-MS). In this instrument the third quadrupole of a triple quadrupole mass spectrometer has been replaced with a TOF tube. This combination gives this instrument the ion selection and tandem MS capabilities of a triple quadrupole MS with the high mass accuracy and resolution capabilities of a TOF.*

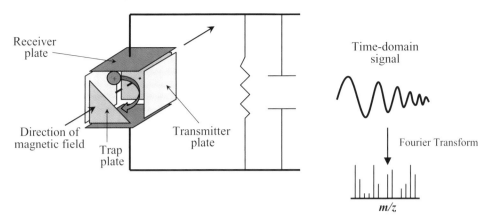

Figure 1.6. *Principles of Fourier transform ion cyclotron resonance (FTICR) mass spectrometry. In FTICR-MS the ion trap is placed in a strong magnetic field. The magnetic field causes ions captured within the trap to resonate at their cyclotron frequency. By applying the appropriate electric field energy, the ions are excited into a larger orbit that can be measured as they pass by detector plates on opposite sides of the trap. Energy can also be applied to dissociate the ions or to eject ions from the trap by accelerating them to a cyclotron radius larger than the radius of the trap. The detector measures the cyclotron frequency of all of the ions in the trap and uses a Fourier transform to convert these frequencies into m/z values.*

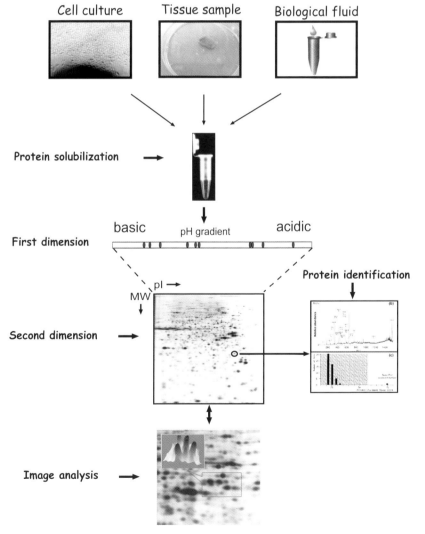

Figure 2.1. *Typical gel-based proteomic project scheme for generating and analyzing 2D protein images.*

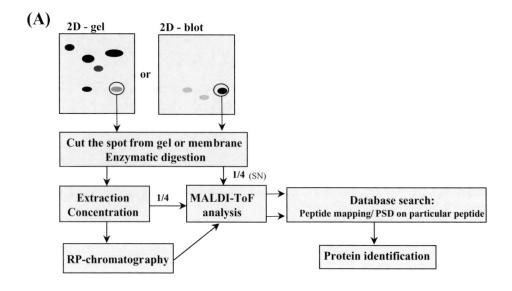

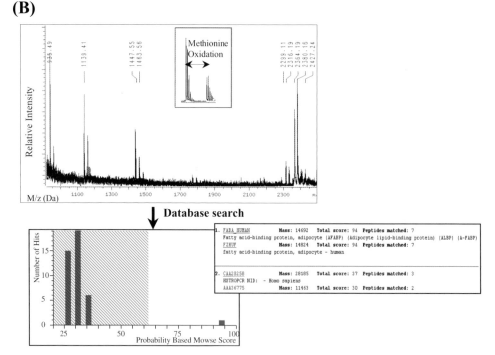

Figure 2.7. (A) Strategy for identifying proteins using MALDI-TOF/MS. (B) Identification of proteins excised from 2D gels followed by in situ digestion and MALDI-TOF analysis. The upper panel shows the MALDI-TOF mass spectrum. The pick list provides the input data for protein identification obtained by searching the MCDB database with the MASCOT software. The lower panels present the identification probability plot from the database search. Probability scores greater than 57 are considered significant ($p < 0.05$).

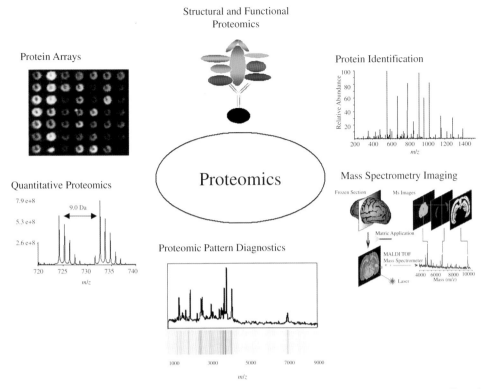

Figure 4.1. A schematic of some of the various solutions used to gain a greater understanding of the proteome. Of these methods, protein identification is the best developed technology; however, exciting strides are being made in the field of diagnostic medicine through the use of proteomic pattern methods and mass spectrometry imaging. Extraordinary progress has been made in structural and functional proteomics that promise greater understanding of protein interactions and three-dimensional structure. While protein arrays are still not in widespread use, this developing technology has a very tangible future in proteome characterization.

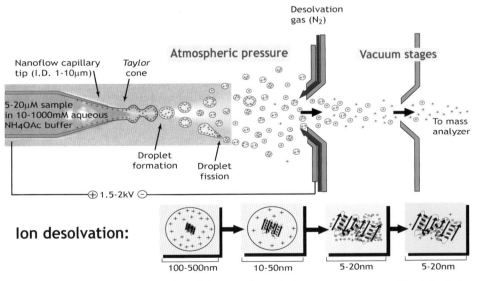

Figure 8.1. Schematic representation of the nanoflow electrospray process used for introducing noncovalent protein RNA complexes. Complexes at concentrations typically in the range 5–20 μM are introduced from ammonium acetate buffer using a nanoflow capillary. A voltage of 1.5–2 kV is typically applied to the capillary and backing pressure is often used to initiate flow. Droplet formation takes place at atmospheric pressure, each droplet calculated to contain one complex molecule, providing the appropriate concentration and needle orifice are employed. With the aid of a counter-current flow of gas, desolvation takes place such that the droplet shrinks from a few hundred nm to ~10 nm, yielding gas-phase ions largely devoid of solvent and buffer molecules.

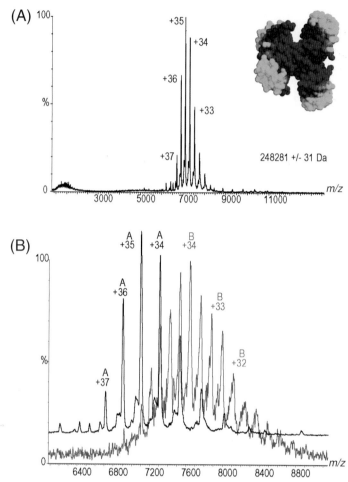

Figure 8.3. *(A) Nanoflow electrospray mass spectrum of the N-terminal domain of RNase E catalytic domain recorded under nondissociating conditions. Charge states assigned to the protein tetramer are labeled. (B) Expansion of the spectrum over the m/z range 6300–9000. The lower spectrum was recorded in the presence of a twofold stoichiometric excess of a substrate RNA analog. The additional charge states correspond in mass to a distribution of RNA molecules binding to the protein tetramer. The major series labeled with charge states corresponds to up to three molecules of RNA binding to the protein tetramer, but higher m/z species can be observed with up to four molecules on the lower charged states. Inset: A possible solution for the N-terminal domain of RNase E based on solution scattering profile and symmetry constraints. The structure was produced with Molscript (Kraulis, 1999) and reproduced with permission from Callaghan et al. (2003).*

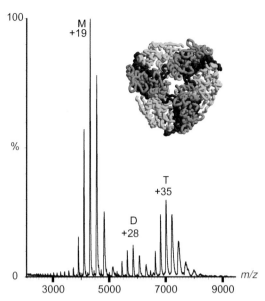

Figure 8.4. *Mass spectrum of PNPase from Streptomyces antibioticus showing the three major components assigned to monomer, dimmer, and trimer. Inset: Crystallographic structure in which subunit is represented in different shades. Adapted from Symmons et al. (2000).*

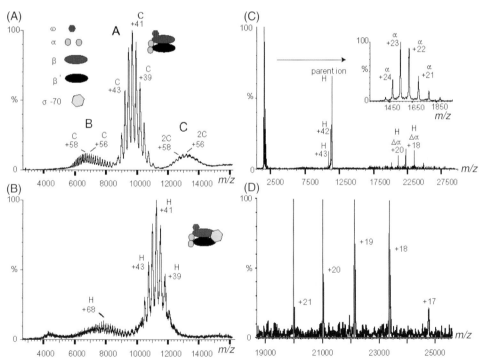

Figure 8.5. *Mass spectra of core and holo forms of E. coli RNA polymerase (A, B) and tandem mass spectrum of the +41 parent ion of the holo complex (C, D). (A) The core enzyme gives rise to a predominant series of peaks labeled A, a more highly charged series labeled B, and a low intensity series corresponding in mass to dimer labeled C. (B) Addition of recombinant s^{70} leads to formation of the holo complex at higher m/z values. Inset: Schematic representation of the subunits adapted from the X-ray structure. (C) Tandem mass spectrum of the +41 charge state of the holo enzyme. (D) Expansion of the high m/z ions formed by loss of an a subunit and comparison with the theoretical isotope simulation of the holoenzyme minus the a subunit. Reproduced with permission from Ilag et al. (2005).*

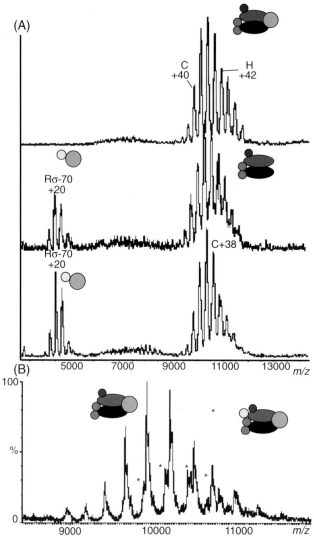

Figure 8.6. Spectra recorded after addition of Rsd to the holo RNA polymerase (A) and acceleration of the resulting high mass complex (B). Upper trace holoenzyme alone, middle trace holoenzyme with equimolar Rsd, and lower trace holoenzyme with twofold excess of Rsd. The series of ions at low m/z corresponds in mass to a 1:1 complex of Rsd with σ^{70}. The complex formed in the presence of a twofold excess of Rsd over that of holoenzyme was accelerated in the collision cell at 150 V, revealing the presence of species labeled * assigned to the core enzyme bound to Rsd. Reproduced with permission from Ilag et al. (2005).

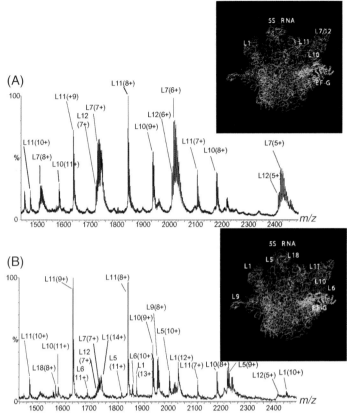

Figure 8.7. Mass spectra of the m/z region 1450–2500 of the ribosome EF-G complex in the presence of (A) fusidic acid and (B) thiostrepton. The two spectra are markedly different. The spectrum recorded in the presence of fusidic acid is similar to that observed for ribosomes in the absence of EF-G under these solution and mass spectrometry conditions. By contrast, the complex inhibited by thiostrepton demonstrates the absence of L7/L12 and the presence of additional proteins L5, L6, and L18. The structure of the 50S subunit was produced using the coordinates from Thermus thermophilus at 5.5 Å resolution PDB file 1GIY (Yusupov et al., 2001). The structure of EF-G (pdb ascension code 1FNM) was fitted according to the structure of Ban et al. (2000). The proteins shaded in the two structures represent those that are released from the two complexes. Reproduced with permission from Hanson et al. (2003).

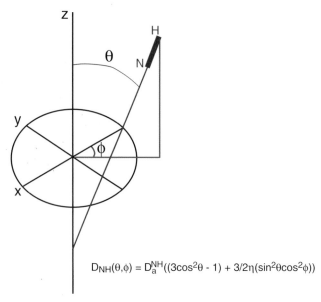

$$D_{NH}(\theta,\phi) = D_a^{NH}((3\cos^2\theta - 1) + 3/2\eta(\sin^2\theta\cos^2\phi))$$

Figure 9.3. *Schematic illustration of orientational information derived from residual dipolar coupling measurements. The observed dipolar coupling, D_{NH}, is dependent on the angle θ between the N-H interatomic vector (shown as the thick line) and the z axis of the tensor, the angle ϕ, which describes the position of the projection of the interatomic vector on the xy plane of the tensor, the magnitude (D_a^{NH}) of the principal component of the tensor, and the rhombicity (η) of the tensor.*

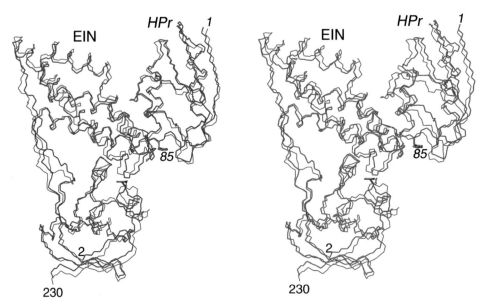

Figure 9.4. *Comparison of the structure of the EIN·HPr complex obtained using the conventional full structure determination approach (red) with that obtained by conjoined rigid body/torsion angle dynamics on the basis of 231 backbone N–H dipolar coupling data and either a full complement (blue) or partial complement (green) of NOE-derived intermolecular interproton distance restraints. The full complement of intermolecular NOEs comprises 109 interproton distance restraints; the partial complement consists of only eight intermolecular methyl proton–NH interproton distance restraints. The relative orientation of the proteins in all three calculated structures is identical. The backbone rms difference between the conventional NMR structure (red) and the structure calculated by docking the X-ray coordinates of the free proteins using the full complement of intermolecular NOEs (blue) only reflects the differences in the NMR and X-ray coordinates of the individual proteins, and these differences are within the uncertainty of the NMR coordinates. Adapted from Clore (2000).*

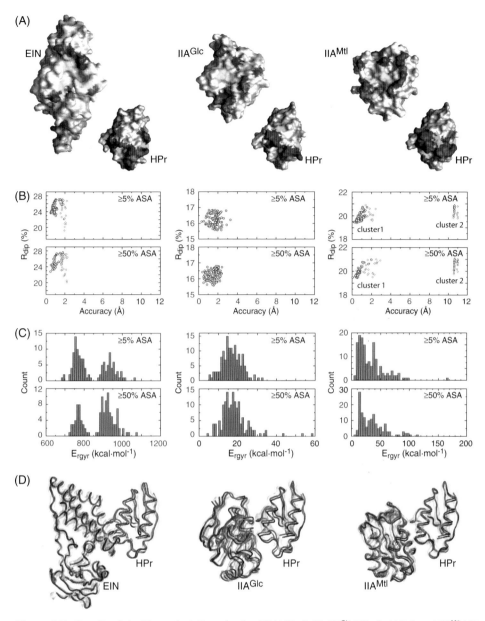

Figure 9.5. *Results of docking calculations for the EIN·HPr (left), IIA^{Glc}·HPr (middle) and IIA^{Mtl}·HPr (right) complexes on the basis of highly ambiguous distance restraints derived from ^{15}N/^{1}H$_N$ chemical shift perturbation maps and backbone N–H dipolar couplings. (A) Interfacial residues (blue/cyan for HPr, red/orange for the three enzymes, and purple for active site histidines) identified by ^{1}H$_N$/^{15}N chemical shift perturbation are displayed on a molecular surface representation of the proteins. (The blue and red colored interfacial residues indicate residues with an accessible surface area (ASA) in the free proteins ≥50% of that in an extended Gly-X-Gly peptide; the cyan and orange colored residues indicate interfacial residues in the free proteins with 5% ≤ ASA < 50%. (B) Plots of the dipolar coupling R-factor (R_{dip}) versus accuracy for the converged structures characterized by no violations >0.5 Å in the highly ambiguous intermolecular distance and $R_{dip} \leq R_{dip}^{median}$. In the case of the EIN·HPr (left panel) and IIA^{Mtl}·HPr complexes (right panel), the circles and diamonds indicate structures in the lower and higher energy populations, respectively, of the radius of gyration energy function (E_{rgyr}) distribution. (C) Histograms of the E_{rgyr} distributions for the converged structures. The E_{rgyr} distribution is unimodal for the IIA^{Glc}·HPr complex (middle), but bimodal for the EIN·HPr (left) and IIA^{Mtl}·HPr (right) complexes. For the bimodal distributions, the lower and higher energy E_{rgyr} populations are colored red and blue, respectively. Note that in the case of the IIA^{Mtl}·HPr complex, all the structures in lower energy E_{rgyr} population reside in the correct cluster 1 ensemble; all the structures in the incorrect cluster 2 ensemble reside in the higher energy E_{rgyr} population. (D) Backbone (depicted as tubes) best-fit superpositions of the average coordinates (red) of the converged structures on the previously determined NMR structures (blue) solved on the basis of intermolecular NOEs and residual dipolar couplings. In the case of the IIA^{Mtl}·HPr complex, the mean coordinates are derived from the cluster 1 ensemble. The ensemble distributions of the docked structures are depicted by isosurfaces of the reweighted atomic density maps. Reproduced from Clore and Schwieters (2003).*

(A) Glucose PTS

(B) EIN·HPr complex

(C) IIA^{Glc}·HPr complex

(D) IIA^{Glc}·IICB^{Glc} complex

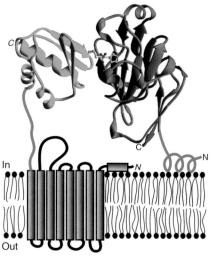

Figure 9.6. *Summary of the glucose arm of the E. coli PTS. (A) Diagrammatic illustration of the PTS cascade illustrating the transfer of phosphorus originating from phosphoenolpyruvate and ending up on glucose through a series of bimolecular protein–protein complexes between phosphoryl donor and acceptor molecules. Ribbon diagrams of the (B) first (EIN·HPr), (C) second (HPr·IIA^{Glc}), and (D) third (IIA^{Glc}·IICB^{Glc}) complexes of the glucose PTS. The N-terminal domain of EI (EIN) is shown in gold, HPr in red, IIA^{Glc} in blue, and the IIB^{Glc} domain of IICB^{Glc} in green. Also shown in yellow are the active site histidine residues of EIN (His189), HPr (His15) and IIA^{Glc} (His90) and the active site cysteine (Cys35) of IIB^{Glc}, together with the pentacoordinate phosphoryl group (red atoms) in the transition states of the complexes. IIB^{Glc} constitutes the C-terminal cytoplasmic domain of IICB^{Glc}. The transmembrane IIC^{Glc} domain of IICB^{Glc} is thought to comprise eight transmembrane helices (shown diagrammatically in black). Note that the N-terminal end of IIA^{Glc} (residues 1–18) is disordered in free solution (C), but upon interaction with a lipid bilayer, residues 2–10 adopt a helical conformation (D), thereby further stabilizing the IIA^{Glc}·IIB^{Glc} complex, by partially anchoring IIA^{Glc} to the lipid membrane. Reproduced from Cai et al. (2003).*

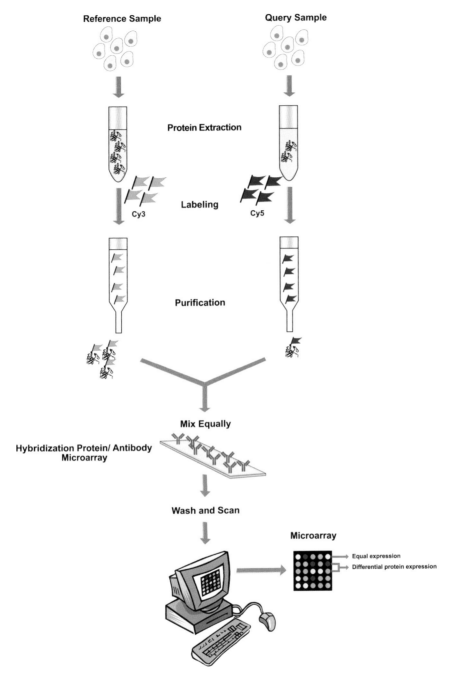

Figure 10.1. Schematic representation of the dual color labeling approach. Proteins are extracted from reference and query samples and labeled with fluorescent dyes (typically Cy3 and Cy5). Following a "clean-up" to remove unincorporated dye, the labeled proteins are mixed and hybridized to the microarray. After washing, the microarray is scanned using a microarray scanner. Hybridization signals are extracted and differential protein levels are determined.

- Target
- Antibody
- Label
- Surface

Figure 10.2. *Common microarray formats. In forward-phase protein arrays, affinity reagent (antibody) is immobilized onto a solid support. In reverse-phase protein arrays, samples to be analyzed (cell extracts or tissues) are immobilized onto a solid surface.*

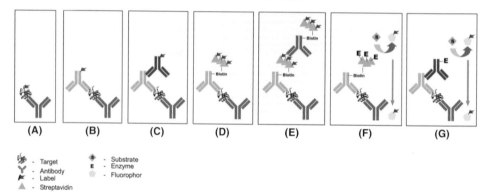

(A)   (B)   (C)   (D)   (E)   (F)   (G)

- Target
- Antibody
- Label
- Streptavidin

- Substrate
E - Enzyme
- Fluorophor

Figure 10.4. *Signal generation schemes used in antibody microarray applications. (A) Direct labeling, (B) and (C) indirect labeling—sandwich assays, (D) and (E) indirect labeling—biotin mediated; (F) and (G) Indirect labeling—enzyme mediated.*

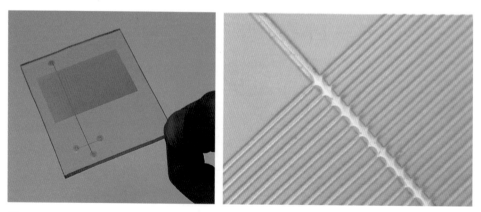

Figure 11.1. *(Left) Fabricated polycarbonate two-dimensional microfluidic chip. (Right) Micrograph showing detail of intersecting microchannels. (Image courtesy of Calibrant Biosystems.*

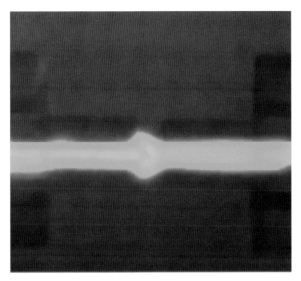

Figure 12.5. *Coaxial sample interface. A fluorescent microscope is used to image the transfer of a plug of fluorescein between two capillaries.*

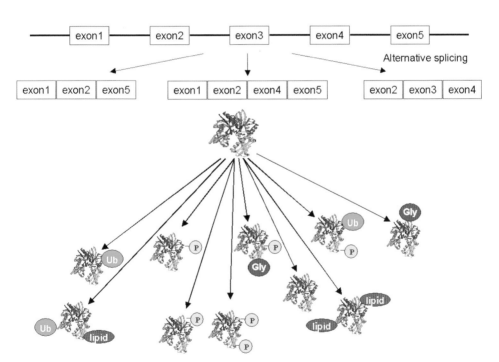

Figure 14.1. *Figure showing the ultimate protein complexity that can arise from the transcription and translation of a single gene.*

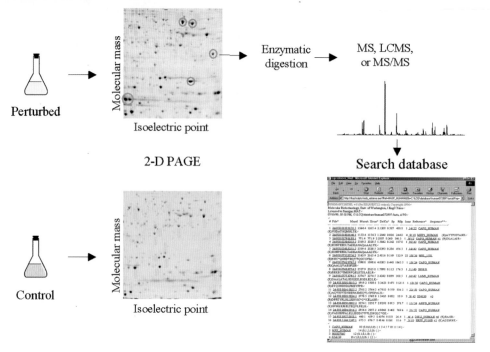

Figure 14.2. Schematic showing general comparative analysis of two proteome samples using two-dimensional polyacrylamide gel electrophoresis (2D-PAGE). Samples are initially separated by 2D-PAGE and protein spots that are of greater intensity on one of the gels are excised, digested into peptides, and analyzed using mass spectrometry (MS). The MS results are searched against an appropriate database to identify the differentially abundant protein.

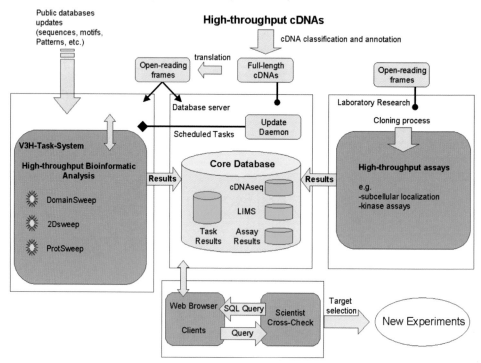

Figure 14.4. Schematic of a data analysis pipeline for an integrated and automated strategy to obtain and administer data from high-throughput investigations of cDNAs. The data input is full-length cDNAs. Verified open reading frames (ORFs) are cloned and expressed for their use in experimental assays to measure such parameters as subcellular location and structure. All identified ORFs move through the pipeline, where they are exhaustively analyzed by running automated tasks such as DomainSweep, 2Dsweep, and ProtSweep. Computational and experimental results are integrated into a relational database (Core Database). The core database contains several single databases allowing researchers to cross-check various protein features in silico through the execution of SQL queries via web browsers or clients.

membrane proteins and can act as recognition sites on different receptors (Bertozzi and Kiessling, 2001).

Many methods have been developed over the past few decades to determine the PTM state of a protein; however, most of these techniques have been applied to the analysis of a single purified protein. Affinity reagents such as anti-phosphoamino acid-specific monoclonal antibodies (mAbs) have been a popular choice to detect phosphoproteins, and lectin binding has widely been utilized to characterize glycoproteins. Such studies suffer, however, unless additional characterization is conducted as these affinity-based detection methods determine only if a protein is modified and are not necessarily designed to identify the specific site of modification. Detailed knowledge of the sites of PTM is important since identical modifications at different sites within the same protein can have widely different effects on protein activity and many proteins may be phosphorylated by many different kinases. For example, TFII-I, a ubiquitously expressed multifunctional transcription factor with broad biological roles in transcription and signal transduction in a variety of cell types, has been shown to contain Src-dependent tyrosine phosphorylation sites; however, the protein is known to interact both physically and functionally with Bruton's tyrosine kinase (Sacristan et al., 2004). In addition, the α-amino-3-hydroxy-5-methyl-4-isoxazolepropionate (AMPA) receptor is phosphorylated by cAMP-dependent protein kinase II as well as protein kinases A and C, with each modification being related to a different AMPA receptor function (McDonald et al., 2001). As mentioned earlier, site-specific monoclonal antibodies can be used to map specific sites of phosphorylation; however, this undertaking is typically laborious and it is unlikely that a useful inventory of site-specific antibodies can be produced for all types of modifications. Antibodies will also continue to be dissected concerning their absolute site specificity.

Presently, the best available technology (and one that has tremendous untapped potential) to identify PTMs within a proteome is MS. The ability to obtain highly accurate masses coupled with the ability to elucidate sequence information of peptides with a high degree of accuracy by MS is directly applicable to the site-specific identification of modifications. As with protein identification, MS has been used primarily to identify specific sites within a protein that are modified, but with the incorporation of the proper sample processing methods it also holds great potential in the determination of the relative extent of modification.

4.2 PHOSPHORYLATION

Arguably the most important and best understood modification used to modulate protein activity and propagate signals for cell homeostasis is phosphorylation (Cohen, 2002). Cell cycle progression, differentiation, development, peptide hormone response, and adaptation are just a few of the processes that are regulated by protein phosphorylation. Historically, the most popular method used to study protein phosphorylation requires radiolabeling with ^{32}P inorganic phosphate ($^{32}P_i$) in vivo or [$\gamma^{32}P$]ATP in vitro (Manning et al., 1980). The radioactive proteins are separated using fractionation procedures such as two-dimensional polyacrylamide gel electrophoresis (2D-PAGE) or high-performance liquid chromatography (HPLC). To determine the amino acid types that are modified, the phosphoprotein can be hydrolyzed completely and the phosphoamino acid content determined. The specific site(s) of phosphorylation is determined by proteolytically digesting the radiolabeled protein, separating and detecting the phosphorylated peptides, and sequencing the radioactive peptide(s) by Edman degradation. These methods require significant amounts of protein, are laborious, and inherently possess all of the difficulties associated with working with radioactive compounds. Quantitative differences in phosphorylation status of specific proteins are conducted by ^{32}P-labeling entire proteomes that are subsequently fractionated by two-dimensional polyacrylamide gel electrophoresis (2D-PAGE) and the relative spot intensities are compared (Chu et al., 2004). Unfortunately, [32]P$_i$-labeling does not lend itself to high-throughput proteome-wide analysis due to issues with handling radioactive compounds and the associated contamination of analytical instrumentation.

Immunostaining is a more universally applicable approach to studying phosphoproteins that is highly sensitive. The general low abundance nature of phosphoproteins, however, often dictates a two-pronged analytical approach for detection and subsequent identification of these proteins. As shown in Figure 4.2, two distinct 2D-PAGE gels of a given sample can be run on both an analytical and a preparative gel, where the analytical gel is used to immunohistochemically detect phosphoproteins after transfer to a PVDF membrane, which serves as a template to guide subsequent spot picking from the preparative gel. The preparative gel is run such that it contains a greater amount of total protein and is stained using a conventional staining method such as Coomassie Blue or silver stain. Any spots that match between the immunoblot and the preparative gel are then cored from the preparative gel and in-gel digested, and the protein is identified by MS or tandem MS.

In its simplest form, MS can be used to obtain an accurate mass measurement of an intact phosphorylated protein, which is then compared to the calculated mass of the unmodified protein or the mass of the protein after phosphatase treatment. The difference (in multiples of 80 Da, the mass of phosphate) in these two masses can then be used to calculate the number of phosphorylated sites within the protein (Han et al., 2002). The measurement of the mass of the intact protein does not provide any specific information related to the site of phosphorylation—a key piece of information since many proteins possess multiple phosphorylation sites and each can have a distinct effect on the protein's function. Analysis of the protein at the peptide level, followed by MS or tandem MS, is required to identify the exact site of phosphorylation (Sakchaisri et al., 2004). When two or more possible phosphorylatable residues are present within a peptide, tandem MS is necessary to establish the specific site of phosphorylation. Since a majority of the identification of phosphorylation sites are made at the peptide level, this chapter will focus on MS analysis of phosphopeptides using what is referred to as a bottom-up approach; however, top-down approaches, while not yet fully matured, are expected to be of increasing importance in the coming years in proteomics.

4.2.1   Identification of Phosphorylated Proteins

Enrichment of Phosphopeptides   In utilizing MS for identifying phosphorylated residues within a protein, the sample is most typically first digested into pep-

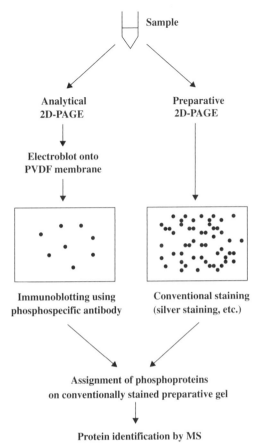

Figure 4.2. *The use of 2D-PAGE and immunoblotting to identify phosphoproteins. Due to the low abundance of most phosphoproteins, the sample, a preparative gel, and the analytical gel of the proteome sample can be run. Obviously, a greater amount of sample is loaded onto the preparative gel. After separation, the analytical gel is immunoblotted using a phospho-specific antibody to reveal the location of phosphoproteins within the gel. Spots visualized in this manner are aligned with their spot on the preparative gel that has been stained using a conventional staining method. The phosphoprotein can then be identified using standard mass spectrometry-based proteome methods.*

tides using either chemical (i.e., cyanogen bromide) or enzymatic (i.e., trypsin) means (Meek and Milne, 2000). At this point in proteomic research, enzymatic digestion, particularly trypsin, is chosen over chemical means by a vast margin. In some cases, however, trypsin may generate peptides that are too small or large to allow accurate identification of the modification, if the site of interest is localized within a lysine (Lys)- or arginine (Arg)-rich or Lys/Arg-poor region; therefore, alternative methods or enzymes are often required. The choice of the correct proteolytic agent is simplified when the site(s) of phosphorylation is suspected or known. Searching for unknown phosphorylated residues may require a digestion strategy that uses different enzymes that produce peptides that allow broad coverage of the protein's primary structure. The ability to identify all possible peptides generated from a phosphoprotein digest is the key limiting factor in the identification of unknown phosphorylation sites since complete coverage of a protein by MS analysis is rarely achieved.

Enrichment of Phosphorylated Peptides   The relatively low abundance and stoichiometry of many phosphorylated proteins is one of the key challenges in their characterization. Success in detecting and characterizing phosphopeptides can be enhanced through methods that reduce the amount of nonphosphorylated peptides within the mixture while concomitantly enriching for the phosphopeptides. A mixture enriched in phosphopeptides not only increases the ability to detect this class of peptides but also aids in the identification since a greater percentage of phosphopeptides can be anticipated and can be used as a constraint in a truly discovery driven, global analysis of a complex proteome sample.

Immunoaffinity Chromatography   Antibodies represent one of the earliest means to enrich a sample for phosphopeptides prior to MS analysis (Pelech, 2004). Phosphorylated protein(s) can be immunoprecipitated from a mixture using antibodies. The extracted phosphoproteins can then be digested and analyzed by MS to identify the phosphoproteins as well as the specific site(s) of phosphorylation. If the protein of interest is known and the goal is to determine at which site(s) it is modified, an antibody directed toward any epitope within the protein may be used. A number of phosphoseryl (pSer)-, phosphothreonyl (pThr)-, and phosphotyrosyl (pTyr)-specific antibodies have been developed and can be used to isolate proteins with such types of modified residues. While the phosphoamino acid is part of the recognition epitope for these phospho-specific antibodies, it is often not the major recognition factor between the antibody and phosphoprotein. The neighboring residues surrounding the phosphorylated residue contribute to the specificity between the antibody and specific phosphoproteins, allowing the phosphorylation status of single proteins within complex mixtures to be monitored.

A more common procedure is to use antibodies whose affinity is more dependent on the phosphorylated residue and not the neighboring primary sequence. These antibodies are generally used to generate mixtures of proteins or peptides that are phosphorylated at a specific residue type. The components within these mixtures can then be identified by MS. Such a strategy was used to identify differences in the phosphorylation state of proteins within resting platelets and those that had been stimulated with thrombin (Marcus et al., 2003). Tyrosine-phosphorylated proteins were immunoprecipitated using an anti-phosphotyrosine antibody and the mixture was then separated by 2D- and 1D-PAGE. Phosphotyrosine-containing proteins were assigned by immunoblotting with an anti-phosphotyrosine antibody. Protein bands were excised from the gel, digested with trypsin, and analyzed by matrix-assisted laser desorption ionization time-of-flight (MALDI-TOF) MS and nano-LC-electrospray ionization (ESI)-MS/MS. Several phosphorylated proteins were identified and some in vivo phosphorylation sites were assigned to specific residues. A similar strategy was used to study changes in the phosphorylation states of proteins extracted from L929 cells (murine fibroblasts) exposed to tumor necrosis factor-α (TNF-α) (Yanagida et al., 2000). Proteins were extracted from L929 cells at various time points after TNF-α treatment and resolved by 2D-PAGE. The 2D-PAGE resolved proteins were either electroblotted onto a PVDF membrane to enable the identification of phosphoproteins as well as quantify any changes in the phosphorylation state of the proteins over time, or silver stained to visualize them for selection for MS identification. Twenty-one different phosphoproteins were identified within the L929 cell lysates, including eight that showed a time-dependent change in their phosphorylation state based on the immunostaining of the PVDF membrane.

While the above examples describe the use of phosphoamino acid specific antibodies to isolate intact phosphoproteins prior to digestion into peptides and MS analysis, these antibodies may also be used to isolate phosphopeptides after digestion of the intact protein. Such a postdigestion strategy was used to identify phosphorylated residues within EphB2. EphB2 is a receptor tyrosine kinase that is involved in neuronal axon guidance, neural crest cell migration, the formation of blood vessels, and the development of facial structures and the inner ear (Kalo et al., 2001). Intact EphB2 protein was first isolated using an anti-EphB2 antibody and was subsequently digested using trypsin. An anti-pTyr antibody was used to isolate the phosphopeptides from this mixture, which were analyzed using MALDI-TOF MS. Eight major peaks were observed in the mass spectrum, representing phosphopeptides containing nine different pTyr residues.

Immobilized Metal Affinity Chromatography   One of the most popular affinity-based methods to extract phosphoproteins or phosphopeptides from complex mixtures is based on immobilized metal affinity chromatography (IMAC) (Porath, 1992). In the application of IMAC, trivalent cations such as Fe^{3+} or Ga^{3+} are chelated by iminodiacetic or nitrilotriacetic acids that are coupled to a chromatographic solid phase support. Enrichment for phosphopeptides in a complex proteome digestate using IMAC can be accomplished because of the greater affinity of the phosphate moiety in phosphopeptides for the transition metal ion. After washing the column to elute nonspecifically bound peptides, the remaining bound species (e.g., phosphopeptides) are eluted using high pH or a phosphate buffer. Unfortunately, IMAC notoriously binds peptides rich in carboxylate groups (e.g., Asp and Glu). A method to reduce this nonspecific binding of nonphosphopeptides to IMAC was reported in which the proteins are digested with trypsin and the resulting peptides are then converted to methyl esters using methanolic·HCL, prior to enrichment using IMAC (Ficarro et al., 2002). This strategy was applied to a whole yeast lysate that was subsequently analyzed by nano-LC-ESI-MS/MS resulting in the detection of more than 1000 phosphopeptides. A total of 216 peptide sequences defining 383 sites of phosphorylation were determined. Of these, 60 were singly phosphorylated, 145 doubly phosphorylated, and 11 triply phosphorylated. Eighteen of these sites were previously identified in other studies, including a doubly phosphorylated motif pTXpY derived from the activation loop of two mitogen-activated protein (MAP) kinases.

The above method was then extended to display and quantify differential expression of phosphoproteins in two different cell systems by using differential isotopic versions of the esterification reagent (i.e., D0-methanolic·HCL and D3-methanolic·HCL) as schematically shown in Figure 4.3 (Ficarro et al., 2003). In this study, the phosphorylation status of proteins within mammalian sperm undergoing capacitation, a process that must occur before fertilization, was studied. Phosphopeptides were isolated from sperm using IMAC, after they had been methyl esterified using methanolic·HCL. More than 60 phosphorylated sequences

(A)

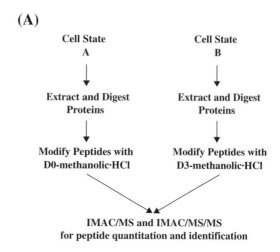

(B)

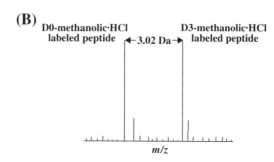

Figure 4.3. (A) Schematic representation of phosphopeptide quantitation using D0- and D3-methanolic HCl derivatization of peptides followed by IMAC enrichment of phosphopeptides and quantitation and identification by MS and MS/MS. (B) Representative MS spectrum of a pair of phosphopeptides derivatized with isotopically different versions of methanolic·HCL.

were identified using MS/MS, including precise sites of tyrosyl and seryl phosphorylation of the sperm tail proteins AKAP-3 and AKAP-4. Differential isotopic labeling was used to then quantify phosphorylation changes that occur during capacitation. In this study, one set of proteins was esterified using D0-methanolic·HCL and the other set with D3-methanolic·HCL prior to IMAC enrichment and MS analysis of the phosphopeptides. Although the exact site of phosphorylation of valosin-containing protein, a key protein involved in sperm capacitation, was not identified, a cross-immunoprecipitation approach did confirm that this protein is tyrosine phosphorylated during capacitation.

The first group to actually demonstrate the incorporation of an isotopic tag to quantitatively measure the phosphorylation status of phosphopeptides from two distinct sources was Chait's laboratory (Oda et al., 1999). The general strategy is shown in Figure 4.4. In this initial demonstration, a wild-type and G1 cyclin Cln2-deficient yeast strain were grown in either natural isotopic abundance or ^{15}N-enriched media, respectively. After combining the cells and extracting the proteins, the Cln2-dependent protein STE20 was isolated and analyzed by ESI-MS. Mass

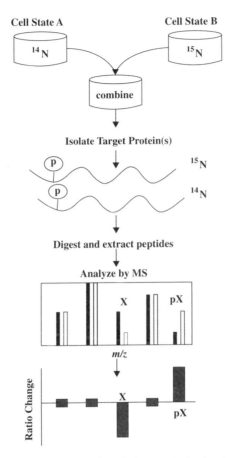

Figure 4.4. *Method for site-specific quantitation of changes in the level of phosphorylation on pro-*
teins. For illustration, peptides that remain unchanged in the two cell pools are assumed to be present
in equal abundance, and the level of phosphorylation of peptide A is assumed to change from 30%
(cell state A) to 70% (cell state B), leading to a decrease in the measured intensity ratio of unphos-
phorylated peptide X and an increase for phosphorylated peptide pX as shown in both the mass
spectrum and graph of the ratio change of each detected pair of isotopically labeled peptides.

spectral measurement of the intensity ratios of the isotopically labeled ($cln2^-$)
versus unlabeled ($CLN2^+$) phosphopeptides showed that at least four sites exhibit
large increases in phosphorylation in the $CLN2^+$ cell pool. These Cln2-dependent
sites appear to be consensus cyclin-dependent pSer/pThr, consistent with direct
phosphorylation of STE20 by Cln2-Cdc28.

Recently, another group put a twist on this approach by using an amino acid
specific isotope label (Ibarrola et al., 2003). In this study, two populations of 293T
cells were adapted to grow either in lysine-$^{12}C_6$- or lysine-$^{13}C_6$-containing media.
The cells were then transfected with FLAG epitope-tagged expression construct
encoding Frigg, a protein that is phosphorylated when the cells are treated with
calyculin A. After treatment for 3 or 30 min, the protein content of the cells was
lysed and equal amounts of protein from each cell sample were combined. The
Frigg protein was enriched from the mixed lysate by immunoprecipitation using an

anti-FLAG antibody. After separating the mixture by 1D-PAGE, the band corresponding to Frigg was excised, digested with trypsin, and analyzed by LC-MS/MS. The resulting MS spectra showed several pairs of peptides differing by 6 Da; the mass difference between singly charged peptide species containing a normal lysyl residue (i.e., lysine-$^{12}C_6$) versus those containing a heavy lysine residue ($^{13}C_6$), and with the same signal intensity indicating total incorporation of [$^{13}C_6$]lysine in the experimental cell population. The intensities of the unmodified peptide pairs, identified as originating from the protein Frigg, were equal based on the extracted ion currents of these molecular ions. Within the spectra, a pair of [M + 3H]$^{3+}$ peptide molecular ions (peaks at m/z 987.78 and 989.78) was found with a signal intensity ratio of 1:9. Tandem MS confirmed both of these signals as arising from a phosphorylated peptide within Frigg (LTRYpSQGDDDGSSSSGGSSVAG-SQSTLFK). A mass difference of 69 Da was observed between the b_4 and b_5 ions, indicating the presence of dehydroalanine, generated by the β-elimination of phosphoric acid (98 Da) from the pSer residue at position 5. The intensity ratio observed in the extracted ion chromatogram of this pair reflected a ninefold increase in the amount of phosphorylation after a 30 min calyculin A treatment as compared to a 3 min treatment. Both approaches demonstrate an effective means by which isotopic labeling directed to the entire protein (and not specifically to the site of phosphorylation) can be used to quantitate changes in the phosphorylation state of proteins within complex mixtures. Unfortunately, these methods suffer from two disadvantages. Since the protein itself is isotopically labeled using heavy isotope-enriched medium, the method is amenable only to organisms that can be labeled metabolically. In addition, this approach doesn't provide any means to specifically enrich a mixture for phosphopeptides beyond the use of IMAC or antibodies as described earlier.

Chemical Modification and Isotopic Labeling of Phosphopeptides

Other methods that are phosphorylation site specific and universally applicable have been developed to enable the specific enrichment *and* quantification of phosphopeptides. Both methods incorporate stable isotopes to differentially label the samples to be compared and employ subsequent MS analysis for the identification and quantitation of the enriched phosphopeptide mixture. Unfortunately, neither of these methods has been applied to a large, discovery driven study designed to quantitate global changes of phosphorylation status in a global proteome sample.

The first strategy to isolate and quantify phosphopeptides was developed concurrently, and independently, by two groups (Goshe et al., 2001; Oda et al., 2001). There are subtle differences in the specific procedures, however, the overall approaches of both methods are quite similar. The reaction scheme is specific for labeling pSer and pThr residues; unfortunately, pTyr residues are refractory to the chemistry involved and only pSer and pThr residues are successfully labeled. In the first step, reactive thiolates of cysteinyl residues are blocked via reductive alkylation or performic acid oxidation. The phosphate moieties are subsequently removed via hydroxide ion-mediated β-elimination from the pSer and pThr residues, resulting in their conversion to dehydroalanyl and β-methyl-dehydroalanyl residues, respectively. While it may seem counterproductive to remove the phosphate groups when the goal is to detect phosphorylation sites, this step is necessary to create the reactive centers required for subsequent modification. The newly formed α,β-unsaturated double bond renders the β-carbon in each of the newly formed residues sensitive to nucleophilic attack. The next modification involves a Michael-type addition of the bifunctional reagent 1,2-ethanedithiol (EDT). The addition of EDT results in the creation of a free thiolate in place of what was formerly a phosphate moiety. This free thiolate can now react with and become covalently modified with iodoacetyl-PEO-biotin. The result is the modification of pSer and pThr residues with a linker molecule containing a terminal biotin group. Relative quantification is achieved by using either the light ($HSCH_2CH_2SH$ or EDT-D_0) or the heavy ($HSCD_2CD_2SH$ or EDT-D_4) isotopic version of EDT during the Michael-type addition. Once the samples have been modified, they are proteolytically digested and the digestate is passed over an immobilized avidin chromatography column to extract the modified peptides. These extracted peptides can then be analyzed using either LC-MS/MS or MALDI-MS.

Two MS strategies are required to identify and quantitate the phosphorylation state of the phosphopeptides. In the MS mode, the mass-to-charge ratios (m/z) and signal intensities of the intact peptides are measured. Signals originating from the modified versions of the phosphopeptides appear as resolved doublets within the mass spectrum separated by the mass difference between the EDT-D_0 and EDT-D_4 labels (i.e., 4.0 Da). The mass spectrum of the phosphorylated peptide FQS^{P}E-EQQQTEDELQDK from β-casein in which the pSer residue has been modified with EDT (D_0 or D_4) and iodoacetyl-PEO-biotin is shown in Figure 4.5A (Oda et al., 2001). The identity of the sequence of the phosphorylated peptide(s) is then derived from the tandem mass spectrum. Two key attributes of this strategy is that the chemical modification remains attached to the residue during MS/MS fragmentation of the peptide and the peptide ionization is not hampered by the presence of the phosphate moiety. In a typical experiment, the phosphate group may dissociate from the peptide during MS/MS, preventing the site-specific assignment of the phosphate modification. The modification strategy described above, however, allows the exact phosphorylation site to be determined by MS/MS as shown in Figure 4.5B (Goshe et al., 2001). In this example, the MS/MS spectra for the EDT-D_0 and EDT-D_4 modified versions of a peptide are shown. The mass difference between the y_{13}^{2+} and y_{14}^{2+} daughter ions of these modified peptides is equal to the mass of a seryl residue modified as described earlier, allowing the specific site of phosphorylation to be clearly identified.

A second labeling method to isolate and quantify phosphopeptides was developed by Aebersold and co-workers (Zhou et al., 2001). This method potentially provides greater enrichment than the method described earlier since the phosphopeptides are covalently linked to a solid support during processing. In the first preparatory step, the peptide amino groups are protected using t-butyl-dicarbonate (tBoc) chemistry to eliminate any potential for intra- and intermolecular condensation. The carboxylate and phosphate groups are then modified via a carbodiimide-catalyzed condensation reaction to form amide and phosphoramidate bonds. Using an acid wash to hydrolyze the phosphoramidate bonds deprotects the phosphate groups, and cystamine is attached to the regenerated phosphate group via a second carbodiimide-catalyzed condensation reaction. A reduction of the internal disulfide of cystamine is used to generate a free sulfhydryl group, enabling the peptides

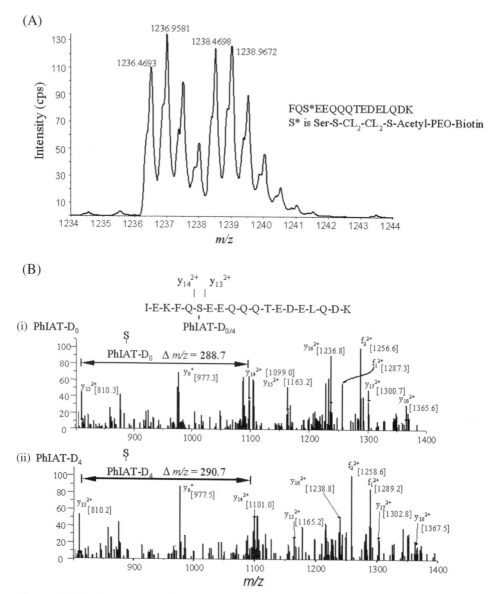

Figure 4.5. (A) Mass spectra of EDT-D_0/D_4 labeled β-casein peptide. The enriched mixture of bioti-
nylated phosphopeptides was analyzed by capillary reverse phase liquid chromatography coupled
directly on-line to a PE Sciex API QStar Pulsar hybrid quadrupole-TOF mass spectrometer. The
$[M + 2H]^{2+}$ ion pair corresponds to the mass of the derivatizde β-casein phosphopeptide
FQS*EEQQQTEDELQDK, where S* has the modified side chain —CH_2—SCL_2CL_2S-acetyl-PEO-
biotin and L is either H (EDT-D_0 label) or D (EDT-D_4 label). (B) Tandem mass spectrometry identification of β-casein phosphopeptides. The tandem MS/MS spectra of a phosphopeptide modified and
affinity isolated using the (i) light and (ii) heavy isotopic versions of EDT and iodoacetyl-PEO-biotin
are shown. Both labeled versions of the phosphopeptide were identified in a single LC data-
dependent MS/MS analysis of the enriched mixture.

to be attached to glass beads coupled with iodoacetyl groups. By covalently attaching the peptides to a solid support, more stringent washing conditions can be used, reducing the number of nonspecifically bound components being recovered. Phosphopeptides are recovered, and the tBoc protection groups removed by acid cleavage, thus regenerating peptides with free amino and phosphate groups. The carboxylate groups, however, remain blocked.

This method yielded mixtures highly enriched in phosphopeptides with minimal contamination from other nonphosphorylated peptides. This strategy does not require the removal of the phosphate groups; therefore, it is equally applicable to pSer-, pThr-, and pTyr-containing peptides. The tandem mass spectra of the modified phosphopeptides were of high quality, allowing the peptides to be identified using sequence database searching. The resulting tandem mass spectra allowed pSer/pThr and pTyr-containing peptides to be discriminated since pSer and pThr often lose a H_3PO_4 group in collision-induced dissociation (CID) (Bennett et al., 2002). Due to the increased stability of pTyr residues in the CID process, they most often do not lose their phosphate group during fragmentation. This strategy was applied for the identification of phosphopeptides isolated from a *Saccharomyces cerevisiae* cell lysate. Analysis of the extracted phosphopeptides showed that greater than 80% of the useful tandem mass spectra identified phosphopeptides. While a direct method to quantitate changes in phosphorylation state was not demonstrated in this initial application, stable isotope tags could be incorporated during the blocking of the carboxylates using either normal isotopic abundance or deuterated ethanolamine (i.e., ethanolamine-d_4).

Another labeling method was developed to identify seryl and threonyl phosphorylation sites by creating a novel trypsin cleavable site within proteins (Knight et al., 2003). In the first step, as shown in Figure 4.6, β-elimination is conducted under basic conditions to convert the pSer and pThr residues to generate dehydroalanine and β-methyldehydroalanine, respectively. In the next step, the dehydroalanine and β-methyldehydroalanine sites are modified with cysteamine via Michael addition to generate an aminoethylcysteine or β-methylaminoethylcysteine residue. The isosteric nature of aminoethylcysteine compared with lysine permits the cleavage of these newly created sites with proteases that recognize lysine (e.g., trypsin, Lys-C, and lysyl endopeptidase). This method has been tested using a standard mixture of seven pSer and two pThr peptides. After modification of the phosphorylated site within each peptide, they were analyzed by MS, resulting in the identification of the expected modification and digestion of the peptide at the newly created lysyl mimic using trypsin. Digestion at the modified pThr site was found to be less efficient than at the pSer site. The efficacy of this method was further tested on two model proteins, α- and β-casein, which contain three and five sites of phosphorylation, respectively. Many of the aminoethylcysteine-containing peptides were identified by LC-MS/MS sequencing. These peptides were found to produce typical peptide MS/MS fragmentation patterns, rendering them easily interpretable. A characteristic y_1 ion at *m/z* 165.1 resulting from a loss of a C-terminal aminoethylcysteine residue appeared as a highly abundant product ion in this MS/MS spectrum and other CID spectra of peptides containing this C-terminal residue. The efficacy of using this method on a global proteomic level has yet to be shown, however, it obviously holds promise.

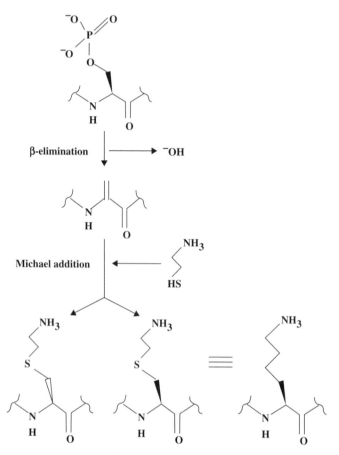

Figure 4.6. *Aminoethylcysteine modification of phosphoserine for the manufacturing of "artificial" trypsin cleavage sites in proteins. In this scheme for transformation, phosphoserine residues and phosphothreonine residues are converted to dehydroalanine and β-methyldehydroalanine residues, respectively. Following their conversion to aminoethylcysteine or β-methylaminoethylcysteine residues, these sites are now susceptible to lysine-specific proteases such as trypsin and Lys-C.*

4.2.2 Mass Spectral Identification of Phosphopeptides

Phosphopeptide Mapping   Once the sample has been prepared, the next step is to identify the phosphopeptides and the specific sites of phosphorylation. A variety of MS-based methods have been designed that either identify a peptide as being phosphorylated or identify the specific site of phosphorylation. The most straightforward method of identifying phosphorylated peptides, termed phosphopeptide mapping, involves enzymatically digesting a purified phosphoprotein and analyzing the resulting fragments by MS (Loughrey et al., 2002). If the protein being studied is known, the phosphopeptide(s) is identified as having an experimental mass shifted by a multiple of 80 Da compared to its predicted mass. If the identity of the protein is unknown, the complement of measured masses can be used to first identify the protein and the phosphopeptides distinguished as described earlier. An additional means to identifying phosphopeptides via peptide mapping involves measuring the masses of the digested protein prior to and after phosphatase treatment. Since treatment with phosphatase enzymatically removes the phosphate groups, peaks representing phosphopeptides in the original mass spectrum will show a decrease of 80 Da after phosphatase treatment.

Collision-Induced Dissociation The most common method of identifying sites of phosphorylation is CID of samples produced by ESI (Ummarino et al., 2003). While phosphopeptide mapping is capable of identifying phosphorylated peptides, it does not provide the specific site of phosphorylation, unless the peptide contains a single phosphorylatable residue. As described later, there are many different CID strategies, however, they all rely on the lability of the phosphoester bonds of the pTyr, pThr, and pSer residues. This bond can easily be fragmented within a collision cell, an ESI ion source, or during post-source decay (PSD) in a MALDI-TOF instrument. The result is the loss of a phosphate that can be identified by several phosphate-specific ion scans. Loss of phosphate as HPO_3 or H_3PO_4 is a favored fragmentation event and usually dominates over the amide backbone cleavages that are useful for sequence determination. The preferred loss of phosphate is particularly dominant when the molecular ion is singly charged. However, even in cases where the loss of phosphate is dominant, important fragmentation information can be obtained provided that sufficient statistics can be accumulated to discern the weaker backbone fragment ions. As a general trend for low-energy CID of phosphopeptides, it has been observed that phosphate tends to be lost from pSer more readily than pThr and from pThr more readily than from pTyr. Phosphate is generally eliminated from shorter phosphopeptides more readily than from longer phosphopeptides, because roughly the same amount of energy for collision is dispersed across fewer bonds. Loss of 98 Da (H_3PO_4) from a phosphorylated fragment ion is indistinguishable from loss of 18 Da (H_2O) from a nonphosphorylated fragment ion, sometimes making the interpretation of a CID spectrum challenging.

Electron Capture Dissociation While the focus of this chapter is not on various fragmentation schemes to identify phosphorylation sites by MS/MS, one such process does require attention based on its potential in phosphorylation site mapping. One of the primary difficulties in identifying phosphorylated sites by tandem MS is the lability of the phosphoester linkage, which typically fragments prior to the amide bonds. A newly developed ion fragmentation technique, electron capture dissociation (ECD), appears promising for these problems as well as having an impact in sequencing large protein fragments via top-down proteomic approaches (Stensballe et al., 2000). In the application of this technique, multiply protonated peptide/protein molecular ions are trapped in the Penning cell of a Fourier transform ion cyclotron resonance (FTICR) MS and energetic electrons are introduced, resulting in the near exclusive fragmentation of the backbone of a peptide or protein, forming c and z ions (Chalmers et al., 2004). Labile modifications such as γ-carboxylation, O-glycosylation (as discussed later), or phosphorylation, however, are retained. In comparison to CID spectra of peptide and protein ions, ECD spectra show far less side-chain losses of PTMs, but rather result in a higher proportion of backbone cleavages providing more extensive sequence information.

The use of ECD to identify the phosphopeptides and sites of phosphorylation has been demonstrated for β-casein, protein kinase A, and many other proteins

(Shi et al., 2001; Chalmers et al., 2004). When fragment ions are generated in the ECD experiment of β-casein, no loss of the phosphate groups is observed. With the sequence known, the ECD-generated fragment ions clearly identify phosphorylation at Ser-15. Three more phosphorylation sites among the cluster of four seryl residues 17, 18, 19, and 22, and Thr-24, and one more between Lys-32 and Val-59, among Ser-35 and Ser-57 and Thr-41 and Thr-55 were identified. In none of the ECD spectra of these phosphorylated peptides was a corresponding nonphosphorylated peptide found. Therefore, ECD should be valuable for quantitative determination of the phosphorylation status of proteins.

Electron Transfer Dissociation Though ECD has shown great promise in its utility for characterizing PTMs, due to the physical constraints of this technique its application is restricted solely to FTICR instruments, which require a great deal of technical expertise to operate and are the most expensive MS instruments. It remains to be seen, therefore, how broadly utilized this technique will be in proteomics. Very recently, Hunt and co-workers have developed an alternative to ECD, where fragmentation is accomplished through a gas phase reaction after coincidentally trapping peptide/protein molecular ions with anions formed by a chemical ionization source (Syka et al., 2004). The entire process can be accomplished on millisecond time scales and utilizes a commercially available and affordable quadrupole linear ion trap (QLT).

In the initial demonstration of the utility of ETD, a synthetic doubly phosphorylated peptide (LPISASHpSpSKTR) was subjected to fragmentation by reaction with anthracene anions in the QLT. After 50 ms of reaction time, all predicted c- and z-type fragment ions were observed, including both of the intact phosphoseryl residues (Syka et al., 2004). With the ability of ETD to be coupled online with nanoflow HPLC separations, a low femtomole mixture of ten synthetic peptides was analyzed in a data-dependent mode with average scan times of 500–600 ms. All peptides were selected for fragmentation by ETD and the resulting fragment ion spectra were easily sequenced from the observed fragment ions, including a doubly phosphorylated peptide present at 1 fmol. In contrast, the same peptide mixture was analyzed by nanoflow HPLC-MS/MS utilizing low energy collision-induced dissociation (CID). In this experiment, the phosphopeptide fragment ion spectrum was dominated by ions corresponding to the loss of phosphoric acid from the seryl residues. The only fragment ions arising from amide bond cleavage were present at <0.5% relative ion abundance (Syka et al., 2004). Hence, ETD promises to reveal much greater information content on a chromatographically relevant time scale with higher efficiencies than CID or ETD, utilizing cost effective instrumentation.

4.3 GLYCOSYLATION

Along with phosphorylation, the post-translational modification of proteins by carbohydrates is one of the most common modifications that occur within the cell (Haltiwanger and Lowe, 2004). Protein glycosylation was thought to be limited to eukaryotes; however, recent investigations have identified many glycoproteins within archaea and bacteria (Spiro, 2002). A significant percentage of proteins, particularly secreted and membrane proteins, have been shown to be glycosylated. There is no well-defined singular purpose for carbohydrate attachment, as the biological activity of a glycoprotein often is not changed as a result of deglycosylation. The most relevant function of the glycosyl groups is to increase the aqueous solubility of a protein and permit specific interactions with other molecules (Dell and Morris, 2001). Consistent with this finding is that carbohydrate groups are commonly found on classes of proteins such as immunoglobulins, proteases, cytokines, hormones, and cell surface receptors that function primarily through interactions with other biological molecules. While unmodified proteins can often be efficiently studied by X-ray crystallography or nuclear magnetic resonance spectroscopy, such methods may not provide structural information about the carbohydrate constituency due to their nonrigid structure, rendering significant challenges to the structural characterization of these proteins.

There are two main types of glycosylation: N linked (covalently attached to the side chain nitrogen atom of Asn) or O linked (attached to the side chain oxygen atoms of primarily Ser and Thr) (Kobata, 2000). There are consensus sequences that can be used to conservatively predict whether a specific Asn residue has the potential to be glycosylated. N-glycosylation generally occurs within the sequence -Asn-Xaa-Ser-, -Asn-Xaa-Thr-, or -Asn-Xaa-Cys- (where Xaa is any residue except Pro). There are no readily apparent consensus sequences to predict sites of O-linked glycosylation. O-linked glycosylation occurs primarily within the Golgi; while N-linked glycosylations occur cotranslationally as the nascent protein emerges from the endoplasmic reticulum.

The importance of characterizing glycosylation sites within proteins and peptides is duly illustrated in a recent publication that describes the identification of a biomarker for interstitial cystitis (Keay et al., 2004). Interstitial cystitis (IC) is a debilitating, chronic, nonbacterial, inflammatory bladder disease that has very painful and distressing symptoms. Despite considerable research, the cause of IC remains a mystery. Simply diagnosing IC is a long, complex process that involves the exclusion of other possible disorders, ultimately leading to a bladder biopsy carried out under anesthesia. The presence of an antiproliferative factor (APF) in the urine of IC patients was first postulated almost ten years ago (Keay et al., 1996). APF was hypothesized to be a small ($M_r < 10 \, kDa$), heat stable peptide that inhibited bladder epithelial growth based on experiments that showed decreased [^{3}H]thymidine incorporation in bladder epithelial cells treated with fractionated urine from IC patients. Although recognized almost a decade ago, the actual structure of APF remained elusive until recently. Using an experimental scheme that combined fractionation with a cellular assay to measure epithelial cell proliferation, investigators were able to isolate APF in a single fraction in urine obtained from patients with IC (Keay et al., 2004). Nano-LC-MS/MS analysis of the isolate allowed APF to be identified as a nine-residue glycosylated peptide, as shown in Figure 4.7A. Specifically, APF was identified as a uniquely modified frizzled 8-related sialoglycopeptide that bears 100% sequence identity to the sixth transmembrane segment of this G protein-coupled Wnt ligand receptor. A synthetic version of this sialoglycopeptide had an antiproliferative effect on human bladder epithelial cells, while the nonglycosylated version did not inhibit cell proliferation (Figure 4.7B). Subsequent Northern blot analysis demonstrated that APF is expressed solely in the bladder epithelium of IC patients, whereas transcript was evident in

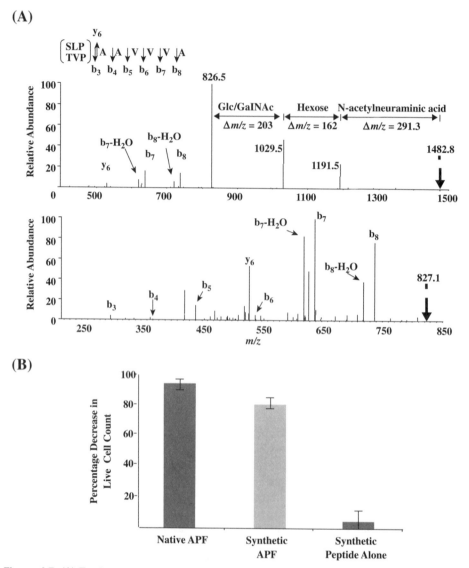

Figure 4.7. (A) Tandem mass spectrometry analysis of antiproliferative factor (APF) showing the identification of the glycosylation group (upper trace) and peptide sequence (lower trace). (B) Inhibition of bladder epithelial T24 cells. Native APF isolated from urine of patients with interstitial cystitis and the synthetic version had a significant negative effect on cell proliferation, while the synthetic peptide alone (i.e., minus the glycosylation modification) had little, if any, effect.

normal human bladder epithelial cells in vitro. Not only does this study demonstrate the key role glycosylation plays in protein activity, but it also illustrates the complexity of proteomic analysis and how complete characterization of proteins (beyond simple sequence) is necessary to fully understand the repertoire of protein functions within cells and biofluids.

The multibranched character of glycans makes their structural determination particularly challenging. This difficulty is compounded by the isomeric nature of monosaccharides and the variety of linkage positions possible between neighboring monosaccharides (Gerwig and Vliegenthart, 2000). The MS analysis of N-linked eukaryotic glycoproteins is somewhat simplified since these groups contain a common pentasaccharide core consisting of three mannose and two *N*-acetylglucosamine (GlcNac) units. Additional sugar residues are attached to this core element via two outer mannose residues; however, other sugar residues can be attached in a variety of different configurations. As the name implies, high-mannose type modifications are comprised of additional mannose residues that are linked to the pentasaccharide core. In the complex type of glycosylation, GlcNac, galactose, sialic acid, and L-fucose residues may be present with the glycan chains typically terminating with sialic acid residues. To further complicate the problem, there are hybrid-type glycans that have features of both high-mannose and complex glycans. O-linked glycans, on the other hand, do not possess a common core structure and can range from monosaccharide modifications to large sulfated polysaccharides.

To characterize glycosylated proteins, various pieces of information are required such as identification of the site(s) of glycosylation, structural characterization of the glycolytic side chain, quantitation of the extent of glycosylation at each site, and identification of the number of different glycoforms of each protein. Mass spectrometry can significantly contribute to the identification of the sites of glycosylation and structural characterization of the glycolytic side chain using both MS and MS/MS measurements. The complete characterization of the different glycoforms is highly dependent on the ability to fractionate the various glycoforms and identify the glycolytic composition of each.

The location of the N- or O-linked modified site(s) is usually performed via peptide mapping of the glycoproteins before and after treatment with a deglycosylation agent (Hagglund et al., 2004). O-linked carbohydrates are typically removed from the glycoproteins by base-catalyzed β-elimination while N-linked groups are cleaved using *N*-glycanase (Fryksdale et al., 2002). Both the native and deglycosylated proteins are proteolytically digested (usually with trypsin) prior to MS analysis. The appearance of new molecular ions along with the disappearance of others enables the identification of those peptides that were originally glycosylated. For N-linked sugars, the carbohydrate attached (i.e., complex, high mannose, or hybrid) can be determined by digesting the glycoprotein with endoglycosidase H, causing the high-mannose and hybrid type of sugar to be released (Henriksson et al., 2004). The resultant glycopeptide contains a GlcNAc attached to the Asn residue. A peptide with a mass 203 Da higher than that of its deglycosylated counterpart is therefore observed in the mass spectrum. As an illustration of this differential mapping of glycopeptides, the mass spectra of endopolygalacturonase (EPG II) before and after endoglycosidase H digestion is shown in Figure 4.8. EPG II has one potential site for N-linked glycosylation that has been modified with eight different high-mannose structures, whose compositions range from $(Man_5GlcNAc_2)$ to $(Man_{12}GlcNAc_2)$ (Colangelo and Orlando, 2001). The MALDI mass spectrum observed after digestion demonstrates the disappearance of the high-mannose structure and the appearance of a new molecular ion at *m/z* 2061.2.

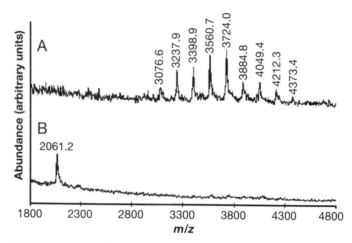

Figure 4.8. *MALDI mass spectra of the glycopeptides from EPG II (A) before and (B) after treatment with Endo H. Reproduced with permission from Colangelo et al. (2001).*

This *m/z* value is representative of the mass of the peptide modified with a single GlcNAc monomer.

Digestion of an N-linked glycosylated peptide with *N*-glycanase converts the Asn residue to which the oligosaccharide is attached to an aspartyl (Asp) residue (Fryksdale et al., 2002). The mass of the deglycosylated peptide is consequently increased by 1 Da relative to the calculated mass of the corresponding unmodified peptide. In addition to identifying the possibility of a glycosylation site through direct measurement of the peptide itself, the presence of low mass ions can be diagnostic of sugar moieties that were bound to the peptide. These low mass ions can be produced in the ESI source region by increasing the potential difference between the orifice and skimmer or within the collision cell of the mass spectrometer. This fragmentation method is useful for producing sugar-specific oxonium ions corresponding to Hex^+ (*m/z* 162), $HexNAc^+$ (*m/z* 203), $NeuAc^+$ (*m/z* 274 and 292), $Hex-HexNAc^+$ (*m/z* 366), and $NeuAc-Hex-HexNAc^+$ (*m/z* 657) (Plummer and Tarentino, 1981).

As mentioned earlier, the complex branching present in many oligosaccharides makes structural elucidation by MS/MS extremely difficult. Cross-ring fragmentation products must be identified to determine the intersaccharide linkages; however, these products are usually of low intensity in tandem mass spectra. In the case of complex glycans, this class of ions may be absent entirely from the CID spectrum, due to the low favorability of multiple-bond cleavage processes when in the presence of highly facile pathways originating at glycosidic bonds. This results in spectra with little information content necessary to map the glycan structure, requiring additional derivatization methods or other analytical techniques to be employed, such as methylation analysis, to establish the missing structural details.

Electron Capture Dissociation   As with phosphoprotein analysis described earlier, ECD holds tremendous promise in the characterization of glycoproteins (Mirgorodskaya et al., 1999). The FTICR mass spectrum of a tryptic digest of the

28 kDa lectin of the coral tree, *Erythrina corallodendron*, was acquired and two of the major peaks were accurately matched by mass to known glycopeptide structures (Hakansson et al., 2001). To confirm these assignments the ECD fragment spectrum was obtained from the N-glycosylated peptide of m/z 1005.5 (residues 100–116) as shown in Figure 4.9B. This fragmentation spectrum is dominated by N-terminal c-type ions, allowing the amino acid sequence and glycan structure of this peptide and sites of dissociation in ECD to be identified.

In the resulting tandem mass spectrum, fragment ions corresponding to cleavages at 10 of 15 peptide backbone amide bonds were observed, as shown in Figure 4.9A. The sites that were not cleaved were located in close proximity to the glycosylated Asn residue, suggesting that the bulky glycan sterically hinders access to the backbone carbonyl oxygens of the neighboring amino acid residues. The carbonyl oxygens are involved in the proposed ECD cleavage mechanism, in which, after electron capture, an energetic hydrogen atom is released and subsequently captured by a site of high hydrogen atom affinity, such as a carbonyl oxygen (Mirgorodskaya et al., 1999). In the ECD spectrum of the lectin glycopeptide, three glycosylated fragment ions are observed. All three fragments contain the entire, complex glycan structure; no carbohydrate loss due to cleavage of glycosidic bonds is observed. The current result extends the applicability of ECD for glycopeptide analysis to N-glycosylated peptides and to peptides containing branched, highly

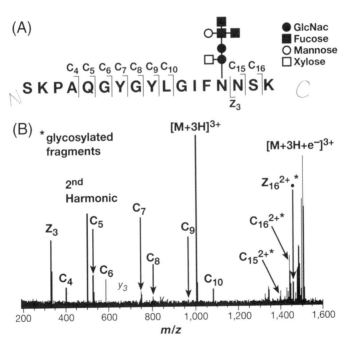

Figure 4.9. *Electron capture dissociation FTICR mass spectrum obtained from the triply protonated N-glycosylated peptide of* m/z *1005.5 (peptide segment 100–116) in the lectin digest (B). The y-axis is magnified 10×. Cleavages at 10 of 15 backbone amide bonds are observed (A). The achieved c ions provide a peptide sequence tag of six amino acids. No fragmentation of the branched, N-linked heptasaccharide was observed. Reproduced with permission from Hakansson et al. (2001).*

substituted glycans. A peptide sequence tag containing six amino acid residues obtained from the achieved ECD fragment ion series was used to retrieve the protein from the database.

Since ECD does not result in the fragmentation of the carbohydrate groups themselves, infrared multiphoton dissociation (IRMPD) (Little et al., 1994) was also employed to characterize the attached monosaccharides (Hakansson et al., 2001). A complex fragmentation pattern is observed; however, several ions were identified as the parent glycopeptide with loss of one or more sugars. In the $900 < m/z < 1100$ range, both $[M + 2H]^{2+}$ and $[M + 3H]^{3+}$ ions are observed. Several correspond to the parent glycopeptide with loss of one or more sugars. Finally, in the $200 < m/z < 900$ range, $[M + 3H]^{3+}$ ions corresponding to the glycopeptide with loss of several sugars are seen. Also, dehydrated sugar $[M + H]^+$ ions are observed. Dissociation at each glycosidic bond was observed, as well as the loss of the entire glycan. The extensive monosaccharide losses are consistent with the presence of multiple branch points in the structure of the glycan. Observation of a doubly protonated fragment at m/z 1097.0 ($[M - Man_3XylGlcNAc + 2H]^{2+}$) made it possible to specify the site of fucosylation as the inner GlcNAc residue. While the IRMPD spectrum provides no information about the peptide sequence, combining this information with that obtained by ECD provided a complete characterization of the glycopeptide.

Proteome-wide Identification of Glycoproteins

Much of the focus in global proteomics has been on the proteome-wide identification and quantitation of proteins or peptides. Recently, there has been much attention directed to characterizing the phosphoproteome. While not as aggressively pursued, characterization of the glycome (i.e., all the glycosylated proteins in a proteome) is also generating much needed attention. One of the major obstacles in characterizing the glycome is the complexity and heterogeneity of this important class of PTM. Presently, the best approaches involve separating a proteome mixture by 2D-PAGE and then using methods to specifically stain or detect the glycosylated proteins. One detection method developed uses the fluorescent hydrazide, Pro-Q Emerald 300 dye, which may be conjugated to glycoproteins by a periodic acid Schiff's (PAS) mechanism (Steinberg et al., 2001). The glycans present in glycoproteins are initially oxidized to aldehydes using periodic acid. The dye then reacts with the aldehydes to generate a highly fluorescent conjugate. The glycoproteins can be detected directly in the gels within 2–4 h of electrophoresis. Gels labeled with the Pro-Q Emerald 300 dye can also be subsequently stained with SYPRO Ruby dye, which allows sequential two-color detection of glycosylated and nonglycosylated proteins. Detection of glycoproteins may be achieved in sodium dodecyl sulfate–polyacrylamide gels, two-dimensional gels, and on polyvinylidene difluoride membranes. Lectin affinity chromatography has also been used to isolate glycopeptides resulting from the tryptic digestion of complex proteome mixtures (Hang et al., 2003). While this method circumvents the use of 2D-PAGE, for complete glycoprotein coverage many different lectins are required due to the heterogeneity of the carbohydrate groups present within a glycome sample. Even when the glycoproteins are fractionated, however, a significant amount of work is still required to characterize both the modified residue and the carbohydrate attached.

4.4   OTHER POST-TRANSLATIONAL MODIFICATIONS

While this chapter has focused on the characterization of two major types of PTMs, there exist well over 200 different types. While many of these may be rare, their presence may nonetheless be critical to the function of the protein. While the characterization of phosphorylated and glycosylated sites dominates the field of proteomics as it pertains to the identification of PTMs, other more specific modifications have also been detected by MS. An elegant method for the identification of ubiquitinated sites has been developed that relies on the detection of a mass shift of lysine residues after digestion of the proteins of interest with trypsin (Peng et al., 2003). While ubiquitin represents too large a modification to be conveniently measured directly, if the ubiquitin-modified protein is digested with trypsin, a residual glycyl–glycyl dipeptide remains covalently attached to the modified lysyl residue of the target protein (Figure 4.10). This residual modification has a

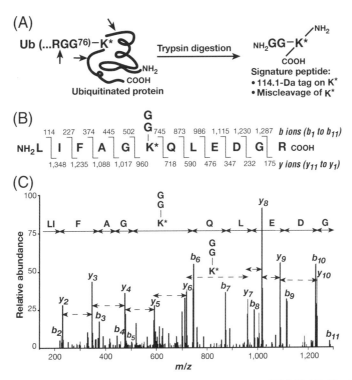

Figure 4.10. *Strategy for identifying the precise sites of ubiquitination by MS/MS. (A) After trypsin diges-tion, ubiquitin-conjugated proteins contain a diglycine remnant of ubiquitin (Ub) covalently attached to a lysine residue that is resistant to trypsin proteolysis. Amino acids are denoted by single-letter code. (B) Example showing the sequence of a signature peptide produced by trypsin proteolysis. Cleavage at the peptide backbone in the tandem mass spectrometer would result in the predicted fragment ion masses shown (b- and y-type ions) with the intact diglycine modification. (C) Fragmentation pattern (MS/MS spectrum) acquired for the peptide shown in (B). Only the singly charged ions are shown for simplicity. Much of the primary amino acid sequence can be determined including the precise site of ubiquitination. The ubiquitin-conjugated protein used in this example was ubiquitin itself and the iso-peptide bond linkage was through Lys-48. Reproduced with permission from Peng et al. (2003).*

signature mass of 114.1 Da and the modified peptide will contain a missed cleavage site since the modified lysyl residue is resistant to trypsin cleavage. The modified tryptic fragments are identifiable by database-searching algorithms, allowing the identification of the peptide (and protein) of interest. This method was initially applied to the discovery of ubiquitinated proteins in yeast. In this study, whole cell lysates isolated from yeast cells expressing hexahistidine (6xHis)-tagged ubiquitin were passed over a Ni-NTA affinity column to isolate proteins that had been modified by this ubiquitin fusion protein. After tryptic digestion, the samples were analyzed by online MS/MS to determine the sequences of the potentially modified peptides. The search was performed using a variable modification of 114.1 Da to each lysyl residue. The net result was the identification of 110 ubiquitination sites from 72 different proteins. About one-third of these proteins were localized to the internal membrane, which supports the proposed role of ubiquitination in the trafficking of membrane proteins, often resulting in their transport to the lysosomes and vacuole where they are internalized and degraded.

The identification of acetylation sites has traditionally followed a straightforward approach in which a modification of 42 Da on lysyl or N-terminal residues is examined in the MS/MS spectra of the parent peptides. A similar strategy is used for the identification of methylated (14 Da tag on lysyl or arginyl residues) or oxidized sites (16 Da tag on methionyl, tryptophanyl, or tyrosyl residues). An excellent example where such a strategy was employed has been presented for the characterization of modified proteins from human lens tissue (MacCoss et al., 2002). In this study, proteins extracted from human lens tissue were digested using three different enzymes: one that cleaves in a site-specific manner and two others that cleave nonspecifically. After analyzing the three peptide mixtures using multidimensional fractionation coupled with MS/MS, the resulting data was analyzed for different types of modifications including acetylation, methylation, and oxidation. Modifications identified in the crystallin proteins included arginyl and lysyl methylation, lysyl acetylation, and methionyl, tyrosyl, and tryptophanyl oxidations as well as phosphorylated seryl, threonyl, and tyrosyl residues. The method developed will be useful in discovering co- and post-translational modifications of proteins.

4.5   CONCLUSION

Ultimately, one of the goals of proteomics is to characterize the protein constituents of various pathways and networks that enable the cell to carry out the processes necessary to maintain life. While the identification of the proteins involved is critical, this information is insufficient to achieve the ultimate goal. Lists of proteins involved in these processes are analogous to knowing that a spark plug, piston, and fuel are used to operate an engine. Without knowing the signal generated by the spark plug that activates the fuel that drives the piston, the entire process would remain a mystery. Fortunately, the advances that have been made in the technology to identify proteins in complex biological mixtures have also benefited the characterization of post-translational modifications. Although characterization of phosphorylated proteins, and even more so glycoproteins, is far from routine, a much greater effort is being focused toward these two classes of proteins than ever before. Unfortunately, we have only begun to scratch the surface based on the number of known, and possibly yet undiscovered, post-translational modifications. Fortunately, there is also much room for improved developments in ways to analyze these modifications and these improvements can and will be made in many different areas, ranging from sample preparation through instrumentation.

ACKNOWLEDGMENTS

This project has been funded in whole or in part with federal funds from the National Cancer Institute, National Institutes of Health, under Contract No. NO1-CO-12400.

By acceptance of this article, the publisher or recipient acknowledges the right of the U.S. Government to retain a nonexclusive, royalty-free license and to any copyright covering the article. The content of this publication does not necessarily reflect the views or policies of the Department of Health and Human Services, nor does mention of trade names, commercial products, or organizations imply endorsement by the U.S. Government.

REFERENCES

Bennett KL, Stensballe A, Podtelejnikov AV, Moniatte M, Jensen ON. 2002. Phosphopeptide detection and sequencing by matrix-assisted laser desorption/ionization quadrupole time-of-flight tandem mass spectrometry. *J Mass Spectrom* 37:179–190.

Bertozzi CR, Kiessling LL. 2001. Chemical glycobiology. *Science* 291:2357–2364.

Chalmers MJ, Hakansson K, Johnson R, Smith R, Shen J, Emmett MR, Marshall AG. 2004. Protein kinase A phosphorylation characterized by tandem Fourier transform ion cyclotron resonance mass spectrometry. *Proteomics* 4:970–981.

Chu G, Egnaczyk GF, Zhao W, Jo SH, Fan GC, Maggio JE, Xiao RP, Kranias EG. 2004. Phosphoproteome analysis of cardiomyocytes subjected to beta-adrenergic stimulation: identification and characterization of a cardiac heat shock protein p20. *Circ Res* 94:184–193.

Cohen P. 2002. The origins of protein phosphorylation. *Nat Cell Biol* 4:E127–E130.

Colangelo J, Orlando R. 2001. On-target endoglycosidase digestion matrix-assisted laser desorption/ionization mass spectrometry of glycopeptides. *Rapid Commun Mass Spectrom* 15:2284–2289.

Conrads TP, Issaq HJ, Hoang VM. 2003. Current strategies for quantitative proteomics. *Adv Protein Chem* 65:133–159.

Dell A, Morris HR. 2001. Glycoprotein structure determination by mass spectrometry. *Science* 291:2351–2356.

Ficarro SB, McCleland ML, Stukenberg PT, Burke DJ, Ross MM, Shabanowitz J, Hunt DF, White FM. 2002. Phosphoproteome analysis by mass spectrometry and its application to *Saccharomyces cerevisiae*. *Nat Biotechnol* 20:301–305.

Ficarro S, Chertihin O, Westbrook VA, White F, Jayes F, Kalab P, Marto JA, Shabanowitz J, Herr JC, Hunt DF, Visconti PE. 2003. Phosphoproteome analysis of capacitated human sperm. Evidence of tyrosine phosphorylation of a kinase-anchoring protein 3 and valosin-containing protein/p97 during capacitation. *J Biol Chem* 278:11579–11589.

Fryksdale BG, Jedrzejewski PT, Wong DL, Gaertner AL, Miller BS. 2002. Impact of deglycosylation methods on two-dimensional gel electrophoresis and matrix assisted laser desorption/ionization–time of flight–mass spectrometry for proteomic analysis. *Electrophoresis* 23:2184–2193.

Gerwig GJ, Vliegenthart JF. Analysis of glycoprotein-derived glycopeptides. *EXS* 88:159–186.

Goshe MB, Conrads TP, Panisko EA, Angell NH, Veenstra TD, Smith RD. 2001. Phosphoprotein isotope-coded affinity tag approach for isolating and quantitating phosphopeptides in proteome-wide analyses. *Anal Chem* 73:2578–2586.

Griffin TJ, Goodlett DR, Aebersold R. 2001. Advances in proteome analysis by mass spectrometry. *Curr Opin Biotechnol* 12:607–612.

Hagglund P, Bunkenborg J, Elortza F, Jensen ON, Roepstorff P. 2004. A new strategy for identification of N-glycosylated proteins and unambiguous assignment of their glycosylation sites using HILIC enrichment and partial deglycosylation. *J Proteome Res* 3:556–566.

Hakansson K, Cooper HJ, Emmett MR, Costello CE, Marshall AG, Nilsson CL. 2001. Electron capture dissociation and infrared multiphoton dissociation MS/MS of an N-glycosylated tryptic peptic to yield complementary sequence information. *Anal Chem* 73:4530–4536.

Haltiwanger RS, Lowe JB. 2004. Role of glycosylation in development. *Annu Rev Biochem* 73:491–537.

Han JM, Kim JH, Lee BD, Lee SD, Kim Y, Jung YW, Lee S, Cho W, Ohba M, Kuroki T, Suh PG, Ryu SH. 2002. Phosphorylation-dependent regulation of phospholipase D2 by protein kinase Cdelta in rat pheochromocytoma PC12 cells. *J Biol Chem* 277:8290–8297.

Hang HC, Yu C, Kato DL, Bertozzi CR. 2003. A metabolic labeling approach toward proteomic analysis of mucin-type O-linked glycosylation. *Proc Natl Acad Sci USA* 100:14846–14851.

Henriksson H, Denman SE, Campuzano ID, Ademark P, Master ER, Teeri TT, Brumer H 3rd. 2004. N-linked glycosylation of native and recombinant cauliflower xyloglucan endotransglycosylase 16A. *Biochem J* 375:61–73.

Ibarrola N, Kalume DE, Gronborg M, Iwahori A, Pandey A. 2003. A proteomic approach for quantitation of phosphorylation using stable isotope labeling in cell culture. *Anal Chem* 75:6043–6049.

Kalo MS, Yu HH, Pasquale EB. 2001. In vivo tyrosine phosphorylation sites of activated ephrin-B1 and ephB2 from neural tissue. *J Biol Chem* 276:38940–38948.

Keay S, Zhang CO, Trifillis AL, Hise MK, Hebel JR, Jacobs SC, Warren JW. 1996. Decreased ^{3}H-thymidine incorporation by human bladder epithelial cells following exposure to urine from interstitial cystitis patients. *J Urol* 156:2073–2078.

Keay S, Szekely Z, Conrads TP, Veenstra TD, Barchi JJ, Zhang CO, Koch KR, Michedja CJ. 2004. An antiproliferative factor from interstitial cystitis patients is a frizzled 8 protein-related sialoglycopeptide. *Proc Natl Acad Sci USA* 101:11803–11808.

Knight ZA, Schilling B, Row RH, Kenski DM, Gibson BW, Shokat KM. 2003. Phospho-specific proteolysis for mapping sites of protein phosphorylation. *Nat Biotechnol* 21:1047–1054.

Kobata A. 2000. A journey to the world of glycobiology. *Glycoconj J* 17:443–464.

Kumar A, Agarwal S, Heyman JA, Matson S, Heidtman M, Piccirillo S, Umansky L, Drawid A, Jansen R, Liu Y, Cheung KH, Miller P, Gerstein M, Roeder GS, Snyder M. 2002. Subcellular localization of the yeast proteome. *Genes Dev* 16:707–719.

Little DP, Speir JP, Senko MW, O'Connor PB, McLafferty FW. 1994. Infrared multiphoton dissociation of large multiply charged ions for biomolecule sequencing. *Anal Chem* 66:2809–2815.

Loughrey Chen S, Huddleston MJ, Shou W, Deshaies RJ, Annan RS, Carr SA. 2002. Mass spectrometry-based methods for phosphorylation site mapping of hyperphosphorylated proteins applied to Net1, a regulator of exit from mitosis in yeast. *Mol Cell Proteomics* 1:186–196.

MacCoss MJ, McDonald WH, Saraf A, Sadygov R, Clark JM, Tasto JJ, Gould KL, Wolters D, Washburn M, Weiss A, Clark JI, Yates JR III. 2002. Shotgun identification of protein modifications from protein complexes and lens tissue. Proc *Natl Acad Sci USA* 99:7900–7905.

McDonald BJ, Chung HJ, Huganir RL. 2001. Identification of protein kinase C phosphorylation sites within the AMPA receptor GluR2 subunit. *Neuropharmacology* 41:672–679.

Manning DR, DiSalvo J, Stull JT. 1980. Protein phosphorylation: quantitative analysis in vivo and in intact cell systems. *Mol Cell Endocrinol* 19:1–19.

Marcus K, Moebius J, Meyer HE. 2003. Differential analysis of phosphorylated proteins in resting and thrombin-stimulated human platelets. *Anal Bioanal Chem* 376:973–993.

Meek DW, Milne DM. 2000. Analysis of multisite phosphorylation of the p53 tumor-suppressor protein by tryptic phosphopeptide mapping. *Methods Mol Biol* 99:447–463.

Mirgorodskaya E, Roepstorff P, Zubarev RA. 1999. Localization of O-glycosylation sites in peptides by electron capture dissociation in a Fourier transform mass spectrometer. *Anal Chem* 71:4431–4436.

Oda Y, Huang K, Cross FR, Cowburn D, Chait BT. 1999. Accurate quantitation of protein expression and site-specific phosphorylation. *Proc Natl Acad Sci USA* 96:6591–6596.

Oda Y, Nagasu T, Chait BT. 2001. Enrichment analysis of phosphorylated proteins as a tool for probing the phosphoproteome. *Nat Biotechnol* 19:379–382.

Pelech S. 2004. Tracking cell signaling protein expression and phosphorylation by innovative proteomic solutions. *Curr Pharm Biotechnol* 5:69–77.

Peng J, Schwartz D, Elias JE, Thoreen CC, Cheng D, Marsischky G, Roelofs J, Finley D, Gygi SP. 2003. A proteomics approach to understanding protein ubiquitination. *Nat Biotechnol* 21:921–926.

Plummer TH Jr, Tarentino AL. 1981. Facile cleavage of complex oligosaccharides from glycopeptides by almond emulsin peptide: *N*-glycosidase. *J Biol Chem* 256:10243–10246.

Porath J. 1992. Immobilized metal ion affinity chromatography. *Protein Expr Purif* 3:263–281.

Sacristan C, Tussie-Luna MI, Logan SM, Roy AL. 2004. Mechanism of Bruton's tyrosine kinase-mediated recruitment and regulation of TFII-I. *J Biol Chem* 279:7147–7158.

Sakchaisri K, Asano S, Yu LR, Shulewitz MJ, Park CJ, Park JE, Cho YW, Veenstra TD, Thorner J, Lee KS. 2004. Coupling morphogenesis to mitotic entry. *Proc Natl Acad Sci USA* 101:4124–4129.

Shi SD, Hemling ME, Carr SA, Horn DM, Lindh I, McLafferty FW. 2001. Phosphopeptide/phosphoprotein mapping by electron capture dissociation mass spectrometry. *Anal Chem* 73:19–22.

Spiro RG. 2002. Protein glycosylation: nature, distribution, enzymatic formation, and disease implications of glycopeptide bonds. *Glycobiology* 12:43R–56R.

Steinberg TH, Pretty On Top K, Berggren KN, Kemper C, Jones L, Diwu Z, Haugland RP, Patton WF. 2001. Rapid and simple single nanogram detection of glycoproteins in polyacrylamide gels and on electroblots. *Proteomics* 1:841–855.

Stensballe A, Jensen ON, Olsen JV, Haselmann KF, Zubarev RA. 2000. Electron capture dissociation of singly and multiply phosphorylated peptides. *Rapid Commun Mass Spectrom* 14(19):1793–1800.

Syka JEP, Coon JJ, Schroeder MJ, Shabanowitz J, and Hunt DF. 2004. Peptide and protein sequence analysis by electron transfer dissociation. *Proc Natl Acad Sci USA* 101:9528–9533.

Ummarino S, Corsaro MM, Lanzetta R, Parrilli M, Peter-Katalinic J. 2003. Determination of phosphorylation sites in lipooligosaccharides from *Pseudoalteromonas haloplanktis* TAC 125 grown at 15 degrees C and 25 degrees C by nano-electrospray ionization quadrupole time-of-flight tandem mass spectrometry. *Rapid Commun Mass Spectrom* 17:2226–2232.

Yanagida M, Miura Y, Yagasaki K, Taoka M, Isobe T, Takahashi N. 2000. Matrix assisted laser desorption/ionization–time of flight–mass spectrometry analysis of proteins detected by anti-phosphotyrosine antibody on two-dimensional-gels of fibrolast cell lysates after tumor necrosis factor-alpha stimulation. *Electrophoresis* 21:1890–1898.

Zhou H, Watts JD, Aebersold R. 2001. A systematic approach to the analysis of protein phosphorylation. *Nat Biotechnol* 19:375–378.

5

Technologies for Large-Scale Proteomic Tandem Mass Spectrometry

David L. Tabb

University of Washington, Seattle, Washington

John R. Yates*

The Scripps Research Institute, La Jolla, California

5.1 INTRODUCTION

Proteomics represents a rearrangement and augmentation of classical protein biochemistry. Where traditional techniques have focused on a few proteins per analysis, proteomics attempts to conduct the comprehensive analysis of complex protein mixtures. Proteomics could be defined as the science of assessing the protein state for a complex mixture, cell, or tissue under a particular set of conditions. Depending on the techniques used, this may entail giving a catalog of proteins present (Washburn et al., 2001), characterizing the post-translational modifications of those proteins (Mann and Jensen, 2003), or evaluating their relative quantities (Gygi et al., 2000; MacCoss et al., 2003; Ranish et al., 2003).

The two general strategies for analyzing proteomic samples differ in approach (VerBerkmoes et al., 2002). In "top-down" proteomics, the accurate masses or fragment ions of intact proteins are used to identify the constituents of a sample (Reid and McLuckey, 2002). In the "bottom-up" approach, peptides are frag-

---

* To whom correspondence should be addressed.

Proteomics for Biological Discovery, edited by Timothy D. Veenstra and John R. Yates.
Copyright © 2006 John Wiley & Sons, Inc.

mented through tandem mass spectrometry to ascertain their identities from sequence-level information (Hunt et al., 1986). The bottom-up approach is sometimes referred to as "shotgun" proteomics because it identifies peptides throughout the proteome in a random sampling process similar to that found in shotgun genomics approaches. These techniques are the focus of this chapter. The tools used for this type of analysis fall into three categories: sample separation, mass analysis, and informatics.

Because proteomics deals with more complex mixtures than most protein biochemistry experiments, the science relies more heavily on the separation of components in protein mixtures. The physical separation of proteins is usually conducted by two-dimensional gel electrophoresis (2D-GE) (Klose and Kobalz, 1995) or through liquid chromatography (LC) (McCormack et al., 1997; Gatlin et al., 1998). Denaturation and enyzmatic digestion are used in bottom-up methods to cleave the intact proteins into peptides. In these protocols, peptides are separated rather than proteins. Section 5.2 describes strategies for spreading a sample in time or space prior to mass analysis.

Mass analysis has been the driving engine of proteomics. Essentially, the adaptation of mass spectrometry to proteins has greatly increased the amount of data resulting from these experiments. Each mass spectrometer has three major elements: ion source, mass analyzer, and detector. Ionization in proteomics is generally handled via matrix-assisted laser desorption ionization (MALDI) (Moyer and Cotter, 2002) or electrospray ionization (ESI) (Fenn et al., 1989). Mass analysis is typically managed through time-of-flight (TOF) (Moyer and Cotter, 2002) or quadrupole analyzers (Miller and Denton, 1986; Jonscher and Yates, 1997). A detector records the ion current resulting from ions exiting the analyzer.

Mass spectrometry produces proteomic data, but proteomic information requires that the spectra be interpreted. Many programs interact to make this possible. First, mass spectrometry is conducted via instrument control software; this control impacts which spectra will be produced and how they will be represented in files. When tandem mass spectrometry has been employed, the spectra must be extracted from the instrument capture files, filtered, and assigned to particular precursor ion charge states. Then the peptides may be identified, typically by sequence database search algorithms. Section 5.4 describes the handling of data after its production in the tandem mass spectrometer.

5.2   SAMPLE SEPARATION

Proteins can vary widely in mass, hydrophobicity, structure, and other characteristics. As a result, many techniques have evolved for separating them (Link, 2002). The primary techniques used for proteomics are gel-based separations and liquid chromatography separations. In this section, gel separations for proteomics are first considered, and then a powerful 2D-LC separation technique is described. These techniques can separate proteins or peptides such that particular portions of a sample can be analyzed independently of the rest.

5.2.1 Gel-Based Proteomic Protocol

Gel separations are common experiments for evaluating the components of protein complexes. Early proteomics efforts were grafted onto existing gel-separation technologies, as in the discovery by Lawrence and co-workers (1999) of a regulatory factor of insulin-like growth factor. The protocol for gel-based proteomics experiments begins with the denaturation of proteins. The proteins are separated by one- or 2D-GE, and bands or spots containing proteins of interest are cut from the gel. Disulfide bridges are reduced and alkylated, and proteolytic enzymes cleave proteins to peptides. These peptides are separated by reverse phase LC en route to a tandem mass spectrometer (see Figure 5.1).

Gel electrophoresis is a standard technique for separating proteins (Klose and Kobalz, 1995). Typically, the proteins are first denatured by a detergent such as SDS. In one-dimensional electrophoresis, they are separated by size by

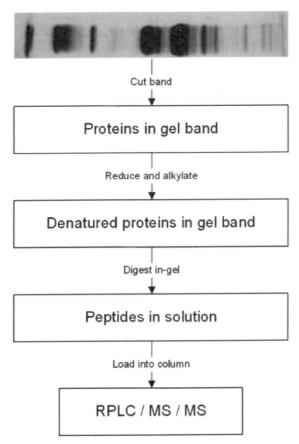

Figure 5.1. *An overview of the steps in gel-based proteomics.*

electrophoretic mobilities through polyacrylamide. Because proteins contain different amino acid residues, the pH values at which they are neutral (called the isoelectric point) differ. In 2D-GE, the proteins are separated by size in one dimension and by isoelectric focusing in another. Such gels have been shown effective for resolving up to 10,000 proteins in a single experiment.

The positions of proteins are marked by applying a dye to the gel (McDonough et al., 2002); Coomassie Blue dye is widely used but only highlights proteins for which at least 30–50 ng is present (this represents a best case scenario). Silver staining, on the other hand, can mark positions for proteins of which at least 1 ng is present. Fluorescent dyes have roughly the same sensitivity as silver staining but may be more easily removed. Once proteins have been separated and positions are marked, gel bands (for 1D gels) or spots (for 2D gels) can be excised for extraction.

Protein tertiary structures may be maintained by disulfide crosslinks between cysteine residues. Reduction of these bridges by dithiothreitol or TCEP can break these links, allowing the protein to be fully denatured. Subsequent alkylation by iodoacetamide blocks off the cysteine side chains and adds 57 Da to their masses. This protocol is useful when the proteins are to be digested to peptides; by breaking the disulfide bonds, reduction and alkylation prevent pairs of peptides from being linked together.

Several enzymatic digestions of proteins are available (Barrett et al., 2003). The most commonly used protocol in proteomics is the trypsin digest. This enzyme, which cleaves proteins after arginine and lysine residues, is available in a form bound to beads for removal from the sample after the digest. EndoK-C has the same cleavage specificity as trypsin but shows better efficiency in the presence of urea and other denaturants; however, use of the enzyme increases experiment cost. Alternative enzymatic cleavages may increase the diversity of peptides produced; subtilisin, elastase, themolysin, and proteinase K can be employed to create peptides covering different portions of a protein sequence.

These digestions yield peptides that can be separated en route to the mass spectrometer by reverse phase LC (McCormack et al., 1997; Gatlin et al., 1998). Columns for these separations are commercially available, but they can also be produced from fused silica capillaries with inner diameters of 100 μm or less. A laser puller can be used to create tips with inner diameters of 5 μm. These columns can then be loaded with 5 μm C_{18}-coated beads, and the sample's peptides are loaded into the column under pressure. A gradient of increasingly hydrophobic solvents elutes the peptides progressively from the column into the mass spectrometer.

This protocol can be effective, but many disadvantages prevent it from being an optimal proteomics strategy. First, gel electrophoresis may fail to retain proteins of extreme p*I*, molecular weight, or hydrophobicity. Second, proteins of low concentration may fail to be selected for removal from the gel if the staining technique is of insufficient sensitivity. Cutting gel bands or spots is a tedious process for which automation has only recently become available. Some reagents for gel electrophoresis of proteins are not compatible with mass spectrometry. On the other hand, these techniques enable existing separation methodologies to be used for proteomic analyses.

5.2.2 MudPIT

A technique created in the Yates laboratory attempts to automate separation and overcome the limitations of gel electrophoresis. The essential modifications in the multidimensional protein identification (MudPIT) experiment are that proteins are reduced, alkylated, and digested to peptides prior to separation and that separation takes place via 2D-LC rather than gel electrophoresis (Link et al., 1999; Washburn et al., 2001). The resulting technique automates the analysis of even very complex samples (such as cellular lysates). See Figure 5.2 for an overview.

The MudPIT separation uses a biphasic column, which elutes directly into a mass spectrometer. The first separation uses strong cation exchange (SCX) material to separate peptides by their charges at acidic pH. The second separation employs hydrophobic C_{18} material for a reverse phase gradient. MudPIT separations proceed in cycles; for each salt concentration in the SCX separation, a separate reverse phase separation is conducted. As sample complexities increase, larger numbers of cycles can be employed.

In a typical MudPIT separation, twelve cycles might be conducted. The peptides are loaded into the biphasic column, and the peptides that initially passed through the SCX material and into the RP material are eluted during the first cycle of the MudPIT. For subsequent cycles, a few minutes' flow of a specified percentage of ammonium acetate moves a subset of the peptides from the SCX material to the reverse phase material. This subset of peptides is then eluted via an acetonitrile gradient. The salinity for each cycle increases throughout the MudPIT. For example, these percentages of 500 mM ammonium acetate were used for separating the proteins of a rat hippocampus (Tabb et al., 2003): 0%, 10%, 15%, 20%, 25%, 30%, 35%, 40%, 45%, 50%, 60%, and 100%. A linear gradient would yield better separation than this type of step gradient, but the intent of the SCX is to segregate

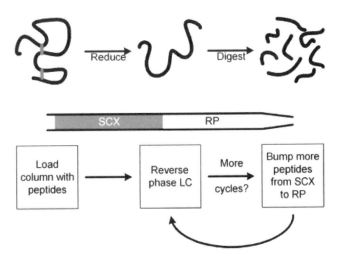

Figure 5.2. *An overview of the steps in MudPIT proteomics.*

the peptides into aliquots for reverse phase separation rather than resolve them in a single dimension.

MudPIT separations help to automate separations for proteomics experiments. Their key advantage is that the eluent from the column flows directly into the mass spectrometer, obviating the need for tedious excision of gel bands or spots. In the next section, the fate of eluting peptides is examined in detail as they move through the mass spectrometer.

5.3   MASS ANALYSIS

The use of mass spectrometry for proteomics yields much more detailed information than does a 2D gel image or UV trace from reverse phase LC. If intact proteins are analyzed, mass spectrometry can give the molecular weights of the proteins to high precision. Peptides, on the other hand, can subsequently be fragmented to reveal primary structure. This section describes the mass analysis of peptides fragmented via collision-induced dissociation.

5.3.1   ESI/MS/MS

Proteins and peptides were not obvious candidates for mass spectrometry due to their high mass, polarity, and ionic nature. The first challenge of proteomic mass spectrometry was to introduce protein or peptide ions to the vaccuum of a mass spectrometer. John Fenn solved this problem by use of the ESI technique (Fenn et al., 1989). Direct injection or LC elutes proteins directly into an electrospray source, where a high voltage is applied, causing the liquid to form a mist of small droplets. Through the application of heat and collisions, the droplets are reduced in size until the ions are desorbed. Pressure and voltage differences then force the charged proteins into the mass analyzer.

The most common type of mass analyzer paired with ESI is the quadrupole (Miller and Denton, 1986). The traditional format of a quadrupole is four linear electrodes arranged in a square to form a channel down which ions fly (see Figure 5.3), but a rearrangement of the electrodes to form the quadrupole ion trap (see

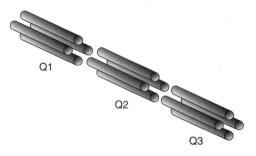

Q1

Q2

Q3

Figure 5.3. *In triple quadrupole CID, mixed peptide ions enter the first quadrupole (Q1) from the ion source. In Q1, peptide ions of a particular sequence are retained while others are filtered out. In Q2, the peptide ions are energized by collision with gas molecules, causing them to fragment. Q3 is employed to scan through the mass range, cataloging the produced fragment ions in a tandem mass spectrum.*

Figure 5.4) can lend extra versatility to the mass analyzer (Jonscher and Yates, 1997). In either case, the way in which voltage is applied to the electrodes controls which ions are stabilized and which are destabilized in the area among the electrodes.

In the traditional quadrupole, a tandem mass analyzer consists of three sets of four electrodes—a triple quadrupole. Collision induced dissociation (CID) selects ions of a particular peptide passing through the first quadrupole by m/z ratio, fragments the ions in the second quadrupole, and scans the fragment ions in the third quadrupole. In this way, a triple quadrupole separates the processes of tandem mass spectrometry in space.

A quadrupole ion trap mass analyzer uses the same set of electrodes for all three processes. First, the trap collects ions of the selected peptide. Next, energy is applied to the peptide ions to cause them to oscillate more rapidly, resulting in more energetic collisions with the noble gas filling the trap and thus fragmentation of the peptide. Finally, the fragment ions resulting from fragmentation are ejected from the trap in order of m/z ratio to constitute the tandem mass spectrometer. The three steps of CID are separated in space for triple quadrupoles but by time in quadrupole ion traps.

5.3.2 Collision Induced Dissociation

Since CID plays such a key role in proteomics research, a closer examination is warranted. While many aspects of these reactions are poorly understood, the basics have been established through experiments spanning the last two decades. The key actors in CID are the surplus protons on each peptide ion and the energy applied through collisions with noble gas molecules.

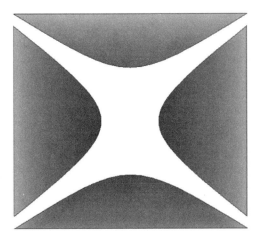

Figure 5.4. *The hyperbolic cross section of the quadrupole is repeated for the ion trap, but in three dimensions rather than two. The left and right electrodes shown above are part of a single "ring" electrode, while the top and bottom electrodes are electrically paired but separate "endcap" electrodes. In quadrupole ion trap CID, mixed peptide ions enter the trap, but only the selected peptide ions are retained. Once enough ions are present, the trap energizes the ions, causing them to collide with gas molecules and dissociate. The trap then scans through the produced fragment ions to produce the tandem mass spectrum.*

Typically, proteomic samples are acidified before separation, resulting in positive charges for peptides. The added proton is most likely to be found at a site of high proton affinity/gas-phase basicity. Among the amino acids, the side chains of arginine, lysine, and histidine show the greatest gas-phase basicity, with arginine the most basic of the three (Afonso et al., 2000; Tabb et al., 2004). In addition, the N terminus of the peptide may play host to protons. Because trypsin is commonly used to digest proteins in proteomic protocols, generated peptides often possess C-terminal arginine or lysine residues, which can retain one proton. Additional protons may be solvated among other side chains or at the N terminus of the peptide.

The energy applied to peptide precursor ions in ion trap CID accelerates their motions within the trap. Noble gas molecules are always present in the trap at low pressure, but at normal energies the peptide collisions with them do not cause fragmentation. When energy is applied in CID, though, these collisions substantially increase the internal energy of the peptides, causing them to adopt unusual conformations and causing the additional protons to sample less accommodating locations within the peptides (as explained in the Mobile Proton Model of peptide fragmentation (Wysocki et al., 2000)).

If a proton comes to rest near the oxygen of a peptide bond carbonyl, it may draw the bond electrons toward the oxygen, leaving the carbon partially positively charged (see Figure 5.5). In some cases, the preceding carbonyl's oxygen may attack the partially positive carbon, forming an unstable ring in the peptide structure. The transference of hydrogens can then lead to a departing C-terminal ion with a normal peptide structure (the *y* ion) and an N-terminal ion terminating in an oxazolone ring (the *b* ion). Once an ion has dissociated, it is no longer energized by ion trap CID, which targets a narrow *m/z* range.

In some cases, alternate mechanisms are at work in forming fragment ions. If a peptide carries only one additional proton and it is immobilized at an arginine side chain, aspartic acid's side chain may attack its carbonyl to produce a dominant cleavage (Huang et al., 2002). If proline is present in a peptide's sequence, cleavage may occur dominantly to its N terminus but almost never to its C terminus (Loo et al., 1993; Breci et al., 2003). The chemical diversity of peptides implies a diverse chemistry for their fragmentation. In addition, fragment ions may subsequently fragment; *b* ions in particular are unstable and may lose carbon monoxide to form *a* ions. Recently, databases of known peptide identifications have been surveyed to find statistical trends in fragmentation (Kapp et al., 2003; Tabb et al., 2003). These and other efforts are beginning to make the process of peptide fragmentation more predictable.

In the final analysis, mass spectrometry of peptides meshes chemical processes (CID) with physical processes (mass analysis) to yield spectra giving structural information for peptides. The data produced, however, require analysis to give proteomic information. In the next section, we examine how the tandem mass spectrum can be identified.

5.4 SPECTRAL ANALYSIS

Initially, tandem mass spectra were matched to peptide sequences manually (Hunt et al., 1986). In contemporary usage, the path from spectrum to identification is

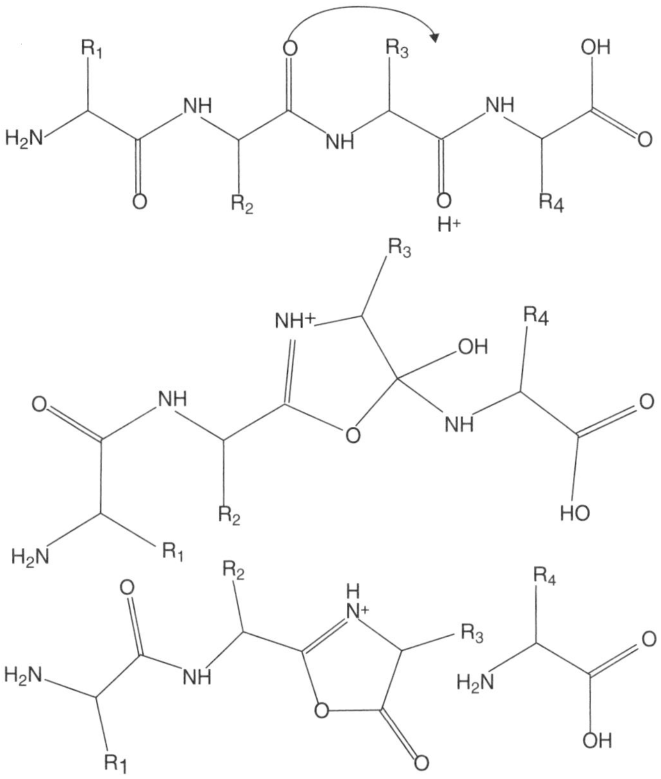

Figure 5.5. Chemistry of collision induced dissociation.

managed by several algorithms. The first stage of this process is the acquisition of the spectra via the instrument control software. Next, the tandem mass spectra are separated from the other data and preprocessed by utility programs. Finally, the spectra are identified, usually by a database search algorithm such as Sequest (Eng et al., 1994). An examination of each of these steps reveals many areas of active research.

5.4.1 Instrument Control Software

The configuration of the software controling the mass spectrometer and pumps can affect the data produced in many ways. It can determine the gradient used in the LC and thus which peptides are eluted when. It determines which peptides will be selected for fragmentation. This software controls the amount of energy applied to accelerate the peptides to begin CID. It sets the number of times each peptide's CID is repeated to be averaged for each spectrum. Importantly, the instrument control software handles data export to other software. A first-rate mass spectrometer can be hamstrung by instrument control software that is unstable or limits export of data to manipulable formats.

Thermo Finnigan's XCalibur instrument control software is the choice of many labs in this field. XCalibur manages the pumps required for producing MudPIT gradients, automates the processes of peptide selection and CID, and exports data via an ActiveX control or to standard file formats. In addition, some mass spectrometers ship with an XCalibur that incorporates the Sequest peptide identification algorithm. Thermo Finnigan was the first to market an ion trap instrument to automate these processes and thus is used widely in the proteomics field.

5.4.2   Separation and Preprocessing of Tandem Mass Spectra

When XCalibur captures data for a tandem mass spectrometry experiment, its capture file contains more than just tandem mass spectra. These data must be extracted to a separate file and then preprocessed. The ExtractMS program (http://fields.scripps.edu/sequest/extractms.html) was created for this purpose, creating a separate "DTA" text file for each tandem mass spectrum. Spectra that contain insufficient information (e.g., spectra that represent failed fragmentation) are not extracted to files. In addition, the software attempts to deduce the charge state of the peptide fragmented for each spectrum. ExtractMS can distinguish spectra from singly charged peptides from those that were multiply charged, but it cannot distinguish between doubly charged and triply charged peptide spectra, so two files are created for each multiply charged spectrum, one representing each of these two charge states.

The 2to3 program (Sadygov et al., 2002) attempts to determine the charge state for the multiply charged spectra in a collection. When the software determines the precursor ion was triply charged, it removes the doubly charged spectrum copy, and vice versa. The software works by finding pairs of fragment ions. The program projects the mass of the precursor ion, first assuming that it is doubly charged and then assuming that it is triply charged. The pairs of fragment ions which sum to each of these masses are counted. If one mass corresponds to a significantly larger number of these pairs, that charge state is selected for that spectrum. The ZSA (Charge State Algorithm) was created in the Lane group at Harvard to achieve the same aim by a different technique (Perez et al., 2003). ZSA conducts eight subtests on each spectrum and then combines the results of these tests by a neural net to determine the charge states of spectra. Inference of precursor charge states for spectra reduces the number of identifications produced for a sample, thus reducing the time required.

5.4.3   Sequest: A Database Search Algorithm

Sequest, originally published in 1994 (Eng et al., 1994), introduced the concept of the database search peptide identification algorithm. Similar algorithms have also been published and made commercially available, notably Matrix Sciences' Mascot (Perkins et al., 1999). Software of this type uses a sequence database as the source of candidate sequences to explain each spectrum.

The algorithms begin by assembling a candidate sequence list. The mass of the peptide that was fragmented to produce a particular spectrum can be derived from the observed m/z value of the precursor ion and the charge state that has been assigned to the spectrum (in some cases, spectral copies at multiple charge states may be present). Since there may be some error in the peptide's calculated mass, an error of a few daltons may be permitted. The database is queried for peptide sequences that have masses that match the precursor peptide's mass within this error range. If a specific enzymatic cleavage is indicated, only those peptides beginning or ending in appropriate amino acid residues may be considered. The collection of peptides that match the precursor peptide's mass constitutes the candidate sequence list for the spectrum.

In Sequest, preliminary scoring is used to prune the candidate list. The positions of fragment ions for each sequence are calculated, and the observed spectrum is inspected for the presence of these ions. A simple scoring function ranks the candidate sequences, and the top 500 are retained. Preliminary scoring plays a more significant role with large databases, when candidate lists may include more than a thousand peptides.

Each of the remaining sequence candidates is compared to the spectrum more rigorously. Once again, the positions of fragment ions for each sequence are calculated. A theoretical spectrum is generated containing peaks at the fragment ion positions, and then each theoretical spectrum is compared to the observed spectrum via cross-correlation. The sequences that generate the closest matching spectra are ranked most highly in Sequest's output.

Algorithms of this type can be differentiated based on the way in which they determine closeness of fit between observed and model spectra. Sequest makes use of the cross-correlation algorithm for its comparisons, while Mascot and others use statistically derived algorithms. Mascot's comparison starts with a few peaks and progressively uses more to refine its analysis, while Sequest uses a fast Fourier transform of the entire spectrum for each cross-correlation. These implementation differences can obscure the fact that, fundamentally, these algorithms are achieving the same thing, comparison of a theoretical spectrum to the experimental one.

Less attention has focused on the way in which the model spectra are constructed. Fragment ion masses are generated from the candidate sequences. Peaks of uniform intensity are inserted into the model spectrum for N-terminal b series and C-terminal y series ions. This technique addresses the horizontal dimension (m/z) of the spectrum adequately, but it represents a simplistic approach to the vertical dimension (intensity). The ions in experimental spectra vary widely in intensity, even with individual ion series. Improving the techniques for intensity modeling could potentially increase the accuracy of database search algorithms.

5.4.4 De Novo Algorithms

Database search algorithms are the most widely used for identifying peptides from shotgun proteomics. Another process for identifying the sequences represented by spectra is de novo sequence inference. In this technique, the fragment ions present in the spectrum are analyzed directly by software to determine the corresponding sequence. Generally, "de novo" is used to describe algorithms that infer the peptide's entire sequence, but the term may also be used to describe software that infers only a portion of the peptide sequence, a so-called "sequence tag" that can be used subsequently to find the entire peptide's sequence in a database.

Bartels's (1990) publication of a peptide sequence inference algorithm based on the concept of a "sequence spectrum" has been influential in shaping subsequent de novo algorithms. Sherenga (Dančik et al., 1999), Lutefisk (Taylor and Johnson, 1997), and SeqMS (Fernandez-de-Cossio et al., 2000) have all used this approach, which transforms the observed spectrum into a directed graph from which the sequence can be deduced. The sequence graph contains nodes that represent collections of peaks in the original spectrum; for example, if a spectrum for a peptide ion of mass 1000.0 Da contains ions at m/z 172.0, 200.0, and 800.0, these may represent a, b, and y ions, respectively, leading to a node in the sequence graph. Sequences that explain the nodes with higest probability are those most likely to correspond to the sequence of the peptide. The "sequence spectrum" approach has not yet been widely adopted for processing real-world samples, in part because these algorithms may require higher quality data than most tandem mass spectrometers can produce.

The sequence tag technique for peptide identification was introduced by Mann and Wilm (1994) in 1994. In this process, a sequence tag for the peptide is inferred directly from the observed spectrum either manually or by software, and then the database is searched to find peptides that include these partial sequences and that match the masses to either side of the partial sequence. Several difficulties have hampered this technique from broader adoption. First, inferring even partial sequences from spectra is error prone; scoring a sequence tag is, in some ways, more challenging that scoring a complete sequence for a spectrum. The low accuracy of automated sequence inference requires that multiple tag sequences be considered for each spectrum. The database search may, in turn, result in multiple peptide sequence hits, and generally the task of evaluating which is the correct sequence is left to the user. A program was recently created in the Yates laboratory that attempts to address these problems (Tabb et al., 2003). This technique may grow in prominence as these difficulties are solved.

5.4.5 Identification Assembly and Data Mining

The identification of proteins in complex mixtures produces a large quantity of data. A key component to turning the data into information is software to assemble and filter database searching results. Tools for rapid evaluation of spectra and to view protein sequence coverage are valuable aids for the analysis of results. We developed a tool, DTASelect, to evaluate the results from shotgun protein identification experiments (Tabb et al., 2002). This software sorts the identified peptides by the proteins they represent. Users can customize a flexible set of rules by which the identifications are filtered, removing identifications that score insufficiently well. For each protein, the algorithm assembles the observed peptides to determine sequence coverage. Each peptide is linked back to the original search results and to the tandem mass spectrum acquired from the peptide. A companion program, Contrast, allows comparison of database searching results from many different experiments. Without the use of software like DTASelect, interpretation of results from shotgun proteomic experiments would be difficult at best. Comparable tools are being developed in other laboratories, such as CHOMPER (Eddes et al., 2002).

Other groups have developed statistically based systems for analyzing identifications. Anderson et al. (2003) reported the creation of a Support Vector Machine algorithm for classifying identifications into true and false matches. Keller et al. (2002) reported on the PeptideProphet algorithm, which uses an expectation maximization algorithm to generate a probability of accuracy for each identification.

Such algorithms have a different aim than DTASelect; instead of emphasizing a biologically oriented presentation of the data meeting specified rules, these programs attempt to separate true from false hits on statistical grounds.

5.5 PROTEOMIC RESEARCH EXAMPLES

To perform physiological processes, cells segregate proteins into compartments. A higher order organization is encompassed by the grouping of proteins into complexes. These molecular machines can contain a small or a large number of proteins. The proteins that participate in the physiological process can change as a function of the cell cycle or cellular state. To determine the networks of proteins involved in physiological processes, efforts have been underway to comprehensively identify the complexes of certain organisms. High-throughput biochemical strategies are fraught with difficulties as isolation of protein complexes is a critical step in biochemical approaches to identify the components of complexes.

5.5.1 Protein Complex Isolation

Most approaches for biochemical characterization of protein complexes use an immunoprecipitation process to isolate the complex. Several strategies have been used to purify protein complexes. Biochemical purifications have been used to isolate protein complexes for decades. A functional assay can be used to follow isolation of the active complex; a specific activity is sought among the fractions of a separation, for example. When one of the components of the complex is known and an antibody is available, the antibody can be used to follow enrichment of that component.

Higher throughput procedures can be used when the organism to be studied can be transformed to incorporate epitope-tagged genes. In *Saccharomyces cerevisiae*, epitope-tagged genes can be inserted using homologous recombination, an approach that allows the use of the endogenous promoters for the gene. Expression levels will then be the typical level for that gene. A popular tag in *S. cerevisiae* to isolate protein complexes is the tandem affinity purification tag. This tag incorporates sequence from the protein A, a protease cleavage site specific for the tobacco etch virus (TEV) protease, and the sequence for the calmodulin binding peptide (CaBP). A complex is isolated by passing cell lysate across an IgG column to capture the tagged protein and associated proteins. The complex is removed by suspending the IgG beads in a solution containing the TEV protease. Cleaving the complex from the column provides a very gentle step, avoiding the release of nonspecific contaminants bound to the column. By passing the solution through a calmodulin column, the complex is recaptured and then released using EGTA to remove calcium from solution. Without calcium in solution, CaBP dissociates from calmodulin. Other methods are also used to incorporate epitope tags into proteins for isolation.

A third approach to isolate interacting proteins uses a protein bound to a column to "fish out" proteins that interact strongly with the protein. An advantage to this approach is the ability to shift the equilibrium for low affinity binders by increasing the amount of cell lysate passed through the column. A disadvantage to this approach is ability to separate signal from noise. Nonspecific interacting proteins can bind to the solid support and other surfaces and thus can be released into solution along with interacting proteins. To circumvent this problem, washes of increasing stringency are used and proteins in each wash step are captured. Individual washes are then analyzed to assess which proteins are specific interactors.

5.5.2 Analysis of the SAGA Complex

The SAGA complex is a collection of proteins engaged in remodeling chromatin during transcription. DNA is organized by histone proteins into nucleosomes. The nucleosomes form repeating structures as part of the chromatin. When DNA is tightly bound in chromatin, transcription is inhibited because transcription factors are unable to bind to DNA. Post-translational modification of the histones is correlated with transcription and is catalyzed by a set of histone acetyltranfersases. Interestingly, acetylation activity is associated with large multisubunit protein complexes. The SAGA complex is a Gcn5-dependent histone acetyltransferase complex. The SAGA complex was enriched by chromatographic separations while following acetyltransferase activity. A final separation of the active 1.8 MDa complex was achieved using gel electrophoresis. Separated protein bands were removed and digested in situ with trypsin. The resulting peptide mixture was analyzed by LC/tandem mass spectrometry. In this 1.8 MDa protein complex, five proteins associated with transcription activation were identified. This data provides a direct connection between a set of TAF proteins and chromatin acetylation or remodeling. Western blot experiments confirmed the copurification of these proteins with the SAGA complex. This independent confirmation of the identities of the proteins provided substantial confidence in the mass spectrometry methods employed. Further functional analyses showed depletion of at least one of the components disrupted acetylation of the nucleosome and transcriptional stimulation, thus providing functional validation of the findings.

These experiments used an established biochemical purification process, chromatographic enrichment of a functional active protein complex. The final separation step prior to mass spectrometry analysis used gel electrophoresis together with silver staining to mark the positions of proteins. This process is very effective in identifying proteins but is hampered by several limitations. First, the process is labor-intensive and the ultimate sensitivity of the method is limited by the ability to efficiently extract digested peptides from the gel matrix. Second, the method is not high-throughput since each component of the complex is analyzed as a separate mass spectrometry sample.

5.5.3 Analysis of the Mitochondrial Nucleoid

To circumvent the limitations imposed by the use of gel electrophoresis, LC/tandem mass spectrometry without prior gel separation was employed to identify the components of the mitochondrial nucleoid. Mitochondrial DNA (mtDNA) is organized into structures within the mitochondria that contain 4–5 copies of mtDNA. mtDNA must undergo replication, repair, segregation, and partitioning. Studies have shown that partitioning of mtDNA is a nonrandom process. By following the purification of Abf2p, a complex of proteins was isolated. The protein mixture was digested using trypsin and then analyzed. The tandem mass spectra obtained were then searched through the *S. cerevisiae* database to identify the proteins present in the mixture. A collection of proteins was identified and Mgm101p was studied in the context of its role in the nucleoid (Meeusen et al., 1999). To test whether Mgm101p localizes to the mitochondrial nucleoid, it was tagged with green fluorescent protein. Cells expressing the construct were viewed via cytofluorescence, and the tagged protein was localized to the nucleoid. Cells deficient in active Mgm101p were observed to be sensitive to oxidative damage to mitochondrial DNA.

5.5.4   Analysis of the Yeast Ribosome

As proteomic research has moved to protein mixtures of ever greater complexity, the need for higher resolution separations has become necessary. An advantage to improved chromatographic separations is an increase in the dynamic range of the analysis of protein mixtures. In some cases, different chromatographic materials can be combined to achieve greater separation than would be achievable by either material alone. Link et al. (1999) combined strong cation exchange and reverse phase chromatographies for the analysis of large protein complexes and cellular lysates. The ribosome from *S. cerevisiae* was isolated using standard density centrifugation. After isolation of the complex, it was digested with the enzyme trypsin. The resulting peptide mixture was then separated on-line with a tandem mass spectrometer and the resulting spectra were searched through the yeast sequence database. Seventy-eight proteins were identified as part of the yeast ribosome. A total of 137 genes encode 78 proteins, with 59 of the proteins encoded by duplicate genes. If the protein complex were analyzed on a standard p*I* gel of 3–10, only two proteins were observed after the gel was stained; most of the polypeptides of the ribosome are basic and thus do not separate on isoelectric focusing gels. If a non-equilibrium pH gradient gel is used, almost all of the proteins are observed. At least one new component of the ribosome was identified, YMR116p. The human homolog of this protein was also observed upon analysis of the human ribosome. By using the higher resolving capability of MudPIT, the components of large protein complexes can be identified.

5.5.5   Analysis of Yeast Whole Cell Lysates

Washburn et al. (2001) extended the method of MudPIT to the analysis of yeast cell lysates for the analysis of proteins expressed in the cell as well as localization of proteins to specific structures of the cell. Components of cells can be fractionated through the use of protein solubility. For example, fractions can be enriched for membrane proteins by extracting proteins from the insoluble pellet produced during cell lysis. By extracting proteins from the pellet with high salt buffers and high pH buffers (sodium carbonate, pH 11), peripheral membrane proteins are removed, leaving a fraction that is enriched in integral membrane proteins. Washburn et al. (2001) lysed yeast cells and extracted the pellet to produce three fractions for analysis by MudPIT. After analysis and careful evaluation of the search results, 1484 proteins were identified. This number of proteins represents approximately one-third of the proteins expected to be expressed in log-phase cell cultures. More importantly, the integral membrane protein fraction was solubilized in formic acid and subjected to cyanogen bromide digestion followed by trypsin digestion. A total of 131 integral membrane proteins were identified with more than a three transmembrane domains predicted. This strategy represented a new approach for the analysis of membrane proteins since it diminished the need to solubilize membrane proteins for separation.

5.6 CONCLUSION

Proteomics may be a relatively new field, but its achievements to date are impressive. Protein identification is no longer limited to single proteins, painstakingly isolated and purified. Proteomic technologies have made it possible to identify the protein content of complexes, organelles, and even whole cells.

Many technological hurdles remain. While identifying common proteins can routinely be achieved, many proteins that perform crucial roles are present at low concentrations, making it far more difficult to identify them, especially in the presence of more ubiquitous proteins. New separation and enrichment methods must evolve to improve the prospects of proteomics for low abundance proteins.

Comprehensive identification of post-translational modifications remains a challenge for proteomics. New algorithms, especially those that use the sequence-tagging approach, are likely to make the identification of modifications more reliable and powerful. In addition, these techniques may make it possible to identify peptides on the basis of related database sequences as well.

Proteomics will move forward, but its motion will not be limited to any one field. Chemical techniques for improving separations and purification are vital. Physical rearrangements of tandem mass analyzers are making significant improvements in the accuracy with which the masses of fragment ions can be known. Insightful software development will improve the accuracy and performance of proteomic algorithms. Perhaps most importantly, proteomics will continue to emerge from specialist laboratories and become a more widespread toolset. As these technologies become ubiquitous, proteomics cannot help but move forward.

REFERENCES

Afonso C, Modeste F, Breton P, Fournier F, Tabet J-C. 2000. Proton affinity of the commonly occuring L-amino acids by using electrospray ionization-ion trap mass spectrometry. *Eur J Mass Spectrom* 6:443–449.

Anderson DC, Li W, Payan DG, Noble WS. 2003. A new algorithm for the evaluation of shotgun peptide sequencing in proteomics: support vector machine classification of peptide MS/MS spectra and SEQUEST scores. *J Proteome Res* 2:137–146.

Barrett AJ, Rawlings ND, Woessner JF. 2003. *Handbook of Proteolytic Enzymes*, Elsevier Academic Press, San Diego, CA.

Bartels C. 1990. Fast algorithm for peptide sequencing by mass spectroscopy. *Biomed Environ Mass Spectrom* 19:363–368.

Breci LA, Tabb DL, Yates JR III, Wysocki VH. 2003. Cleavage N-terminal to proline: analysis of a database of peptide tandem mass spectra. *Anal Chem* 75:1963–1971.

Dančík V, Addona TA, Clauser KR, Vath JE, Pevzner PA. 1999. De novo peptide sequencing via tandem mass spectrometry. *J Computation Biol* 6:327–342.

Eddes JS, Kapp EA, Frecklington DF, Connolly LM, Layton MJ, Moritz RL, Simpson RJ. 2002. CHOMPER: a bioinformatic tool for rapid validation of tandem mass spectrometry search results associated with high-throughput proteomic strategies. *Proteomics* 2:1097–1103.

Eng JK, McCormack AL, Yates JR III. 1994. An approach to correlate tandem mass spectral data of peptides with amino acid sequences in a protein database. *J Am Soc Mass Spectrom* 5:976–989.

Fenn JB, Mann M, Meng CK, Wong SF, Whitehouse CM. 1989. Electrospray ionization for mass spectrometry of large biomolecules. *Science* 246:64–71.

Fernandez-de-Cossio J, Gonzalez J, Satomi Y, Shima T, Okumura N, Besada V, Betancourt L, Padron G, Shimonishi Y, Takao T. 2000. Automated interpretation of low-energy collision-induced dissociation spectra by SeqMS, a software aid for de novo sequencing by tandem mass spectrometry. *Electrophoresis* 21:1694–1699.

Gatlin CL, Kleeman GR, Hays LG, Link AJ, Yates JR III. 1998. Protein identification at the low femtomole level from silver-stained gels using a new fritless electrospray interface for liquid chromatography-microspray and nanospray mass spectrometry. *Anal Biochem* 263:93–101.

Gygi SP, Rist B, Gerber SA, Turecek F, Gelb MH. 2000. Quantitative analysis of complex protein mixtures using isotope-coded affinity tags. *Nat Biotechnol* 17:994–999.

Huang Y, Wysocki VH, Tabb DL, Yates JR III. 2002. The influence of histidine on cleavage C-terminal to acidic residues in doubly protonated tryptic peptides. *Int J Mass Spectrom* 219:233–244.

Hunt DF, Yates JR III, Shabanowitz J, Winston S, Hauer CR. 1986. Protein sequencing by tandem mass spectrometry. *Proc Natl Acad Sci USA* 83:6233–6237.

Jonscher KR, Yates JR III. 1997. The quadrupole ion trap mass spectrometer—a small solution to a big challenge. *Anal Biochem* 244:1–15.

Kapp EA, Schutz F, Reid GE, Eddes JS, Moritz RL, O'Hair RA, Speed TP, Simpson RJ. 2003. Mining a tandem mass spectrometry database to determine the trends and global factors influencing peptide fragmentation. *Anal Chem* 75:6251–6264.

Keller A, Nesvizhskii AI, Kolker E, Aebersold R. 2002. Empirical statistical model to estimate the accuracy of peptide identifications made by MS/MS and database search. *Anal Chem* 74:5383–5392.

Klose J, Kobalz U. 1995. Two-dimensional electrophoresis of proteins: an updated protocol and implications for a functional analysis of the genome. *Electrophoresis* 16:1034–1059.

Lawrence JB, Oxvig C, Overgaard MT, Sottrup-Jensen L, Gleich GJ, Hays LG, Yates JR III, Conover CA. 1999. The insulin-like growth factor (IGF)-dependent IGF binding protein-4 protease secreted by human fibroblasts is pregnancy-associated plasma protein-A. *Proc Natl Acad Sci USA* 96:3149–3153.

Link AJ. 2002. Multidimensional peptide separations in proteomics. *Trends Biotechnol* 20: S8–S13.

Link AJ, Eng J, Schieltz DM, Carmack E, Mize GJ, Morris DR, Garvik BM, Yates JR III. 1999. Direct analysis of protein complexes using mass spectrometry. *Nat Biotechnol* 17:676–682.

Loo JA, Edmonds CG, Smith RD. 1993. Tandem mass spectrometry of very large molecules. 2. Dissociation of multiply charged proline-containing proteins from electrospray ionization. *Anal Chem* 65:425–438.

McCormack AL, Schieltz DM, Goode B, Yang S, Barnes G, Drubin D, Yates JR III. 1997. Direct analysis and identification of proteins in mixtures by LC/MS/MS and database searching at the low-femtomole level. *Anal Chem* 69:767–776.

MacCoss MJ, Wu CC, Liu H, Sadygov R, Yates JR III. 2003. A correlation algorithm for the automated quantitative analysis of shotgun proteomics data. *Anal Chem* 75:6912–6921.

McDonough JL, Neverova I, Van Eyk JE. 2002. Proteomic analysis of human biopsy samples by single two-dimensional electrophoresis: Coomassie, silver, mass spectrometry, and Western blotting. *Proteomics* 2:978–987.

Mann M, Jensen ON. 2003. Proteomic analysis of post-translational modification. *Nat Biotechnol* 21:255–261.

Mann M, Wilm M. 1994. Error-tolerant identification of peptides in sequence databases by peptide sequence tags. *Anal Chem* 66:4390–4399.

Meeusen S, Tieu Q, Wong E, Weiss E, Schieltz D, Yates JR III, Nunnari J. 1999. Mgm101p is a novel component of the mitochondrial nucleoid that binds DNA and is required for the repair of oxidatively damaged mitochondrial DNA. *J Cell Biol* 145:291–304.

Miller PE, Denton MB. 1986. The quadrupole mass filter: basic operating concepts. *J Chem Educ* 63:617–622.

Moyer SC, Cotter RJ. 2002. Atmospheric pressure MALDI. *Anal Chem* 74:468A–476A.

Perez RE, Asara JM, Lane WS. 2003. Peptide precursor charge state determination directly from ion trap MS/MS spectra. *American Society of Mass Spectrometry Annual Conference Proceedings* MPK-336.

Perkins DN, Pappin JC, Creasy DM, Cottrell JS. 1999. Probability-based protein identification by searching sequence databases using mass spectrometry data. *Electrophoresis* 20:3551–3567.

Ranish JA, Leslie DM, Purvine SO, Goodlett DR, Eng J, Aebersold R. 2003. The study of macromolecular complexes by quantitative proteomics. *Nat Genet* 33:349–355.

Reid GE, McLuckey SA. 2002. "Top down" protein characterization via tandem mass spectrometry. *J Mass Spectrom* 37:663–675.

Sadygov RG, Eng J, Durr E, Saraf A, McDonald H, MacCoss MJ, Yates JR III. 2002. Code developments to improve the efficiency of automated MS/MS spectra interpretation. *J Proteome Res* 1:211–215.

Tabb DL, McDonald WH, Yates JR III. 2002. DTASelect and Contrast: tools for assembling and comparing protein identifications from shotgun proteomics. *J Proteome Res* 1:21–26.

Tabb DL, MacCoss MJ, Wu CC, Anderson SD, Yates JR III. 2003. Similarity among tandem mass spectra from proteomic experiments: detection, significance, and utility. *Anal Chem* 75:2470–2477.

Tabb DL, Saraf A, Yates JR III. 2003. GutenTag: high-throughput sequence tagging via an empirically derived fragmentation model. *Anal Chem* 75:6415–6421.

Tabb DL, Smith LL, Breci LA, Wysocki VH, Lin D, Yates JR III. 2003. Statistical characterization of ion trap tandem mass spectra from doubly charged tryptic peptides. *Anal Chem* 75:1155–1163.

Tabb DL, Huang Y, Wysocki VH, Yates JR III. 2004. Influence of basic residue content on fragment ion peak intensities in low-energy collision-induced dissociation spectra of peptides. *Anal Chem* 76:1243–1248.

Taylor JA, Johnson RS. 1997. Sequence database searches via de novo peptide sequencing by tandem mass spectrometry. *Rapid Commun Mass Spectrom* 11:1067–1075.

VerBerkmoes NC, Bundy JL, Hauser L, Asano KG, Razumovskaya J, Larimer F, Hettich RL, Stephenson JL Jr. 2002. Integrating "top-down" and "bottom-up" mass spectrometric approaches for proteomic analysis of *Shewanella oneidensis*. *J Proteome Res* 1:239–252.

Washburn MP, Wolters D, Yates JR III. 2001. Large-scale analysis of the yeast proteome by multidimensional protein identification technology. *Nat Biotechnol* 19:242–247.

Wysocki VH, Tsaprailis G, Smith LL, Breci LA. 2000. Mobile and localized protons: a framework for understanding peptide dissociation. *J Mass Spectrom* 35:1399–1406.

6

Protein Fractionation Methods for Proteomics

Thierry Rabilloud

DRDC/BECP, CEA-Grenoble, Grenoble, France

6.1 INTRODUCTION: PROTEINS, THIS OBSCURE OBJECT OF DESIRE (IN PROTEOMICS)

It might seem trivial to say that proteins are the analytes in proteomics. However, the implied consequences of this statement strongly orient the problems encountered in proteomics, making proteomics much more difficult than genomics and transcriptomics. One common problem to most of the "omics" is the number of objects (genes, transcripts, proteins, etc.) to be analyzed. These objects are often counted in thousands, even for the simplest bacteria. When it comes to upper eukaryotes, the number of objects to be analyzed comes into tens of thousands. While the complexity of complex genomes has been revised downward (from over 100,000 to less than 40,000), the estimates of the number of different proteins present in a human cell now largely exceed the 10,000 number put forward some twenty years ago (Duncan and McConkey, 1982). On top of the high number of genes expressed in a single cell, the added complexity due to post-translational modifications now drives proteomists to propose numbers in tens of thousands of different protein forms per complex eukaryotic cell. Apart from glycosylation, which is probably the post-translational modification introducing most heterogeneity in proteins, recent work examining in detail protein post-translational modifications shows the heterogeneity induced by previously overlooked modifications such as methylation or proteolytic cleavage.

In addition to this number problem, proteomics is complicated by features that are specific to proteins compared to nucleic acids. The first of these features is the dynamic range problem. In a mammalian cell, the quantitative ratio between the less abundant proteins (100 molecules per cell) and the most abundant ones (100 million molecules per cell) builds a dynamic range of 10^6. The situation is much worse in biological fluids, where the dynamic range is probably around 10^{12}. The second problematic feature specific to proteins as opposed to nucleic acids is their chemical heterogeneity. Due to the chemical diversity of the 20 amino acids, and to the incredible number of combinations they afford, proteins differ widely in many physicochemical characteristics such as molecular weight, surface charge, isoelectric point, and hydrophobicity.

All these features (number of proteins, dynamic range, and chemical diversity) combine to make the study of even a simple proteome largely out of reach of any single analytical method. This enormous complexity imposes the necessity to cut the proteome into smaller parts, which should be closer to the range and scope afforded by current analytical methods. This need for complexity reduction means in turn that protein fractionation is, one way or another, a good way for a better mining of proteomes.

6.2 PROTEINS AS ANALYTES OR THE BASES OF FRACTIONATION

Due to the fact that they are composed from a complex combination of 20 chemically diverse amino acids, proteins are chemically diverse and complex analytes. This diversity means that a number of physicochemical parameters differ from one protein to another and can therefore be used as critical parameters to design methods for separating one protein from another. Apart from the protein size, which is related to the length of the protein, many other parameters depend heavily on the sequence. Among those are the surface charge at a given pH and the isoelectric point, which are not equivalent. Proteins of equal p*I* can show different charges at a given pH, and vice versa. A good demonstration of these situations is given by titration curve experiments, which allow access to both parameters (Gianazza et al., 1980). Within the general structural parameters, hydrophobicity is also variable from one protein to another. It must be noted that these parameters depend not only on the primary sequence of the proteins but also on their post-translational modifications. For example, phosphorylation alters the p*I* and surface charge, while glycosylation can alter the protein size, p*I*, charge, and hydrophobicity.

Apart from these parameters, which are closely related to the primary structure, proteins exhibit features that are often linked to their tridimensional structures. Most proteins, if not all, are able to bind, generally on a noncovalent basis, various molecular objects of various size and structure. These ligands can be substrates, cofactors, other proteins, lipids, sugars, or molecular mimics of these chemicals. This molecular binding process can be used to fractionate proteins on the basis of the existence and strength of binding. While this binding ability is most times dependent on the precise structure of the amino acids making a protein, some binding processes are driven mainly by post-translational modifiers of these amino acids. Examples of this type include binding to lectins (induced by sugars) or to hydroxyapatite (largely modulated by phosphate groups).

This binding ability has an important consequence. In the cells, proteins are not molecular objects dispersed in the solvent. They are usually bound to other proteins to form macromolecular complexes. Since these complexes are held together by noncovalent forces, they are very often quite labile and difficult to isolate unless special precautions are taken. However, some complexes are held together relatively tightly and can be isolated as such. Examples of such complexes are the nuclear pores, the cytoskeleton or the ribosomes. This macromolecular complex level, however, is not the last level of organization. Higher order structures are commonly found, especially in eukaryotic cells. Such structures can be a super-assembly of complexes (e.g., nucleolus) or can even be limited by a lipid membrane, as all organelles are.

This organization of proteins into superstructures can be used as a fractionation method, in the sense that such structures are a subset of the complete cell. If these subcellular structures can be purified as such, this purification technique will constitute a very efficient and functionally oriented method of fractionation of the complex protein population found in a proteome.

6.3 SUBCELLULAR FRACTIONATION

A first strategy to fractionate proteins is to make use of the native subcellular compartmentalization of a cell to isolate subcellular structures. Because subcellular structures are by far less complex than the total parent cell, the classical proteomics toolbox can provide more in-depth analysis of these simpler structures. Classical subcellular fractionation therefore receives more and more interest in proteomics.

The rationale for subcellular fractionation is to open the cells of the sample of interest in such a way that the subcellular component of interest receives minimal damage. This lysis is usually done by mechanical homogenization. This process must be carefully controlled and varies from one sample to another in order to achieve the best compromise between total lysis of cells (and thus a high yield) and minimal damage to the intracellular component(s) of interest. This homogenization process is the only one applicable to most organelles, that is, intracellular structures limited by a lipid membrane, which are very sensitive to osmotic pressure and detergents. When the integrity of the component of interest is not solely dependent on the presence of a lipid membrane (e.g., nuclei, cytoskeleton), detergents can be used to solubilize all lipid membranes present in the cells, both the plasma membrane and those limiting the organelles (Hymer and Kuff, 1964). This detergent addition ensures thorough cell lysis and minimal contamination of the components of interest by lipid-delimited organelles.

Once the cells have been opened by a suitable means, the component(s) of interest must be separated from other cellular structures. This separation is usually done by centrifugation, using the fact that different components have a different size and therefore a distinct sedimentation velocity or buoyant density. The separation based on sedimentation velocity separates mainly four components: nuclei, mitochondria, microsomes, and cytosol. While nuclei can be isolated in a relatively pure form using detergents, the other fractions are usually cross-contaminated. Mitochondria preparations are usually contaminated by endosomes, lysosomes, and aggregated microsomes, while microsomes are contaminated by lysosomes and mitochondrial fragments. Furthermore, microsomes are a mix of vesicles arising from plasma membrane, Golgi complex, and endoplasmic reticulum, and it is often desirable to separate these components.

Further separation is therefore based on other parameters. The use of buoyant density is historically the favored method and allows further purification of contaminated organelles such as mitochondria. In addition, fine-tuning of the density provides a finer separation and allows the separation of microsomes into subcomponents. Various media have been used for this purpose (Rickwood et al., 1983), ranging from sucrose to colloidal silica (Percoll) and to iodinated media. Despite their higher cost, the latter media afford greater flexibility and are now more frequently used (Graham et al., 1983; Wattiaux and Wattiaux-de Coninck, 1983).

Another fractionation method for fine purification of subcellular components is free-flow electrophoresis. In this method, the surface charge of the particles is the leading parameter. This surface charge is brought by the membrane proteins and the ionic lipids (e.g., phosphatidyl serine) and differs from one type of organelle to another. While this technique requires a dedicated and rather complex apparatus, it can afford exquisite purification of organelles, which has a dramatic impact on the proteomics analysis downstream. This impact has recently been demonstrated for mitochondria (Zischka et al., 2003). It must be stated, however, that most of these techniques are not able to yield pure organelle preparations, and that contamination is most often a problem. For example, most organelle preparations are usually contaminated with cytoskeletal proteins. This contamination is due to the fact that organelles are in fact bound to the cytoskeleton, which ensures their trafficking. As cytoskeleton is a rather stable assembly, some cytoskeleton remnants are almost always carried along with the organelles to be purified. Apart from this specific problem, membrane vesicles of various origins (i.e., plasma membrane, reticulum, endosomes, and Golgi) are very difficult to separate from each other. In some instances, special purification schemes can be designed to extract with greater purity one of these components. Examples for endosomes include culture of the cells with a special density agent (Triton WR-1339), which is captured by the cells in the endolysosomes and modifies their buoyant density to a zone that is normally free of other organelles (Wattiaux and Wattiaux-De Coninck, 1967). Alternatively, cells can be cultured in the presence of magnetite-dextran, which is absorbed in the endolysosomes. Magnetic purification of the organelles afford endolysosomes with a higher purity (Adessi et al., 1995).

As for the plasma membrane, various techniques have been put forward with variable success. Lectin affinity of the microsomes, which uses the fact that plasma membrane-derived vesicles have an outside-out orientation, while vesicles derived from the secretory pathway are inside-out, has been proposed, but has shown limited success (Clemetson et al., 1977). Other techniques use culture labeling of plasma membrane proteins with nonpermeant biotin-containing tags. After cell lysis, purification of the vesicles deriving from plasma membrane is afforded by avidin chromatography (Zhang et al., 2003). The main problem with this technique is encountered when samples with a compromised cell viability are used. This decrease in cell viability is the case for cancer samples, which often contain necrotic cells. In this case, the so-called nonpermeant labeling molecule can enter dead cells and label intracellular proteins, thereby increasing the contamination of the membrane proteins with major intracellular proteins.

Despite these relatively poor performances in terms of real (and always questionable) purity of the isolated organelles, subcellular fractionation receives a lot of attention in proteomics. This interest is due to the fact that, in addition to the simplification of the protein mixture to be analyzed, it provides very important information (i.e., the subcellular localization of the proteins of interest). This result holds true even when the localization is modulated by a post-translational modification or by a ligand-induced dissocation of a protein complex. Furthermore, "abnormal" localization can have a real biological meaning, and the recent demonstration of the implication of endoplasmic reticulum in phagocytosis (Gagnon et al., 2002) starts from a proteomics study of the phagosomes repeatedly showing reticulum proteins.

6.4 FRACTIONATION OF PROTEIN COMPLEXES

Another approach for protein fractionation is to use lesser order structures, and to try to purify protein complexes. This fractionation is based on the fact that the association of protein into complexes appears more and more as a general phenomenon. While this phenomenon is generally true in vivo, the major problem is to keep these complexes structurally similar in vitro compared to their existence in vivo. Maintaining this structural consistency is a double-edged sword: complexes can lose components in vitro or, conversely, can accumulate some nonspecific components as artifacts of the sample preparation procedure. The magnitude of the problem depends on the cohesion of the complexes.

Some biological complexes are extremely strong and therefore easy to purify. One of the best examples is the ribosome, which has been studied comprehensively for decades, before the word proteomics even existed. As the ribosomal proteins have rather extreme pI values, specialized methods were devised to separate as many ribosomal proteins as possible on special two-dimensional gels run without isoelectric focusing (Madjar et al., 1979). The use of IEF-based two-dimensional gels was made possible only when pH gradients extending in the basic pH range were practicable (Sinha et al., 1990; Gorg et al., 1997). Furthermore, non-gel proteomics also used ribosomes as a case study to show its performance (Link et al., 1999). Other strong protein complexes, such as the nuclear pore, were also good targets for proteomics approaches of various types (Rout et al., 2000).

However, these successes were made possible by the existence of isolation procedures derived from the ones used for organelles to purify the complexes adequately. For other less abundant and more fragile complexes, multistep isolation processes give either too low a yield or destroy the in vivo complex. To limit these problems, more straightforward approaches are needed.

When dealing with natural complexes, a good example of a straightforward and elegant purification is afforded by the one-step isolation of mitochondrial complexes using the blue native electrophoresis approach (Schagger and Von Jagow, 1991). In this approach, the respiratory complexes are separated on the basis of their size on a native acrylamide gel run at neutral pH in a special buffer system. As these complexes are membrane associated, a nonionic, nondenaturing detergent is used to solubilize them. Solubilization is helped by the addition of aminocaproic acid at high concentration. Last but not least, Coomassie Blue is added after initial solubilization to give all complexes a negative charge. This charge addition improves the migration in the gel and limits the aggregation phenomena. With this approach, intact and enzymatically active complexes have been isolated in a single step (Schagger and Von Jagow, 1991). They can then be separated into their subunits by denaturing electrophoresis in the presence of SDS, and the subunits can be analyzed by mass spectrometry. Even hydrophobic subunits can be analyzed with this strategy (Devreese et al., 2002), due to the fact that complexes usually show an averaging effect of all subunits and are generally much more soluble than some of their individual subunits. When the protein complexes of interest are amenable to this approach, and this is not limited to mitochondrial respiratory complexes (Singh et al., 2000), either complete complexes or subcomplexes can be isolated, depending on how the experimental conditions are optimized (Neff and Dencher, 1999).

The basic principle of electrophoretic separation of native complexes is not limited to the blue native electrophoresis setup. The rationale is to be able to perform a mild solubilization of the complexes, to promote separation and solubility by increasing their electrical charge above their natural one by an anionic binder, and then to separate them by electrophoresis at neutral pH. In the blue native electrophoresis system, the solubilizer is a nonionic detergent and the ionic binder is Coomassie Blue. In another setup, the solubilizer and the anionic binder are combined in the form of a specially designed anionic detergent (Hisabori et al., 1991). Although not as popular as the blue native system, the nondenaturing anionic detergent offers some flexibility through the design of the detergent and should afford separation possibilities in cases resistant to the blue native system.

Despite its appealing performances, this native electrophoresis approach has received rather limited use. It is possible that only the strongest complexes survive these separation steps, and that most complexes are dissociated either by Coomassie Blue binding or during electrophoresis. Furthermore, this fractionation approach is based on the size of the complexes and therefore requires that the complexes to be separated show a larger size than the bulk of proteins (i.e., much above 100 kDa).

Another approach for the isolation of protein complexes is based on affinity purification via one known member of the protein complex of interest. The most classical scheme is immunopurification, and this technique has been used for innumerable applications. However, this approach cannot be considered as a generic one, as it requires one to have specific antibodies. Moreover, artifactual binding of some proteins to antibodies is commonplace, and conditions used to decrease this spurious binding (e.g., high ionic strength, presence of detergents) can also lead to partial or total dissociation of the complex of interest. Related, "targeted" purifications of protein complexes also use affinity chromatography schemes. Instead of using antibodies as binders, they use chemical derivatives of known ligands of at least one member of the complex to be purified. One elegant strategy uses a ligand coupled chemically to oligo dA and allows purification of even membrane protein-containing complexes under very mild conditions (oligo dT chromatography) for binding and elution (Roos et al., 1998). Although requiring careful optimization, this approach can be of interest for answering specific questions using the proteomics toolbox.

With the progress of the tagging strategies allowed by gene engineering techniques, other complex purification schemes based on chimeric proteins have been devised. Here again, numerous tags have been used, ranging from the poorly specific, but small size, hexahistidine tag, to fusion proteins with domains such as glutathione transferase (Smith and Johnson, 1988), to maltose binding protein (Maina et al., 1988). Extensive use of these techniques, although successful, has pointed out problems linked to nonspecific binding of proteins to the supports used to purify the tagged proteins. Here again, the conditions used to decrease this nonspecific binding often dissociate the protein complexes. There was thus a need for a specific and generic approach to prepare protein complexes. The best solution up to now is based on the use of two independent tags and is known as the tandem affinity purification (TAP)-TAG approach (Rigaut et al., 1999). In this elegant approach, two tags separated by a specific protease cleavage site are fused in tandem. By combining the specificities of the two tags, nonspecific adsorption is reduced to a minimum, while retaining the capacity of a generic method for the exploration of protein complexes at a proteome-wide scale (Gavin et al., 2002; Janin and Seraphin, 2003).

One of the important caveats in all these strategies based on isolation of protein complexes is the likelihood of the isolated complex to mimic, at least in part, the complexes existing in vivo. This requirement poses the problem of choosing extraction and purification conditions that look as similar as possible to those prevailing in a cell, while still being practicable. It must be recalled at this point that a cytosolic or a mitochondrial matrix is a 10% by weight solution of proteins, that is, a milieu of very high viscosity and of quite unknown chemical parameters (dielectric constant, ionic strength, etc.). It is therefore quite obvious that weakly buffered water solutions of low ionic strength do not represent a good mimic of the intracellular medium, and that problems with the half-lives of complexes in such media are most likely to occur. Furthermore, increasing the ionic strength is probably not a good solution, as this is known to extract many subunits from complexes, at least those held by means of electrostatic interactions. Here again, the seminal work of Schägger and Von Jagow (1991) provides an interesting scheme to follow. In their work, it is shown that the addition of high concentrations of a dielectric compound (in this case aminocaproic acid) dramatically enhances the extraction of protein complexes. Thus, dielectric compounds could show some of the beneficial effects of salts (salting in effects), without showing their dissociating effects. However, aminocaproic acid has a relatively low dipolar moment and might not be the ideal dielectric compound. Chemicals with stronger dipolar moments such as sulfobetaines (Vuillard et al., 1995a) have been shown to increase protein solubility and may prove useful for isolation of protein complexes. Their efficiency as salt mimics has been shown for the purification of halophilic proteins (Vuillard et al., 1995b). This example has also shown the efficiency of betaine in this process (Vuillard et al., 1995b), which is linked with the presence of a high concentration of betaine in bacterial cytosol during osmotic stress (Larsen et al., 1987). Thus, dielectric compounds with varying dipolar moments of hydrophobic moieties may be worth testing to enhance the stability of protein complexes during their fractionation either by electrophoresis or chromatographic techniques.

6.5 FRACTIONATION OF INDIVIDUAL PROTEINS

6.5.1 The Problem of Protein Solubility

Because of the difficulties in maintaining the integrity of protein complexes, it appears much easier to fractionate individual proteins. When such a process is to be performed on a proteome-wide scale, the bottleneck problem turns out to be protein solubility. In a cell, proteins are dissolved in a water-based solvent containing some 10% proteins and some salts. Such proteins are thus usually soluble in water-based solvents containing some salts and/or neutral chemicals such as sugars or polyols. This general rule has of course some exceptions, such as halophilic proteins, which require high salt concentrations (above 0.2 M) to be soluble and active. However, a special, general, and very important case is the one represented by membrane proteins. Membrane proteins usually contain domains protruding in the extracellular environment or in the cytosol. These domains behave as classical protein domains and are therefore stable in a water-based solvent. However, membrane proteins also contain domains that are embedded more or less deeply in the cell's lipid bilayer. Consequently, these domains are stable in a hydrocarbon-like environment. If a membrane protein is to be solubilized prior to its purification, this means that the solvent used for this purpose must possess both water- and hydrocarbon-like properties. A water-based solvent will induce aggregation of the protein's membrane domains via hydrophobic interactions, causing protein precipitation. However, an organic solvent will denature their water-soluble domains, causing them to coalesce, and will result in precipitation of the protein. Some very hydrophobic proteins, however, are soluble in organic solvents (Molloy et al., 1999; Blonder et al., 2003). This solubility is due to the reduced size of their nonmembrane domains, and to the fact that all water-soluble protein domains contain a hydrophobic core, which can be soluble in organic solvents. In some cases, these positive solubilization forces can overcome the precipitation-driving forces and make the protein soluble in organic solvents.

This solubility issue, however, is not a general case. The typical rule for membrane proteins is that they require a solvent that possesses both strong water- and hydrocarbon-like properties. This property is not found within mixed solvents (i.e., mixtures of water and water-miscible solvents), which only offer average properties and not a combination of both properties. Such a combination is only offered by a stable dispersion of hydrocarbon chains in a water-based solvent. This combination is the definition of lipid membranes, but such assemblies are very difficult to handle throughout a purification process. Hopefully, this definition also applies to detergent micelles, which are much easier to handle throughout a purification process. These constraints explain why detergents are universally successful for the disruption of cell membranes and for the solubilization of membrane proteins.

The micelle-forming ability of detergents is driven by their chemical structure, which combines a hydrophobic part (the tail) that promotes aggregation of the molecules into the micelles, linked to a hydrophilic one (the head) that promotes water solubility of the individual molecules and the micelles. By using different structures of heads and tails and combining various tails with various heads, an almost infinite range of detergents can be generated. Their protein and lipid solubilization properties are of course strongly dependent on the chemical properties of the heads and tails. As a rule of thumb, rigid tails (e.g., steroid-based) and weakly polar heads (e.g., glycosides, oligoethylene glycol) generally lead to "mild" detergents, that is, chemicals that have good lipid solubilization properties but weak protein dissociation properties. While these detergents do not denature proteins, they can prove inadequate for preventing protein–protein interactions that promote precipitation. Consequently, their protein solubilization properties are highly variable, and a lengthy optimization process for choosing the best detergent is usually needed for various membrane proteins.

Conversely, detergents having a flexible tail and a strongly polar (i.e., ionic) head are viewed as "strong" detergents. In addition to their lipid solubilizing properties, these detergents are able to disrupt the hydrophobic interactions maintaining the structure of the proteins. They are therefore denaturing. However, because of their ionic nature, the bound detergent imparts a net electrical charge to the denatured proteins and induces a strong electrostatic repulsion between protein molecules. Thus, even denatured proteins can no longer aggregate, and these ionic detergents are very powerful protein solubilizing agents.

These protein solubilization conditions have a key impact on the protein fractionation that can be carried out afterward, in the sense that they will restrict the choice to techniques with which they are compatible. As an example, ionic detergents are not compatible with any technique using protein charge or pI as the fractionation parameter.

6.5.2   Principles Used for Protein Fractionation

Being complex analytes, proteins offer a variety of physicochemical features that can be used as separation parameters for fractionation. Some of these parameters are not strongly dependent on the integrity of the three-dimensional structure of the proteins. These "structural" parameters include the protein's size, electrical charge at a given pH, isoelectric point, and hydrophobicity. Some parameters are more dependent on special amino acid sequences or structural properties and will be examined later in the description of the specific separation modes. Conversely, some parameters depend strongly on the three-dimensional structure of the proteins. These "functional" parameters are generally represented by the ability of the proteins to bind various molecules, and this binding process can be used as a way to separate proteins. These separation parameters can be implemented in a variety of protein separation setups, which can be grouped according to the type of implementation.

6.5.3   Chromatographic Techniques

In the first, major type of implementation, the separation parameter is used to drive the interaction of proteins to a solid support (the chromatographic phase) and then release the proteins from this support to recover them in solution. This implementation is clearly the most versatile and has amply shown its abilities in proteomics research. However, nothing being perfect, a totally neutral support is not available. Consequently, there are always spurious interactions between the support and the proteins, leading to nonspecific adsorption problems and protein losses. As an example, dextran-based supports are ideal to design gel filtration media, in which the protein size is the separation parameter. However, dextran is not electrically neutral, and dextran-based supports contain immobilized anionic groups. These, in turn, induce binding via electrostatic interactions of cationic proteins, unless a buffer of sufficient ionic strength is used to prevent this ionic binding. A general consequence of these extra-binding properties, whatever they are based on, is the fact that binding of proteins to the support is generally not the limiting process. The limiting process is clearly the release of proteins from the support to recover them in the liquid phase. The chromatographic processes that have been used for protein fractionation can be typed according to the separation parameter used and are outlined below.

Fractionation According to Size: Gel Filtration   Gel filtration is a process that has been used for decades to separate analytes, whether they are proteins or not, on the basis on their hydrodynamic size. In this process, the analyte percolates through a porous and beaded support. When the analyte is too big to enter the pores, the analyte travels around the pores and emerges first from the column. With a decreasing size, the analytes can enter deeper into the pores of the separating material and will be more retarded. In this process, the order of the elution is the order of decreasing size. When applied to proteins, assuming that they have roughly a globular shape, this size order corresponds roughly to a molecular weight scale.

Typical porous supports are based on complex sugars, generally chemically crosslinked, or on crosslinked polyacrylamide gels. As the separation parameter is not dependent on the surface functionalities of the proteins to be separated, this means that this separation process is a fairly general one and can be carried out in a variety of conditions—for example, on native or denatured proteins or in the presence or absence of a variety of detergents. The major drawback of this technique is the obvious zone broadening due to the diffusion-based separation process, leading to a rather low resolution separation.

Fractionation According to Electrical Charge: Ion Exchange   In this process, the analytes are bound to the chromatographic support through electrostatic interactions between the electrical charges present on the analytes and those of opposite charge present on the chromatographic support. As all solutions are electrically neutral, this binding is a competition between the ions of interest and, for example, the counterions neutralizing the charges present on the support or the other ions present in the sample solution. Conversely, elution is carried out either by increasing the ionic strength of the solution, and thereby increasing this competition in favor of the bulk ions present in the solvents, and/or by changing the pH and thus the electrical charge present either on the support or on the analytes. When applied to proteins, which generally behave as polyelectrolytes, the binding process is generally quite efficient. Elution is generally carried out by increasing the ionic strength and not by changing the pH, as the pH stability range of proteins is generally limited.

This setup means that proteins must be loaded on the support at low ionic strength and elute at a high ionic strength. When reaching the proteomics scale, that is, when trying to apply this approach to every protein or protein assembly, both can be a problem in some cases. Some proteins (e.g., halophilic proteins) do not tolerate low ionic strength buffers, while the high ionic strength required for elution may dissociate some multisubunit proteins or protein complexes. However, ion exchange can be carried out on native proteins as well as on denatured ones, and in the presence or absence of detergents, provided that the molecules present in solution do not interfere with the ion exchange process. These requirements disqualify charged denaturants, such as guanidinium or thiocyanate salts, which are used at such a concentration that they will prevent any binding to ion exchange supports. Charged detergents may also be excluded, but not on the same basis. In fact, most ionic detergents bind strongly and massively to proteins, masking their native charge and thus destroying the very basis of the separation process.

However, despite these limitations, ion exchange is a technique with a higher resolution than gel filtration. Its resolution, however, is not sufficient to resolve the post-translationally modified forms of the proteins from their unmodified counterparts due to the fact that post-translational modifications generally alter the charge of the proteins by only one unit. As ion exchange is generally most efficient with proteins that are multicharged, a change of a single unit is not sufficient to alter the elution diagram to induce peak separation. Nevertheless, ion exchange is a useful and important technique for initial fractionation of complex protein mixtures.

Fractionation According to Isoelectric Point: Chromatofocusing In this process, the support bears a polyamine functionality designed in such a way that the amino groups have spaced pK_a values. The proteins are loaded at a high pH (higher than the pI range of interest) so that they have a negative charge and bind to the positively charged support. A pH gradient is then made with a low pH eluent containing carrier ampholytes, in order to establish a pH gradient in the column. Due to the geometry in the column format, this means that the pH gradient is a spatial gradient in the column moving downward with time. When the pI of a given protein is reached at a given region, this protein elutes from the column and binds at a lower region of the column, until the pI zone reaches the end of the column and provokes the final elution. This process leads to an important concentration effect, hence the name chromatofocusing.

The most interesting feature of chromatofocusing is its excellent resolution, which is due in part to the concentration effect. In addition, chromatofocusing is often able to separate post-translational variants of proteins. While the change induced by post-translational modifications is often by one electrical charge, the pI change induced by this modification is often sufficient to be resolved by chromatofocusing techniques. This technique has not gained widespread use, however, because of its limitations. When applied to native proteins, it requires the proteins to be stable over a wide pH range, from the basic, loading pH to the pI itself. This broad pH stability is not often the case, and this technique is more successful with denatured proteins. Working with denatured proteins does not, however, change the most critical point, which is that the proteins must be eluted at their pI, which is their solubility minimum. Solubility problems are the real bottleneck of chromatofocusing and are of course more acute with poorly soluble proteins such as membrane proteins. In addition, only electrically neutral additives (e.g., detergents) may be used, as charged additives have adverse effects. Charged detergents completely blur the real pI of the proteins by their binding, and salts prevent the attachment of the proteins to the resin above their pI, which takes place by an ion exchange mechanism. The solubility problem is aggravated by the fact that spurious adsorption can take place on the support itself and further decrease protein solubility.

Fractionation According to Hydophobicity: Hydrophobic Interaction Chromatography   Opposite to the processes discussed previously, which can be carried out on native or on denatured proteins, hydrophobic interaction chromatography (HIC) is carried out almost exclusively on native proteins. It takes advantage of the existence of hydrophobic patches at the surface of many proteins that can interact with a suitable adsorbent, that is, a hydrophilic polymer grafted with hydrophobic groups. The rationale of this interaction is exactly the one of reverse phase, in the sense that interaction is favored by salting out environments (i.e., salt solutions) and weakened in water or, even more, in water–organic solvent mixtures. However, when using water–solvent eluents, great care must be taken not to reach conditions that denature the proteins to be eluted. Upon denaturation, the hydrophobic core of the proteins is exposed to the solvent and will induce a strong binding to the support, as this hydrophobic core is much larger in size than the hydrophobic surface patches present on the native protein. In addition, this method cannot be applied on protein preparations containing detergents, as the detergent will bind most preferentially to the hydrophobic support and will thus be dissociated from the protein.

Thus, ironically enough, this method is best suited for rather hydrophilic proteins and not for membrane proteins. It is therefore not very suitable for proteomic analyses on complete cell extracts but can be used successfully on biological fluids, which contain mainly hydrophilic proteins. However, it has been used successfully for nuclear proteins that are slightly hydrophobic (Schafer-Nielsen and Rose, 1982). It can be used very efficiently as a followup of ion exchange, as it will deionize the sample. It is also highly complementary to the electrophoretic methods, as it uses a separation parameter that cannot be easily implemented in electrophoresis setups.

Fractionation According to Structural Determinants: Metal Chelate Chromatography   In this process, first developed in Porath's group (Porath and Olin, 1983), the immobilized binding moiety is a preformed complex between a metal ion (generally Zn, Cu, or Ni) and an immobilized iminodiacetic function. This preformed complex has the ability to bind additional metal binding compounds. Among those, the imidazole and thiol groups can be found at the surface of the proteins. Thus, the binding of proteins to metal chelates will depend on the presence, at the surface of the proteins, of available imidazole and thiol groups in a geometry that is compatible with binding to the metal chelate. While this induces strong structural constraints when working with native proteins, denatured proteins can also be used. Since denatured proteins are more flexible, they can bind more easily to the chelate. In addition, detergent-containing solutions can be used, provided that the detergent does not bind to the column itself. This association is the case for most detergents except those that are anionic. Moreover, salts can be used, except those that strongly bind heavy multivalent cations (e.g. citrate, tartrate, EDTA). These constraints make this method highly specific for some proteins, even in a complex sample. While this specificity has been exploited extensively on genetically engineered (His-tagged) proteins for purification purposes (Janknecht et al., 1991), this selectivity is more difficult to use with a complex mixture of natural proteins, as is the case in proteomics.

Fractionation According to Binding Capacities: Affinity Chromatography   In this purification scheme, a ligand able to be selectively bound by some proteins is immobilized on the support. The protein mixture to be purified is then passed on this support, and selective binding to the ligand occurs. After washing steps intended to remove nonspecifically bound proteins, elution is performed either by competition with an excess of soluble ligand or by destabilization of the ligand–protein complex, generally by salts or with slightly chaotropic solutions.

Key features in affinity chromatography include the structure and binding geometry of the immobilized ligand to the support. Variants of natural ligands are often used, for example, to provide some spacing between the ligand and the support. By decreasing support-induced steric hindrance, this helps to provide adequate binding of proteins to the support. The second key feature is the elution process, which generally must be mild enough to avoid protein denaturation, but still must be efficient for maximal yields. Innumerable applications of affinity chromatography have been described in the literature. However, in standard biochemistry practice, affinity chromatography is generally designed to be as specific as possible. This specificity is not desirable in many proteomics studies, where the aim is to simplify a very complex extract and thus to purify a class of proteins. However, proteomics studies put an extreme emphasis on robustness. For example, variations in the fine composition of the initial extracts should not overtly perturb the binding process. Alternatively, it is often not possible to try to optimize the ligand–support couple to ensure optimal binding. Thus, for a given protein class to be purified, the choice of the immobilized ligand will not be dictated by the absolute affinity for this ligand, but rather by the ability of the ligand to bind as many members of the protein class as possible. This priority change compared to classical affinity chromatography has probably delayed the application of this technique in proteomics.

A typical example of these problems is given by immobilized dyes (e.g., Cibacron Blue). While this system is quite popular for the purification of some proteins, including the nucleotide-binding proteins, it is not very well suited for proteomics applications. This incompatibility is due to the fact that the basis of the binding is rather unclear. Thus, while it is known that some nucleotide-binding proteins bind to immobilized dyes, not all do so, and some proteins that are known not to bind nucleotides are able to bind immobilized dyes. Therefore, it will be difficult to find a rationale in the proteins retained on immobilized dyes from a complex extract. Moreover, competition phenomena between the proteins of a complex extract for the binding means that some minor changes in the initial extract may lead to important changes in the bound proteins, resulting in a very difficult interpretation of the data.

Thus, affinity methods to be used in proteomics must rely on a very precise, well-known, and controllable binding mechanism. An example of this controlled binding includes lectin chromatography for glycoproteins, although this affinity chromatography has been most often applied to glycopeptides and not to entire proteins. To keep with rather classical types of affinity chromatography, selection of functional classes of proteins (e.g., the nucleotide-binding proteins) could be of interest in proteomics, and examples of these approaches have begun to appear in the literature (Kim and Park, 2003). The biotin–avidin system is also widely used, but generally after in-solution covalent tagging of the proteins of interest within the complex mixture with a biotin-containing reactive probe. Examples of this approach have been shown on serine or cysteine hydrolases (Liu et al., 1999; Kocks et al., 2003) but can probably be extended to other classes of proteins. In some circumstances, however, affinity chromatography is used in its more classical way of providing the highest possible specificity. This specificity, however, is generally used to purify specific protein complexes, and the reader should refer to the previous section for more details.

6.5.4   Electrophoresis Techniques

In this other major type of protein separation method, proteins are made to migrate in an electric field, using the fact that they are charged analytes. The various electrophoretic setups make use of some structural parameters of the proteins to induce a separation of these proteins. Unlike chromatography, there is theoretically no absolute need for a support, and electrophoresis can be carried out in a liquid vein. However, many electrophoresis techniques perform much better when a gelified support is available. This gelified support combines mechanical strength and easy handling of solids with the liquid phase needed for protein separation. Among these, polyacrylamide gels are the most frequently used, as they combine a good separation versatility with a very low adsorption of proteins.

Because of the movement of ions in the electric field used for driving the separation process, most electrophoretic techniques generate an appreciable amount of Joule heat. This Joule heat induces important thermal and convection problems, which considerably limit the possible scale-up of electrophoresis techniques. Thus, most electrophoretic techniques have been optimized as analytical more than preparative techniques. However, with the recent progress of protein microanalysis techniques and the concomitant decrease in the amount of proteins needed for analysis, the capacity of electrophoretic techniques is now more and more able to deliver the required microquantities of proteins, and so-called "micropreparative" electrophoretic techniques are now more frequently used in proteomics. One of the best examples of this evolution is two-dimensional electrophoresis of proteins, as described in an earlier chapter.

Among the many parameters usable for separating proteins, only three are used in electrophoresis: the protein's charge, size, and isoelectric point. The other parameters (e.g., hydrophobicity) require a reasonable scale of adsorption phenomenon to occur between the protein and some solid support. When adsorption takes place in an electrophoretic process, this induces a strong tailing of the separated zones and thus decreases resolution dramatically. This phenomenon is the plague of capillary electrophoresis, where adsorption of proteins on the walls of the capillary prevents this technique from being widely used as a separation tool at the proteomics scale.

Electric Charge as a Separation Parameter in Electrophoresis   The basic equations for the electrophoretic movement of analytes in an electric field show that the mobility is proportional to the electric charge of the analyte and to the electric field, and limited by its friction in the separation medium. Thus, the electric charge of proteins could be an interesting separation parameter. However, this is mainly true in liquid media. In gel-based media, the additional friction induced by the gel material makes the friction parameter dominant over the native charge. This means that charge differences result in a smaller separation than size differences. Thus, electric charge is used almost exclusively in capillary electrophoresis and in free-flow electrophoresis. These methods, however, are not very popular at the proteomics scale because of their rather difficult implementation and rather low resolution for proteins. In addition to this role as a driving parameter in electrophoresis, the electric charge of the proteins also plays a role in preventing spurious aggregation during separation. Proteins of the same type of charge repel each other, and this electrostatic repulsion is very important, although often unnoticed, in the performances of high-resolution electrophoretic techniques, as will be seen later. Conversely, proteins with opposite charges tend to aggregate and this considerably limits the scope of charge-based electrophoresis techniques.

Size as the Separation Parameter in Electrophoresis: Zone Electrophoresis in Gels   As mentioned earlier, size is the dominant separation parameter in electrophoretic separations of proteins in gel-based media. It is thus very easy to implement as a separation parameter in electrophoresis techniques. However, the simplest setup (i.e., sample solubilized in the same buffer as the one used for the gels and for the electrode reservoirs and so-called continuous electrophoresis) is a rather low resolution technique due to the fact that diffusion and, to a lesser extent, convection are at play during the migration of the proteins in the electric field. Thus, the thinness of the protein zones at the end of the separation process can only be wider than the one at the beginning of the process. Because it is often difficult to reach important concentrations in protein solutions, the amounts needed for most purposes result in important volumes compared to the scale of the process and thus to wide zones.

The solution to this vexatious problem is the so-called discontinuous electrophoresis setup, which was described by Ornstein and Davis in the mid 1960's (Davis, 1964) and theoretized by Jovin in the early 1970s (Jovin, 1973). This process uses the isotachophoresis process, which states that under proper conditions, ions rank themselves under an electric field by their decreasing mobility with so-called moving boundaries between the various ionic species. When combined with the use of ions whose mobility is strongly pH dependent and can thus be easily modulated by the pH of the gel, and with the flexibility afforded by gelified media, it is possible to devise a setup in which two different steps take place successively. In the first step, called the stacking step, the isotachophoretic conditions between the fast (leading) and slow (trailing) ions are set with a very low mobility. Proteins therefore concentrate at the moving boundary between the leading and trailing ions. This ensures a very strong concentration effect, and thus a very high resolution afterward. In the second step, the speed of the moving boundary is increased, generally by a change in the pH, to a point where proteins are no longer mobile enough to remain in the moving boundary and are therefore separated. While the resolution of the system is extremely good, its initial versions were plagued by massive aggregation and precipitation phenomena. This effect should not be surprising, as protein concentrations in tens of milligrams per milliliter are commonplace in the stacking step. At such concentrations, the electrostatic repulsion arising from the natural electrical charge of the proteins is no longer sufficient to prevent aggregation.

The straightforward answer to this new problem is to use additives that can bind to proteins and increase their native charge. One example is the Coomassie Blue used in the BN-Page technique for separation of protein complexes (see Section 6.4 of this chapter). However, the most popular additives are charged detergents, especially SDS. SDS has the special property of binding proteins with a quasi-uniform stoichiometry of 1.4 g SDS per g protein (Reynolds and Tanford, 1970). This very important binding has two favorable consequences. The first one is to make every protein strongly negatively charged. This induces a very important electrostatic repulsion and prevents almost all aggregation phenomena during electrophoresis. The second consequence is to mask the native electrical charge of the protein below the charges brought by SDS binding. This masking almost nullifies the influence of the native charge of the proteins and gives rise to an electrophoretic system where protein size is the sole separation parameter.

These positive features explain the very wide popularity of SDS electrophoresis as a protein separation tool. In addition to its resolution, which is quite good for a protein separation tool, since proteins differing in their mass by 2% can be separated, its popularity is bolstered by the robustness of the technique, which is applicable to almost any type of protein. The price to pay for this robustness is the denaturation of the proteins by the binding of SDS, and this technique cannot be applied to the separation of native proteins.

There are, however, a few exceptions of proteins that do not behave properly in SDS electrophoresis, most often due to poor binding of the detergent. The most classical example is glycoproteins, but very basic proteins such as histones or halophilic proteins also show an aberrant migration via SDS electrophoresis. The other drawback of SDS electrophoresis is that it performs best at alkaline pH, where some native post-translational modifications are labile (e.g., esterification of carboxyl groups) and where artifactual modifications (e.g., acrylamide adducts) can arise.

To counteract these problems, systems running with cationic detergents at low pH have been proposed. Although they fulfill the requirements devoted to wide-scope systems (MacFarlane, 1983), they have never achieved broad popularity. This lack of use is due to the fact that polymerization of acrylamide at low pH is difficult and uses rather erratic initiators (MacFarlane, 1983), and also because the resolution is not equivalent to the one that can be reached with SDS electrophoresis and is often not strictly correlated to molecular weight (Lopez et al., 1991). One positive side effect of cationic detergent electrophoresis is that it can be compatible with some preservation of protein activity (Akins et al., 1992). The only system that has reached some popularity in proteomics is the so-called BAC-SDS system (MacFarlane, 1989), which combines a cationic detergent electrophoresis in the first dimension with SDS electrophoresis in the second dimension. Because of the deviations from ideality mentioned above for cationic detergent electrophoresis, the separation is not diagonal but rather cone-shaped. Although the overall resolution is rather low, the system has shown its ability to separate hydrophobic membrane proteins (Hartinger et al., 1996), which are often very difficult to separate with other, higher resolution, electrophoresis systems.

Apart from these minor problems, the main drawbacks of SDS electrophoresis are its inability to separate protein forms differing by a very low mass increment (typically post-translationally modified proteins) and the difficulties encountered for the recovery of whole proteins from the gels after separation. The only really versatile system is blotting onto an adsorptive membrane, but this just replaces entrapment in a gel by adsorption on a surface and does not warrant recovery in a liquid phase. Recovery in a liquid phase, either by passive elution, or by electroelution of excised bands, or by continuous elution in a buffer after electrophoresis, is often associated with variable yields and significant dilution of the proteins or is thus not widely applicable in proteomics.

Isoelectric Point as the Separation Parameter: Isoelectric Focusing The isoelectric point is very easily implemented as a separation parameter in electrophoresis setups. The only necessary condition is to be able to create a stable pH gradient in the separation medium. When proteins are placed in this pH gradient and submitted to an electric field, they migrate up to their isoelectric point. At this very pH, their charge nullifies and they stop migrating. This technique is called isoelectric focusing because a concentration effect takes place at the isoelectric point. Any protein molecule that goes away from its pI (e.g., through diffusion) acquires an electrical charge and migrates back again to its isoelectric point. This ensures very sharp protein zones and a steady-state separation, which is as stable as the pH gradient or the supporting medium itself.

The generation of a stable pH gradient is therefore a critical point in isoelectric focusing. Two technologies are of wide use for this purpose—the carrier ampholytes technology and the immobilized pH gradient technology. Carrier ampholytes are molecules that have a pI, a buffering capacity at their pI, and some conductivity at this pI. Some amino acids (e.g., aspartic and glutamic acids, histidine, lysine, and arginine) are correct carrier ampholytes, while others (e.g., glycine) are not because they show a miserable buffering capacity at their pI. Proteins also behave as carrier ampholytes but are of almost no practical interest. When a mixture of carrier ampholytes is placed between an acidic anodic solution and a basic cathodic solution under an electric field, the carrier ampholyte molecules rank themselves in the order of their pI values and stabilize a pH gradient ranging from the lowest pI species to the highest pI species. The pH gradient is therefore a multistep gradient where each step is represented by a given molecule of a given pI. Each step is as high as the pI increment between the closest neighbors (in pI) and as wide as the concentration of each molecule. Commercial carrier ampholytes are poorly controlled mixtures containing hundreds, maybe thousands, of individual carrier ampholyte molecules. This ensures the production of smooth pH gradients. It must be noted that the pH gradient produced with carrier ampholytes is completely independent from the support but is dependent on the existence of the electric field. However, the fine and long-term reproducibility of pH gradients is difficult to achieve, as different production batches of ampholytes cannot reproduce exactly the same composition and thus exactly the same gradient. Furthermore, the gradient generated by carrier ampholytes is not absolutely stable and decays from the basic side by a phenomenon called cathodic drift. These drawbacks have pushed the development of a new technology, called immobilized pH gradients. As the name implies, pH gradients are generated in this case by an acidobasic titration of buffering monomers, which are copolymerized with the gel-producing monomers. This copolymerization restricts the use of immobilized pH gradients to polyacrylamide gels. However, the drawbacks associated with carrier ampholytes (reproducibility problems, cathodic drift) are completely abolished with this technique, which also allows real pH gradient engineering.

Isoelectric focusing is probably the most resolutive protein separation technology. On a routine basis, proteins differing in their pI by less than 0.1 pH unit are easily separated, meaning that IEF is able to separate some allelic variants, provided that they differ in their charged amino acids composition, or even some post-translationally modified variants (e.g., phosphorylated forms). However, several constraints inherent to the isoelectric focusing process limit the scope of this separation method. The first constraint arises from the high electric field needed in IEF. When proteins are close to their pI, their net charge is low and they are thus poorly mobile in an electric field. A high field is thus required to ensure a correct mobility of proteins close to their pI, and also to maximize the focusing effect. A direct consequence is that low ionic strength media are mandatory for isoelectric focusing, which often limits protein solubility. Furthermore, proteins are at their pI at the end of the IEF process, and the pI is the solubility minimum for a protein. Solubility problems are very important in IEF, especially with native proteins, and therefore IEF is best performed with denatured proteins. However, even in this case, many proteins such as membrane proteins are poorly soluble under IEF conditions. Compared with detergent-based electrophoresis, this problem is further enhanced by the fact that only electrically neutral detergents can be used. Ionic detergents would modify the pI of the proteins and thus prevent any correct separation through this parameter, thereby preventing the beneficial electrostatic repulsion effect used in IEF and leaving proteins under conditions in which their solubility is minimal.

Despite these problems, isoelectric focusing is widely used in proteomics, because of its intrinsic high resolving power. Its most frequent use is through analytical or micropreparative two-dimensional electrophoresis of proteins, which is not within the scope of this chapter. However, with the increased recognition of the complexity of proteomes and thus the increased need for prefractionation techniques at the protein level before the final micropreparative techniques, IEF is more frequently used as a prefractionation tool. It can be used in its very old preparative variants, in liquid vein (e.g., free-flow electrophoresis or the Rotofor® apparatus), or in a semisolid medium made of Sephadex beads (Radola, 1975), as recently repopularized (Gorg et al., 2002). It can also be used in a variant of immobilized pH gradients called multicompartment electrolysis (MCE). This technique was first described in the late 1980s as a preparative tool for the final purification of native proteins (Righetti et al., 1990) but has recently been adapted to crude fractionation of complex mixtures of denatured proteins (Herbert and Righetti, 2000; Zuo and Speicher, 2000).

Because of the low ionic strengths used in IEF, Joule heating is minimal in this process, which greatly facilitates adaptation of IEF to medium-scale purifications needed for prefractionation of complex extracts. This phenomenon explains why more descriptions are found of such preparative setups in the IEF mode rather than in the size-based zone electrophoresis mode.

6.6 PRESENT AND FUTURE TRENDS IN PROTEIN FRACTIONATION IN PROTEOMICS

Most of the methods described in this chapter are quite familiar to many life scientists. As a matter of fact, the protein separation methods described have been optimized over several decades and form the core techniques for classical biochemical protein purification. This long-term optimization means that this field is quite mature, and that it is rather unlikely that revolutionary breakthrough will take place in this field in the near future. There is, however, a major difference in the use of protein separation methods in classical biochemistry and proteomics. In classical biochemistry, the various separation methods are chained to solve a single problem each time, which is the purification of a single protein. This is no longer the case in proteomics. In this field, protein separation is most often used as a prefractionation tool before the proteomic analysis per se. In many cases, the proteomic analysis is also based on protein separation methods, mainly by SDS electrophoresis or by two-dimensional electrophoresis. Despite the performances of these analytical techniques, the complexity of proteomes is so high that prefractionation is now viewed as an absolute requisite if any in-depth proteomic analysis is to be performed.

A further constraint imposed by proteomics is robustness. As proteomic analyses aim toward comprehensiveness, they should be ideally applicable to any type of protein. Owing to the amazing diversity of proteins, it is clear that no separation method can be considered as applicable to every protein. Thus, the usefulness of protein separation methods in proteomics will depend mainly on how they fulfill this robustness requirement and complement the proteomics technique used. As an example, size fractionation of denatured proteins is not attractive as a prefractionation method, as many proteomics setups use that type of fractionation as a core component of the proteomics toolbox. However, prefractionation by preparative IEF is more useful, as it can be used in conjunction with very high resolution 2D electrophoresis to enhance the coverage of proteomes (Herbert et al., 2001). However, as preparative IEF is by definition plagued by the same solubility problems as those encountered in analytical IEF or two-dimensional electrophoresis, the major challenge is one of protein solubilization under IEF conditions, and not IEF per se. This challenge returns to the general protein solubility problem, and the solutions to this problem are to be found in general protein-solubilizing agents such as chaotropes (Rabilloud et al., 1997) or detergents (Chevallet et al., 1998; Luche et al., 2003).

When seeing protein fractionation through the prism of complementation of core proteomics techniques, one of the most interesting added values of protein fractionation is to provide a technique that adds some functional information. This parameter explains why subcellular fractionation is becoming more and more popular in proteomics. It is also quite clear that progress in this field, mainly by increasing the purity of the products of subcellular fractionation, will have a major impact on the quality of the subsequent proteomics analysis (Zischka et al., 2003). Along this line, it is clear that methods aiming at the purification of protein classes (e.g., on the basis of their activity), as exemplified with hydrolases (Liu et al., 1999), or aiming at the purification of protein complexes are of great potential value in proteomics. There is considerable room for progress in the design of affinity-based purification schemes, which should be more robust than most of the schemes used in classical affinity chromatography aiming at the purification of a single protein. In the field of protein complexes, robustness is also a major constraint that must now be taken into account. In addition, the field of protein complexes poses the problem of in vitro mimicking of the in vivo conditions, which is far from being simple or obvious. Ironically, the main progress to be expected in this field is not likely to take place in the protein separation field per se, which is fairly mature, but in the related fields mentioned earlier, which have been much less explored in the past and are now becoming limiting in proteomics.

REFERENCES

Adessi C, Chapel A, Vincon M, Rabilloud T, Klein G, Satre M, Garin J. 1995. Identification of major proteins associated with *Dictyostelium discoideum* endocytic vesicles. *J Cell Sci* 108:3331–3337.

Akins RE, Levin PM, Tuan RS. 1992. Cetyltrimethylammonium bromide discontinuous gel electrophoresis: Mr-based separation of proteins with retention of enzymatic activity. *Anal Biochem* 202:172–178.

Blonder J, Conrads TP, Yu LR, Terumuma A, Janini GM, Issaq HJ, Vogel J, Veenstra TD. 2003. A detergent- and cyanogen bromide-free method for integral membrane proteomics: application to *Halobacterium* purple membranes and human epidermis. *Proteomics* 4:31–35.

Chevallet M, Santoni V, Poinas A, Rouquie D, Fuchs A, Kieffer S, Rossignol M, Lunardi J, Garin J, Rabilloud T. 1998. New zwitterionic detergents improve the analysis of membrane proteins by two-dimensional electrophoresis. *Electrophoresis* 19:1901–1909.

Clemetson KJ, Pfueller SL, Luscher EF, Jenkins CS. 1977. Isolation of the membrane glycoproteins of human blood platelets by lectin affinity chromatography. *Biochim Biophys Acta* 464:493–508.

Davis BJ. 1964. Disc electrophoresis. II. Method and application to human serum proteins. *Ann N Y Acad Sci* 121:404–427.

Devreese B, Vanrobaeys F, Smet J, Van Beeumen J, Van Coster R. 2002. Mass spectrometric identification of mitochondrial oxidative phosphorylation subunits separated by two-dimensional blue-native polyacrylamide gel electrophoresis. *Electrophoresis* 23:2525–2533.

Duncan R, McConkey EH. 1982. How many proteins are there in a typical mammalian cell? *Clin Chem* 28:749–755.

Gagnon E, Duclos S, Rondeau C, Chevet E, Cameron PH, Steele-Mortimer O, Paiement J, Bergeron JJ, Desjardins M. 2002. Endoplasmic reticulum-mediated phagocytosis is a mechanism of entry into macrophages. *Cell* 110:119–131.

Gavin AC, Bosche M, Krause R, Grandi P, Marzioch M, Bauer A, Schultz J, Rick JM, Michon AM, Cruciat CM, Remor M, Hofert C, Schelder M, Brajenovic M, Ruffner H, Merino A, Klein K, Hudak M, Dickson D, Rudi T, Gnau V, Bauch A, Bastuck S, Huhse B, Leutwein C, Heurtier MA, Copley RR, Edelmann A, Querfurth E, Rybin V, Drewes G, Raida M, Bouwmeester T, Bork P, Seraphin B, Kuster B, Neubauer G, Superti-Furga G. 2002. Functional organization of the yeast proteome by systematic analysis of protein complexes. *Nature* 415:141–147.

Gianazza E, Gelfi C, Righetti PG. 1980. Isoelectric focusing followed by electrophoresis of proteins for visualizing their titration curves by zymogram and immunofixation. *J Biochem Biophys Methods* 3:65–75.

Gorg A, Obermaier C, Boguth G, Csordas A, Diaz JJ, Madjar JJ. 1997. Very alkaline immobilized pH gradients for two-dimensional electrophoresis of ribosomal and nuclear proteins. *Electrophoresis* 18:328–337.

Gorg A, Boguth G, Kopf A, Reil G, Parlar H, Weiss W. 2002. Sample prefractionation with Sephadex isoelectric focusing prior to narrow pH range two-dimensional gels. *Proteomics* 2:1652–1657.

Graham J, Bailey D, Wall J, Patel K, Wagner S. 1983. The fractionation and subfractionation of cell membranes. In: Rickwood D (Ed), *Iodinated Density Gradient Media, A Practical Approach*, IRL Press, Oxford, UK, pp 91–118.

Hartinger J, Stenius K, Hogemann D, Jahn R. 1996. 16-BAC/SDS-PAGE: a two-dimensional gel electrophoresis system suitable for the separation of integral membrane proteins. *Anal Biochem* 124:126–133.

Herbert B, Righetti PG. 2000. A turning point in proteome analysis: sample prefractionation via multicompartment electrolyzers with isoelectric membranes. *Electrophoresis* 21:3639–3648.

Herbert BR, Harry JL, Packer NH, Gooley AA, Pedersen SK, Williams KL. 2001. What place for polyacrylamide in proteomics? *Trends Biotechnol* 19:S3–S9.

Hisabori T, Inoue K, Akabane Y, Iwakami S, Manabe K. 1991. Two-dimensional gel electrophoresis of the membrane-bound protein complexes, including photosystem I, of thylakoid membranes in the presence of sodium oligooxyethylene alkyl ether sulfate/dimethyl dodecylamine oxide and sodium dodecyl sulfate. *J Biochem Biophys Methods* 22:253–260.

Hymer WC, Kuff EL. 1964. Isolation of nuclei from mammalian tissues through the use of Triton X-100. *J Histochem Cytochem* 12:359–363.

Janin J, Seraphin B. 2003. Genome-wide studies of protein–protein interaction. *Curr Opin Struct Biol* 13:383–388.

Janknecht R, de Martynoff G, Lou J, Hipskind RA, Nordheim A, Stunnenberg HG. 1991. Rapid and efficient purification of native histidine-tagged protein expressed by recombinant vaccinia virus. *Proc Natl Acad Sci USA* 88:8972–8976.

Jovin TM. 1973. Multiphasic zone electrophoresis. I. Steady-state moving-boundary systems formed by different electrolyte combinations. *Biochemistry* 2:871–879.

Kim E, Park JM. 2003. Identification of novel target proteins of cyclic GMP signaling pathways using chemical proteomics. *J Biochem Mol Biol* 36:299–304.

Kocks C, Maehr R, Overkleeft HS, Wang EW, Iyer LK, Lennon-Dumenil AM, Ploegh HL, Kessler BM. 2003. Functional proteomics of the active cysteine protease content in *Drosophila* S2 cells. *Mol Cell Proteomics* 2:1188–1197.

Larsen PI, Sydnes LK, Landfald B, Strom AR. 1987. Osmoregulation in *Escherichia coli* by accumulation of organic osmolytes: betaines, glutamic acid, and trehalose. *Arch Microbiol* 147:1–7.

Link AJ, Eng J, Schieltz DM, Carmack E, Mize GJ, Morris DR, Garvik BM, Yates JR III. 1999. Direct analysis of protein complexes using mass spectrometry. *Nat Biotechnol* 17:676–682.

Liu Y, Patricelli MP, Cravatt BF. 1999. Activity-based protein profiling: the serine hydrolases. *Proc Natl Acad Sci USA* 96:14694–14699.

Lopez MF, Patton WF, Utterback BL, Chung-Welch N, Barry P, Skea WM, Cambria RP. 1991. Effect of various detergents on protein migration in the second dimension of two-dimensional gels. *Anal Biochem* 199:35–44.

Luche S, Santoni V, Rabilloud T. 2003. Evaluation of nonionic and zwitterionic detergents as membrane protein solubilizers in two-dimensional electrophoresis. *Proteomics* 3:249–253.

MacFarlane DE. 1983. Use of benzyldimethyl-*n*-hexadecylammonium chloride ("16-BAC"), a cationic detergent, in an acidic polyacrylamide gel electrophoresis system to detect base labile protein methylation in intact cells. *Anal Biochem* 32:231–235.

MacFarlane DE. 1989. Two dimensional benzyldimethyl-*n*-hexadecylammonium chloride–sodium dodecyl sulfate preparative polyacrylamide gel electrophoresis: a high capacity high resolution technique for the purification of proteins from complex mixtures. *Anal Biochem* 176:457–463.

Madjar JJ, Arpin M, Buisson M, Reboud JP. 1979. Spot position of rat liver ribosomal proteins by four different two-dimensional electrophoreses in polyacrylamide gel. *Mol Gen Genet* 171:121–134.

Maina CV, Riggs PD, Grandea AG 3rd, Slatko BE, Moran LS, Tagliamonte JA, McReynolds LA, Guan CD. 1988. An *Escherichia coli* vector to express and purify foreign proteins by fusion to and separation from maltose-binding protein. *Gene* 74: 365–373.

Molloy MP, Herbert BR, Williams KL, Gooley AA. 1999. Extraction of *Escherichia coli* proteins with organic solvents prior to two-dimensional electrophoresis. *Electrophoresis* 20:701–704.

Neff D, Dencher NA. 1999. Purification of multisubunit membrane protein complexes: isolation of chloroplast FoF1-ATP synthase, CFo and CF1 by blue native electrophoresis. *Biochem Biophys Res Commun* 259:569–575.

Porath J, Olin B. 1983. Immobilized metal ion affinity adsorption and immobilized metal ion affinity chromatography of biomaterials. Serum protein affinities for gel-immobilized iron and nickel ions. *Biochemistry* 22:1621–1630.

Rabilloud T, Adessi C, Giraudel A, Lunardi J. 1997. Improvement of the solubilization of proteins in two-dimensional electrophoresis with immobilized pH gradients. *Electrophoresis* 18:307–316.

Radola BJ. 1975. Isoelectric focusing in layers of granulated gels. II. Preparative isoelectric focusing. *Biochim Biophys Acta* 386:181–195.

Reynolds JA, Tanford C. 1970. Binding of dodecyl sulfate to proteins at high binding ratios. Possible implications for the state of proteins in biological membranes. *Proc Natl Acad Sci USA* 66:1002–1007.

Rickwood D, Wattiaux R, Wattiaux-de Coninck S. 1983. Choice of media for centrifugal separations. In: Rickwood D (Ed), *Centrifugation, A Practical Approach*, IRL Press, Oxford, UK, pp 15–32.

Rigaut G, Shevchenko A, Rutz B, Wilm M, Mann M, Seraphin B. 1999. A generic protein purification method for protein complex characterization and proteome exploration. *Nat Biotechnol* 17:1030–1032.

Righetti PG, Wenisch E, Jungbauer A, Katinger H, Faupel M. 1990. Preparative purification of human monoclonal antibody isoforms in a multi-compartment electrolyser with immobiline membranes. *J Chromatogr* 500:681–696.

Roos M, Soskic V, Poznanovic S, Godovac-Zimmermann J. 1998. Post-translational modifications of endothelin receptor B from bovine lungs analyzed by mass spectrometry. *J Biol Chem* 273:924–931.

Rout MP, Aitchison JD, Suprapto A, Hjertaas K, Zhao Y, Chait BT. 2000. The yeast nuclear pore complex: composition, architecture, and transport mechanism. *J Cell Biol* 148: 635–651.

Schafer-Nielsen C, Rose C. 1982. Separation of nucleic acids and chromatin proteins by hydrophobic interaction chromatography. *Biochim Biophys Acta* 696:323–331.

Schagger H, von Jagow G. 1991. Blue native electrophoresis for isolation of membrane protein complexes in enzymatically active form. *Anal Biochem* 199:223–231.

Singh P, Jansch L, Braun HP, Schmitz UK. 2000. Resolution of mitochondrial and chloroplast membrane protein complexes from green leaves of potato on blue-native polyacrylamide gels. *Indian J Biochem Biophys* 37:59–66.

Sinha PK, Praus M, Kottgen E, Gianazza E, Righetti PG. 1990. Two-dimensional maps in the most extended (pH 2.5–11) immobilized pH gradient interval. *J Biochem Biophys Methods* 21:173–179.

Smith DB, Johnson KS. 1988. Single-step purification of polypeptides expressed in *Escherichia coli* as fusions with glutathione S-transferase. *Gene* 67: 31–40.

Vuillard L, Braun-Breton C, Rabilloud T. 1995a. Non-detergent sulphobetaines: a new class of mild solubilization agents for protein purification. *Biochem J* 305:337–343.

Vuillard L, Madern D, Franzetti B, Rabilloud T. 1995b. Halophilic protein stabilization by the mild solubilizing agents nondetergent sulfobetaines. *Anal Biochem* 230:290–294.

Wattiaux R, Wattiaux-de Coninck S. 1967. Influence of the injection of "Triton WR-1339" on lysosomes of a rat transplantable hepatoma. *Nature* 216:1132–1133.

Wattiaux R, Wattiaux-de Coninck S. 1983. Separation of cell organelles. In: Rickwood D (Ed), *Iodinated Density Gradient Media, A Practical Approach*, IRL Press, Oxford, UK, pp 119–138.

Zhang W, Zhou G, Zhao Y, White MA, Zhao Y. 2003. Affinity enrichment of plasma membrane for proteomics analysis. *Electrophoresis* 24:2855–2863.

Zischka H, Weber G, Weber PJ, Posch A, Braun RJ, Buhringer D, Schneider U, Nissum M, Meitinger T, Ueffing M, Eckerskorn C. 2003. Improved proteome analysis of *Saccharomyces cerevisiae* mitochondria by free-flow electrophoresis. *Proteomics* 3:906–916.

Zuo X, Speicher DW. 2000. A method for global analysis of complex proteomes using sample prefractionation by solution isoelectrofocusing prior to two-dimensional electrophoresis. *Anal Biochem* 284:266–278.

Part II

Functional Proteomics

7

Protein Localization by Cell Imaging

Eric G. D. Muller and Trisha N. Davis

University of Washington, Seattle, Washington

7.1 INTRODUCTION

By mapping the jungle of interconnections among proteins, proteomics searches for simplifications. What are the protein connections that define a complex biological function? What are the pairwise interactions between proteins within a complex? How do stresses and biological cues affect the flow of information? The task is daunting. Among the relatively small number of open reading frames in the yeast *Saccharomyces cerevisiae*, 6355, the online database "the GRID" recognized 13,819 unique interactions as of December 2002 (http://biodata.mshri.on/ca/grid/servlet/index) (Breitkreutz et al., 2003). The GRID database is coupled to a tool for network visualization, Osprey (http://biodata.mshri.on.ca/osprey/servlet/Index). By setting stringent connectivity parameters, one can identify the approximately 20 most highly connected complexes. With less stringent connectivity parameters, however, the shear number of protein interactions can overwhelm attempts at comprehension. Further undermining understanding are the presence of false positives in the interaction data sets.

Protein localization via live cell microscopy offers unique advantages for exploring a proteome. Under ideal circumstances, localization reveals not only where a protein is found, but when it is found there. Dynamic movements from one location to another can be followed as the cell proceeds through the cell cycle or responds to environmental stresses or internal signaling pathways. Of the many techniques applied to study proteins, microscopy alone views the whole intact cell. This global

---

Proteomics for Biological Discovery, edited by Timothy D. Veenstra and John R. Yates.
Copyright © 2006 John Wiley & Sons, Inc.

perspective can confirm whether localization of a protein in vivo is physically compatible with the interaction maps drawn from other proteomic methods.

This chapter focuses on the use of wide field fluorescence microscopy to examine live cells using the color variants of GFP as protein tags. Although the approaches described are generally applicable to most biological systems, we take our examples from studies on the yeast *Sacchacromyces cerevisiae* as the model organism. Two applications are described in detail: the simple imaging of a cell tagged with fluorescent proteins and the use of FRET (fluorescence resonance energy transfer) to assess the proximity of one protein to another.

Recently, several pertinent reviews have appeared on fluorescent protein markers. Lippincott-Schwartz and Patterson (2003) have written an excellent overview discussing GFP and related techniques to look at the kinetics of protein movement. They present an introduction to the use of FRAP (fluorescence recovery after photobleaching), FLIP (fluorescence loss in photobleaching), and photoactivatable GFP to monitor protein mobility. Zhang et al. (2002) examine not only the GFP family of fluorescent tags, but the wealth of other fluorescent probes available for tracking proteins. They also summarize the vast number of techniques that use these probes as spatial and temporal markers. Zhang et al. (2002), van Roessel and Brand (2002), Sekar and Periasamy (2003), and Truong and Ikura (2001) introduce the use of FRET to indicate changes in protein activity, ion changes, protein–protein distances, and other cellular parameters. Finally, Molecular Expressions (http://micro.magnet.fsu.edu/primer/) is an excellent resource describing optical microscopy, digital imaging, and photomicrography as they apply to the fluorescent imaging of proteins.

7.2 PROTEIN LOCALIZATION WITH FLUORESCENT PROTEINS

7.2.1 Equipment

The choice of microscope must be coupled to (i) the specimen characteristics (i.e., thickness and size), (ii) signal strengths of the proteins of interest when they are coupled to the fluorescent tags, (iii) the temporal and spatial resolution required for the experimental process under study, (iv) the excitation and emission wavelengths of the fluorescent probes, and (v) the capabilities of the software and the processing needs of the researcher. There is no perfect system for all uses and therefore no recommended "best buy" system. In general, the optical and mechanical components in modern microscopes are of comparable high quality. Whether a particular system meets a researcher's needs is a scientific judgment best evaluated by empirical testing of different systems with representative experimental samples.

In surveying microscopes for fluorescence imaging of yeast, two components are of particular importance: the light source and the camera. Fluorescence microscopy demands very bright sample illumination to generate enough excitation energy to make the fluorescence emission detectable. Typically, mercury or xenon arc lamps are used. Mercury lamps have the advantage of being significantly brighter, thus permitting the detection of weaker fluorescent signals. On the other hand, xenon burners have a more even intensity across the visible spectrum, suffer from slightly less flicker, and are longer lived. For our studies in yeast, an HBO100 mercury lamp was essential for the detection of our samples with relatively weak signals. Illumination by lasers using confocal systems is also possible, but here the choice of fluorescent tags is limited by the emission wavelength of the laser.

Whatever the choice of illumination, a photodetector to monitor the output of the lamp is critical. An occasional lamp may come fresh out of the box dimmer or brighter than the norm. The effective life of a bulb depends to a large extent on usage, but also seems to be an inherent property of a bulb. Thus, simply counting the number of hours of usage is not an adequate test of bulb properties. In proteomic studies where a large number of proteins are processed over a period of time, accurate determination of bulb intensity ensures reproducible image quality and fosters more meaningful image comparisons. Finally, in FRET applications the intensity of illumination influences correction factors (Muller et al., 2005).

At the other end of the optical train, the choice of camera can make the difference between capturing and missing an elusive weak signal. There are many factors that come into play in the choice of a camera. We will not review them all here. The primary question is whether the desired signal being measured is greater than the noise level from the system. If the signal is bright, and the object being studied is relatively large and immobile, then image capture will be straightforward. However, dim, small and/or fast moving objects demand the highest signal-to-noise ratio.

To increase the strength of the signal from fluorescent proteins, the quantum efficiency of the chip in the charge coupled device (CCD) camera should be highest in the part of the spectrum that coincides with the emission spectra of these proteins. The fluorescent proteins CFP, GFP, YFP, and DsRed have peaks of emission at approximately 480, 510, 535, and 583 nm, respectively. One currently popular chip is the Sony ICX285, which has a quantum efficiency of 50–60% from 450 to 650 nm.

To decrease the read noise, adjoining pixels can be binned or summed on most chips. This binning is very useful when signals are low. Binning increases the signal-to-noise ratio by a factor equal to the binning factor. A small pixel size on the CCD will maintain relatively high spatial resolution even after binning. For imaging yeast, we use the Roper CoolSnap HQ camera, which incorporates the Sony ICX285 chip with a pixel size of $6.45 \times 6.45\,\mu m$. Images are typically captured with a 100X Plan Apochromat objective (1.35 numerical aperture) using 2×2 binning. This binning factor furnishes a fourfold increase in signal to noise. Finally, to decrease the dark current noise of a camera requires cooling the CCD. Dark current noise is not reduced by binning. Most high end cameras include thermoelectric cooling to temperatures below $-25\,°C$.

7.2.2 Common Fluorescent Protein Tags

GFP is a relatively small protein (MW 27 kDa) comprised of just 238 amino acids. The chromophore matures spontaneously from an autocatalytic cyclization, dehydration, and oxidation of the folded protein and requires no specialized cofactor. Thus, GFP usually fluoresces wherever it is expressed. The chromophore sits protected within the center of a rigid, inert 11-stranded β-barrel. The N and C termini dangle from one end of the β-barrel and can be genetically fused to either the N

TABLE 7.1. Sample of Fluorescent Proteins Used in Microscopy

Fluorescent Protein	ABS	EM	Omega Optical®		ChromaTechnology®	
			EX	EM	EX140	EM
BFP	382	448	387AF28	450AF58	380/30	460/50
CFP	458	480	440AF21	480AF30	435/20	480/40
GFP	470	508	475AF40	535AF45	480/40	535/50
YFP	513	527	500AF25	545AF35	500/20	535/30
DsRed	558	583	540AF30	585ALP	545/30	620/60

or C terminus of a target protein. Importantly for the use of this system to study protein localization is that a GFP–protein fusion usually has the fluorescence properties of GFP and the biochemical activity and cellular localization of the parent protein.

Mutagenesis of GFP has created a family of color variants. The peak absorption and emission wavelengths of the four primary classes, BFP, CFP, GFP, and YFP, are listed in Table 7.1. However, the excitation and emission spectra of each variant are quite broad as indicated by the spectral properties of the optical filters used for microscopy (Table 7.1). Each color class has a number of different forms within each class that can differ in their folding properties, quantum yields, sensitivity to ions and pH, and precise shift in spectral properties. For most purposes the "enhanced" forms of CFP and GFP are used. For YFP we typically use a variant with reduced sensitivity to pH that contains the following changes: S65G, V68L, Q69K, S72A, T203Y (Miyawaki et al., 1999). We have also investigated the behavior of the "Venus" form in yeast. The "Venus" form of YFP was engineered to enhance the quantum yield and rate of chromophore maturation, although it is more sensitive to photobleaching than YFP (Nagai et al., 2002).

Also listed in Table 7.1 is the protein DsRed. Whereas GFP was isolated from the jellyfish *Aequorea victoria*, the fluorescent protein DsRed was isolated from the sea corral of the genus *Discosoma*. The ongoing generation of DsRed mutants continues to improve the poor rate of folding and oligomerization of the native protein (Zhang et al., 2002). For the studies described in this chapter, we have used the "T1" form (Bevis and Glick, 2002).

To compare the different fluorescent proteins as tags, Figure 7.1 shows the localization of the yeast spindle pole body (SPB). Using standard genetic techniques, the SPB component Spc110 was tagged with each of the fluorescent proteins by the insertion of the corresponding gene at the 3′ end of *SPC110*. The fusion is the sole copy of the gene in the cell. The haploid strains are otherwise isogenic. The SPB is a dynamic structure, growing, duplicating, and exchanging components as the cell progresses through the cell cycle (Yoder et al., 2003). For this reason the cells were compared at the same stage in the cell cycle, with duplicated SPBs and a short spindle.

Due to the high autofluorescence in the UV, coupled with a low quantum yield of BFP, we were unable to detect a signal from the Spc110-BFP labeled cell (not shown). All the other protein tags gave bright signals with good signal to noise and similar background subtracted values (Figure 7.1). Subtle differences, however, are revealed in the images. Autofluorescence is greatest for CFP, but the signal is

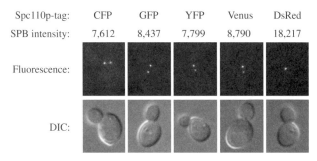

Spc110p-tag:	CFP	GFP	YFP	Venus	DsRed
SPB intensity:	7,612	8,437	7,799	8,790	18,217

Fluorescence:

DIC:

Figure 7.1. *A comparison of fluorescent proteins. The wild-type copy of the essential gene SPC110 was replaced with fusions of SPC110 linked to sequences encoding the indicated fluorescent proteins. Live cells were imaged using the filter sets from Omega Optical® listed in Table 7.1. Large budded cells with duplicated SPBs and a short spindle were selected for analysis. The sum of intensities in a 5 × 5 pixel square was collected for the SPB closest to the bud. An adjacent square was used to measure background for correction. For each strain, data was averaged from 16 to 26 SPBs. The standard deviation was about 15% of the mean. Representative images are shown scaled to maximum intensity.*

strong. GFP labeled both SPBs uniformly, with low background and high signal. YFP-Venus was modestly brighter than our standard YFP. Although not shown here, we have compared the labeling of proteins with high rates of turnover and seen that YFP-Venus does a better job of visualizing these proteins than YFP (unpublished). YFP-Venus also does not diminish in fluorescence at 37 °C, whereas the brightness of our standard YFP drops by 35% (Tess Yoder and Bryan Sundin, unpublished results). DsRed yielded both the brightest and dimmest SPBs. Following SPB duplication, the old pole is inherited by the newly budded cell. In these images of rapidly dividing cells, the newly synthesized Spc110-DsRed has not had sufficient time to mature the chromophore. Thus, the new pole is just barely visible (beneath and to the right of the bright pole in Figure 7.1) with a signal strength of less than one-sixth of the old pole. On the other hand, the old pole labeled with DsRed was twice as bright as the signals from the other GFP variants. The YFP labeled poles also had a slight tendency to label the old pole more brightly, but the difference was much less.

The biggest difference between the tags is the rate of photobleaching (Figure 7.2). Cells were repetitively imaged with short exposure times (0.25 second for Spc110-CFP, and 0.5 second for Spc110-GFP and Spc110-YFP). The signal intensities at the SPB were summed and plotted as a function of the cumulative exposure time. The exponential decay of CFP, YFP, and GFP occurred with half-lives of 1.5, 6.75, and 26 seconds, respectively. DsRed yielded a multiphasic decay (data not shown). For practical purposes, the results show that GFP is the most useful for long-term imaging of a process that requires successive imaging over time. In addition, to quantify the total amount of signal in a cell requires the capture of multiple (typically 10–15) optical planes, followed by deconvolution to remove the out-of-focus light in each plane. Rates of photobleaching preclude the use of CFP for three-dimensional deconvolution and YFP exposure times must be kept to less than 0.3 second in order to maintain relatively uniform signal strength.

Despite their problems with photobleaching, the combination of CFP and YFP are generally still the most useful for colocalization and FRET studies. Their exci-

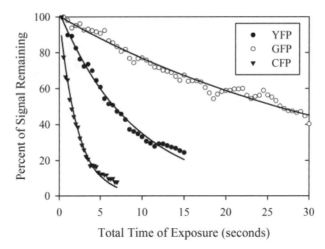

Figure 7.2. Decay of fluorescence from photobleaching. Spc110 was tagged with CFP, GFP, and YFP and examined as described in Figure 7.1. Cells were repetitively imaged with short exposure times, 0.25 second for Spc110-CFP, and 0.5 second for Spc110-GFP and Spc110-YFP. The signal intensities at the SPB were summed and plotted as a function of the cumulative exposure time.

tation and emission spectra show good separation and components colabeled with CFP and YFP are easily resolved. In the future, once DsRed is fully optimized, GFP and DsRed hold great promise as a FRET pair.

7.2.3 Localization as a Tool in Proteomics

This section illustrates how the subcellular localization of a protein complements data from proteomic methods that identify protein–protein interactions. As an example, we use the protein calmodulin (Cmd1). The major functions of Cmd1 are well described and its localization is known (Cyert, 2001). Moreover, it has been characterized by large-scale proteomics, and thus it serves as a good model for how protein localization can aid the interpretation of proteomic data.

The 30 proteins identified as interacting with Cmd1 by proteomic methods (derived from the GRID database, see Section 7.1) are shown in Figure 7.3. Of the 30 proteins, 28 were identified in a proteomic-wide, high-throughput mass spectrometric identification of protein complexes (Ho et al., 2002) and two (YML095c and YMR111c) were identified by two-hybrid screening (Uetz et al., 2000). We have encircled proteins with shared biological process and annotated the circle with the GO process term. (The GO terms are a product of the Gene Ontology Consortium (http://geneontology.org), an organization that seeks to define a unifying vocabulary for the description of all gene products.)

There are three interpretations of this data. First, Cmd1 is part of a single, large complex that exerts its influence on almost every aspect of cellular activity. As a center for comprehensive regulation, Cmd1 could be characterized as the capital of the cell, the seat of cellular government. A second interpretation is that instead of one complex there are 30 complexes with Cmd1 interacting with each individually. Calmodulin could be a promiscuous structural element of many different multimeric proteins. Given the groupings by GO terms, one could logically

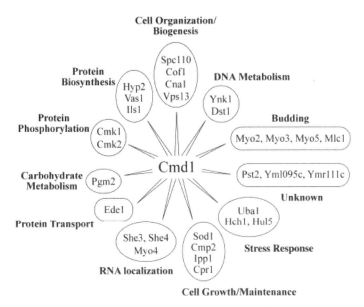

Figure 7.3. *Predicted protein partners of calmodulin (Cmd1). Proteins shown to interact with Cmd1 based on high-throughput proteomic methods were assembled using the online GRID database.*

simplify and conclude that Cmd1 is a member of 12 complexes. However, the main point is that there is no unified central complex. Calmodulin might still direct an array of activity, but Cmd1 would be demoted to governing from a dozen outlying seats. A third interpretation is that the data set has many false positives. The quality of the output from mass spectrometry depends on the purity of the starting material. In large-scale proteomic screens, a standard set of purification conditions is used and it may not be optimal for purification of Cmd1. In addition, the ever increasing sensitivity of the instrumentation enables detection of proteins present in trace amounts. Drawing the distinction between valuable information and rubbish is a challenge in the analysis of large-scale proteomic data.

Protein localization is an effective tool to narrow the possibilities. The localization of Cmd1 might not confirm any particular interaction or validate a particular interpretation. However, localization in live cells highlights where Cmd1 is most abundant and/or concentrated, and if it resides in different compartments during the cell cycle.

A C-terminal fusion of Cmd1-YFP was created using standard techniques. YFP was chosen to allow for colocalization with candidate protein partners fused to CFP. For unequivocal results, the gene encoding Cmd1-YFP replaced *CMD1*, so Cmd1-YFP was the only Cmd1 protein present in the cell. With some fusions a wild-type protein will outcompete a GFP fusion for binding sites on other proteins. To maintain normal levels of expression, the gene encoding the fusion was regulated by the native *CMD1* promoter in its natural context in the genome. Finally, since Cmd1 is essential for the viability of yeast, the growth of the strain ensures that Cmd1-YFP is a functional replacement for Cmd1.

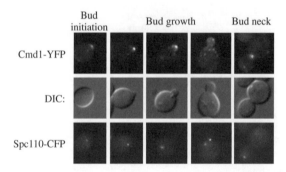

Figure 7.4. *Localization of calmodulin. As described in the text and Figure 7.1, Cmd1 and Spc110 were tagged with YFP and CFP, respectively, and live cells were imaged.*

The striking localizations of Cmd1 as the cell proceeds through the cell cycle are shown in Figure 7.4. Without any knowledge of yeast cell biology, it is clear that there is not one complex of Cmd1 in the cell. Multiple areas of concentration that change in a cell cycle dependent manner imply multiple complexes. At the end of G1 just before bud emergence, Cmd1-YFP concentrates at the incipient bud site. Calmodulin stays concentrated at the bud tip until G2 and then appears at the bud neck in late mitosis. Throughout the cell cycle Cmd1 localizes to actin patches and to the SPB (compare Cmd1-YFP to Spc110-CFP, a component of the SPB).

The localization data focuses attention on the cytoskeleton. Looking for cytoskeletal targets identified in the large-scale proteomic screens, one quickly notices Spc110 at the SPB, cofilin (Cof1), the myosins (Myo2, Myo3, Myo4, and Myo5), and myosin light chain (Mlc1) (Figure 7.3). We know from previously published work that Cmd1 binds to Spc110 at the SPB, binds Myo2 and Myo4 at sites of cell growth, and binds Myo5 in actin patches (Geiser et al., 1993; Brockerhoff et al., 1994; Geli et al., 1998; Stevens and Davis, 1998). Cofilin and Mlc1 associate with the known Cmd1 targets Myo5 and Myo2, respectively (Stevens and Davis, 1998; Idrissi et al., 2002). Thus, the localization data narrowed the field to correct targets and their partners. The localization data did not point to all the Cmd1 targets; the Cmd1-dependent protein kinases, Cmk1 and Cmk2, and calcineurin, Cna1 and Cmp2 (Cyert et al., 1991), are verified Cmd1 targets (Pausch et al., 1991); however, they were not inferred as Cmd1 binding partners from the localization data. These proteins may not have distinctive localizations.

7.3 ARCHITECTURE OF PROTEIN COMPLEXES AS REVEALED BY FRET

7.3.1 FRET Theory

Fluorescence resonance energy transfer (FRET) is the transfer of excitation energy from one fluorophore, referred to as the donor, to a second fluorophore, the acceptor. It can occur when the wave properties of the electrons in the excited state of the donor overlap the wave properties of the ground state electrons of the acceptor. The two electronic states resonate with the excitation energy moving back and forth between the two chromophores. Some of the energy transferred to the acceptor decays to the ground state, emitting light. The donor, by transferring its energy to the acceptor, returns to the ground state without the radiation of light.

Förster described how the efficiency of transfer depends on a number of spectral and spatial properties of the donor and acceptor (Stryer, 1978). For the purpose of using FRET as a probe for the structure of a protein complex, two terms defined by Förster are particularly relevant: J, the overlap integral, and r, the distance between the two chromophores.

J is a measure of the extent of overlap between the emission spectrum of the donor and the excitation spectrum of the acceptor. The greater the overlap, the greater the efficiency of energy transfer. Among the common fluorescent protein tags, the emission and excitation spectra of BFP and GFP, CFP and YFP, and GFP and DsRed make them suitable for FRET (see Table 7.1), and all three pairs have been used successfully experimentally. Given the limitations of BFP and DsRed, however, the most versatile pair is CFP and YFP.

The power of FRET comes from its strong dependence on r, the intermolecular distance between the chromophores. As described by Förster,

$$E = \frac{R_0^6}{R_0^6 + r^6}$$

where E is the efficiency of transfer, and R_0 is the distance at which the efficiency is 50%, referred to as the Förster distance (Stryer, 1978). For the CFP–YFP pair, R_0 is estimated to be 48–52 Å (Heim, 1999).

The calculated efficiency of transfer between CFP and YFP as a function of distance is plotted in Figure 7.5. Clearly, FRET between CFP and YFP is

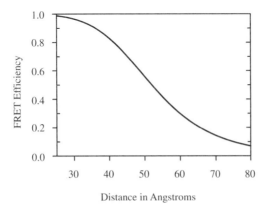

Figure 7.5. FRET efficiency between the chromophores of CFP and YFP as a function of distance. Using Förster's equation for FRET efficiency described in the text, FRET efficiency was plotted as a function of r, the intermolecular distance between CFP and YFP.

exquisitely responsive to changes in distance between 40 and 60 Å. The calculated efficiency for distances closer than 25 Å is not shown since the structure of GFP precludes a proximity of less than 25 Å (Ormo et al., 1996). We find a 10% standard deviation in a given FRET signal (Muller et al., 2005); therefore, the limit of FRET detection above the noise is 75–80 Å, the distance at which FRET occurs with 10% efficiency. We conclude that in exploring the structure of proteins within a complex, if the ends of two proteins are within 80 Å, then CFP and YFP linked to those ends will display detectable energy transfer.

One other spatial dependency should be mentioned. FRET is a function of the orientation of the two chromophores. The flexibility and many vibrational and rotational states of biological molecules led to the conclusion, supported experimentally, that the orientation is effectively random (Stryer, 1978). For rigor, this assumption can be tested by fluorescence anisotropy imaging microscopy (Rocheleau et al., 2003).

7.3.2 FRET Measurements to Determine the Proximity of Proteins to Each Other

In principle, the use of FRET to determine if two proteins are adjacent is fairly straightforward. One protein is labeled with CFP. One protein is labeled with YFP. The proteins are expressed in cells and examined by fluorescence microscopy with the appropriate filter sets. If the ends of the two proteins are within 80 Å, energy transfer will occur. By illuminating the cell with light corresponding to the excitation spectrum of CFP, and detecting the fluorescence that is emitted at the wavelengths corresponding to the emission spectrum of YFP, one should be able to observe the transfer of energy from donor to acceptor.

In practice, the spectral properties of the GFP related proteins, along with the vagaries of fluorescence microscopy, make detection of FRET problematic. Many alternative methods for measuring FRET have been proposed. They fall into three broad categories: those based on the decrease of the fluorescence lifetime of the donor, those that assess the decrease in the donor fluorescence in the presence of the acceptor, and those that measure the enhanced emission from the acceptor in the presence of donor. In addition, within each category, there are sometimes several mathematical models for analyzing the data.

Fluorescence lifetime measurements are the most direct measure of energy transfer and do not depend on the concentration of either donor or acceptor. The lifetime of the excited state of the donor is shorter in the presence of the acceptor since energy transfer to the acceptor provides an additional path to the ground state. The technique shows great promise (see Harpur et al., 2001; Periasamy et al., 2002). Its primary drawback is the expense of the specialized equipment required to measure fluorescence lifetimes.

The second method, to measure the decrease in donor's fluorescence in the presence of acceptor, usually requires taking measurements before and after photobleaching of the acceptor, because physically removing the acceptor is often impossible for in vivo experiments. Bleaching the acceptor in a manner that is not toxic to the cell is difficult and inefficient, particularly with CFP and YFP as the donor–acceptor pair (unpublished results). Still the approach has been used successfully in a number of cases (reviewed in Wouters et al., 2001).

The most prevalent method for measuring FRET takes advantage of the change in emission intensity from both the donor and acceptor. The emission from the acceptor increases and the donor's emission decreases following transfer of energy from the excited state of the donor to the ground state of the acceptor. This increase in acceptor emission is often referred to as sensitized emission. The advantage of this approach is that it is very sensitive to subtle changes in FRET signal, and no additional equipment is necessary other than what is already in use for digital fluorescence microscopy. Given the widespread accessibility for measuring FRET via sensitized emission and our extensive experience with this method, we limit further discussion to this approach.

7.4 FRET-BASED SENSORS

By way of introducing our methodology, we begin by discussing FRET-based biosensors. Of all the uses for FRET microscopy, the FRET-based biosensors have had the greatest successes. FRET sensors have been exploited to detect changes in protease activity, kinase and phosphatase activity, the concentration of calcium ions, the concentration of cGMP, and Ran GTP gradients (see Section 7.1 for reviews listing primary references and Kalab et al., 2002; Miyawaki, 2003).

FRET sensors are built on a common framework in which the distance between the donor and acceptor changes in response to a biochemical activity or stimulus. In one conformation the donor and acceptor are positioned close enough for energy transfer. In the other conformation they are not. The movement of the donor and acceptor is directed by a protein sequence of known structure and function that is positioned between the donor and acceptor, covalently linking them to each other. It is the intervening protein sequence that changes conformation in response to a change in the environmental signal. This change in conformation repositions the donor and acceptor, thereby changing the FRET signal.

The FRET-based biosensors have four strengths that contribute to the success of their use. First, the structure of the intervening sequence or response element is known, and thus the expected changes in the FRET signal in response to a stimulus are also known. Second, the constant 1:1 stoichiometry between the covalently linked donor and acceptor means that changes in the FRET signal can only arise from changes in the distance between the donor and the acceptor and not from changes in their relative abundance. Third, FRET sensors are generally binary in nature and have just two configurations, "on" and "off." The fluorescent signal in the off position is used to calibrate the system and establish a baseline. Deflections of as little as 3% in the FRET signal can be interpreted as meaningful when compared to a stable baseline (Miyawaki et al., 1999). Finally, FRET sensors typically use a very straightforward method for quantification of the FRET signal. The state of FRET is commonly evaluated by simply following the ratio of the acceptor fluorescence to the donor fluorescence under conditions of donor excitation.

7.5 FRET-BASED INDEXES OF PROTEIN PROXIMITIES

The challenge of using FRET to judge protein–protein distance in a protein complex is that not one of the four strengths of the biosensors is present. Unlike FRET sensors, the structure of the protein complex is unknown and, in fact, is usually the point of the investigation. Since the proteins are not covalently bound to each other, the relative concentration of the proteins can vary. Moreover, the stoichiometry of the proteins in the complex is unknown. Both unknowns can significantly handicap FRET measurements.

Finally, the most difficult challenge is the lack of a true baseline. Unlike the FRET sensors, there is usually no "off" condition by which to establish the signal that represents "no FRET." This lack of a null signal is a problem because the spectral properties of the donor and acceptor, for example, CFP and YFP, are not ideal. Some of the light emitted from the fluorescence of CFP extends into the emission wavelengths of YFP. Also, the excitation spectrum of YFP extends slightly into the excitation spectrum of CFP. The result is that even in the absence of energy transfer there is an apparent FRET signal that results from this spillover of fluorescence from CFP and YFP into the FRET channel.

To illustrate the baseline problem, Table 7.2 outlines typical microscope filter settings to examine FRET between two proteins tagged with CFP and YFP. As shown, the CFP and YFP channels do not display any crosstalk between CFP and YFP fluorescence. In addition to the fluorescence from the fluorophores, the only additional sources of light in the CFP and YFP channels come from instrument noise and cellular autofluorescence. On the other hand, the FRET channel combines information not only from the amount of energy transfer but also the amount of CFP, YFP, noise, and autofluorescence in the sample!

For analyzing data from a FRET sensor, a baseline is first established by dividing the FRET channel by the CFP channel under conditions in which CFP and YFP are too distant for transfer. Incorporated into this baseline is all the information about the spillover of YFP and CFP into the FRET channel and the other sources of noise. Experimental conditions are changed to see if CFP and YFP are now brought close enough together for FRET to occur. The contaminating light in the FRET channel is assumed to remain constant throughout the experiment. Thus, an increase above the baseline FRET ratio can be attributed to the results of energy transfer, sensitized emission from YFP, and a decrease in the fluorescence from CFP.

To use FRET to determine whether the distance between the ends of two proteins tagged with CFP and YFP is within 80 Å, the contaminating light in the FRET channel must be dealt with more directly. Since the amount of CFP and

TABLE 7.2. Four-Channel FRET Experiment

Channel	EX	EM	Purpose	Detects
YFP	500AF25	545AF35	Measure YFP	YFP, autofluorescence, noise
FRET	440AF21	545AF35	Measure FRET	FRET, CFP, YFP, autofluorescence, noise
CFP	440AF21	480AF30	Measure CFP	CFP, autofluorescence, noise
DIC[a]			Image the cell	Phase differences as visible light passes through specimen

[a] Differential interference contrast.

YFP fluorescence that spills over into the FRET channel is proportional to the amount of CFP and YFP present, the CFP and YFP channels can be used to approximate the extent of spillover. The factors that will be used to calculate the spillover in the FRET experiment must first be determined.

The CFP spillover factor (*CSF*) represents the amount of CFP emission that spills over into FRET channel normalized to the amount of CFP. *CSF* is calculated by imaging cells expressing only the CFP-tagged protein and then dividing the signal in the FRET channel by the signal in the CFP channel:

$$CSF = \frac{FRET_{CFPalone}}{CFP_{channel}}$$

Likewise, the YFP spillover factor (*YSF*) represents the amount of YFP emission that spills over into FRET channel normalized to the amount of YFP. *YSF* is calculated by imaging cells expressing only the YFP-tagged protein and then dividing the signal in the FRET channel by the signal in the YFP channel:

$$YSF = \frac{FRET_{YFPalone}}{YFP_{channel}}$$

After establishing the constants *CSF* and *YSF*, one returns to the FRET experiment in which two proteins, one tagged with CFP and the other with YFP, are coexpressed. The YFP, FRET, and CFP channels are acquired, in this order, to avoid photobleaching the YFP by repeated exposure to the CFP excitation. The spillover from CFP and YFP into the FRET channel is then calculated as

$$Spillover = (CFP_{channel} \times CSF) + (YFP_{channel} \times YSF)$$

With the value of the spillover in hand, Youvan et al. (1997) were the first to propose that a measure of FRET could be calculated as

$$FRET_{Youvan} = FRET_{channel} - Spillover$$

(Note that in all our equations we have changed the symbols used by the original authors for the sake of clarity.) With this approach, the baseline is essentially set to zero and any positive value for FRET is considered the result of energy transfer.

In all cases to date where live cell FRET investigated the assembly and arrangement of protein complexes (for some recent examples see Damelin and Silver, 2002; Overton and Blumer, 2002; Amiri et al., 2003), the subtraction of spillover from the FRET channel is incorporated into the formula for calculating the FRET signal. There are many other variations that have evolved from this formula in the literature (Berney and Danuser, 2003); however, we introduce our own improvement below.

The first to recognize the shortcomings of $FRET_{Youvan}$ was Gordon et al. (1998). In an often cited paper, they showed that $FRET_{Youvan}$ was not normalized. At a constant FRET efficiency, the FRET signal would be different if the absolute amounts of CFP- and YFP-tagged proteins are different. Thus, the intensity in the FRET channel could increase just because there was more CFP and YFP present, even if *r* was fixed. Gordon et al. (1998) proposed adjusting the FRET value to compensate for the influence of the CFP and YFP signal strength by the following formula:

$$FRET_{Gordon} = \frac{FRET_{channel} - Spillover}{CFP_{channel} \times YFP_{channel}}$$

In this formula the baseline still remains zero, but now all values are divided by the amount of CFP and YFP that is present.

7.6   BIAS OF THE CURRENT FRET INDEXES

Recently, we have uncovered limitations in these and other methods used to measure FRET. We have used FRET to explore the structure of the yeast SPB (Muller et al., 2005). Five proteins that comprise the core of the SPB were labeled at either end using CFP or YFP. These tagged proteins were coexpressed in yeast to label the SPB and images of live cells were acquired as described by Hailey et al. (2002). Incorporating both *CSF* and *YSF* into our calculations, FRET measurements were made to determine the distance between the proteins. Over 30 combinations of proteins were examined in a data set that contains over 3000 data points from individual SPBs. These results will be the subject of a future publication; however, for this review we highlight the major conclusions concerning FRET methodology.

Those conclusions are illustrated by a detailed look at the positive and negative FRET controls. For a positive control the SPB component Spc110 was tagged at its carboxyl end with YFP and CFP in tandem. YFP and CFP were separated by a short linker sequence, so FRET efficiency is predicted to be close to maximum. For a negative control Spc110 was tagged at its amino terminus with YFP and at its carboxyl end with CFP. Between the amino and carboxyl domains, Spc110 has a coiled coil domain that extends for greater than 700 Å, a distance too great for FRET to occur.

As Gordon et al. (1998) and others (Berney and Danuser, 2003) have indicated, the method of Youvan et al. (1997) is biased toward delivering higher FRET values for brighter CFP and YFP. This bias is shown in Figure 7.6A. An increase in CFP intensity leads to a linear increase in $FRET_{Youvan}$ values for the positive control. Measurements were made of 95 SPBs imaged from an asynchronous culture. The yeast SPB grows laterally twofold during the course of a cell cycle, thus leading to a twofold range in the CFP (Figure 7.6) and YFP (not shown) intensity values. As the CFP intensity varied, so did the value of $FRET_{Youvan}$.

The behavior of $FRET_{Gordon}$ though was surprising. Using the same raw data to calculate $FRET_{Gordon}$, we found a strong bias in the FRET values (Figure 7.6B). The best fit of the data was an inverse first-order polynomial. As the CFP intensity values increased, the FRET output flattened out, and as the CFP decreased, the

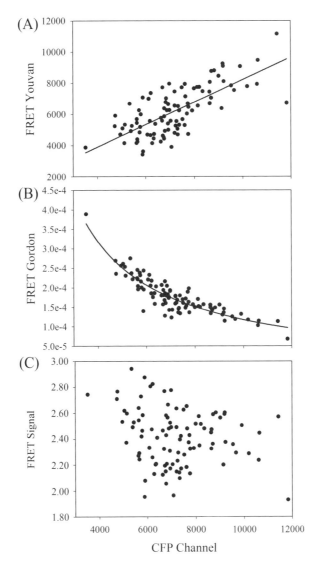

Figure 7.6. *Comparison of three indexes to measure FRET. Spc110-YFP-CFP replaced the wild-type copy of Spc110. Live cells were imaged with the filter sets described in Table 7.1 with the method described in Hailey et al. (2002). The total intensity at 95 individual SPBs was determined as described in Figure 7.1 for each channel described in Table 7.2. The background subtracted values for YFP, FRET, and CFP were then used to calculate the indicated FRET values at each SPB. The FRET values were then plotted versus the corresponding CFP value.*

FRET value curved upward. This effect was even more dramatic when the entire data set was plotted (not shown).

Analysis of the effect revealed that $FRET_{Gordon}$ is markedly influenced by the denominator, CFP × YFP. Thus, at high CFP and YFP intensity values, the index approaches zero, and at very low values it asymptotically approaches infinity (data not shown). The hierarchy of FRET values still tracks with FRET efficiency if the range of CFP and YFP intensities is narrow. Our study of the SPB clearly showed, however, that a new index, which was more independent of CFP and YFP intensities, was needed.

7.7 DEFINING A NEW FRET INDEX

To derive a new index we turned to the model of the FRET-based sensors. We realized that the spillover could provide a baseline much like FRET sensors in the off position. The intensity values of CFP and YFP are already incorporated into the value of the spillover. Indeed, spillover is the intensity in the FRET channel that would be expected in the absence of FRET. Thus, instead of treating spillover as a background that should be subtracted from the observed FRET, we treated spillover as a baseline to be used to scale the extent of FRET. We defined $FRET_{Relative}$ as

$$FRET_{Relative} = \frac{FRET_{Channel}}{Spillover}$$

This new index, $FRET_{Relative}$, measures the fold difference between the observed FRET and the baseline established from the spillover. It normalizes FRET to the spillover instead of the CFP and YFP intensities.

$FRET_{Relative}$ is independent of the CFP (Figure 7.6C) and YFP (not shown) intensities. Whether CFP and YFP are linked to highly expressed proteins, yielding a high baseline, or poorly expressed proteins, yielding a low baseline, $FRET_{Relative}$ indicates the fold increase above the baseline irrespectively. As shown in Table 7.3, for the positive SPB control, $FRET_{Relative}$ was 2.4, meaning the intensity of the FRET channel was 2.4-fold greater than the expected value for no energy transfer. The negative control had a $FRET_{Relative}$ of 1.0, in other words, equivalent to the baseline established by the spillover.

This leads to another advantage of $FRET_{Relative}$: it is standardized to 1 instead of 0 for the value corresponding to the absence of energy transfer. This standardiza-

TABLE 7.3. Generating Values for FRET Signals at the Yeast SPB from Digital Images

Strain[a]	YFP SPB -Bkg	FRET SPB -Bkg	CFP SPB -Bkg	Spillover YFP × YSF + CFP × CSF	FRET Youvan FRET − Spillover	FRET Gordon FRET Youvan/ CFP × YFP	FRET Signal FRET/ Spillover
DHY150	5,094	10,587	7,194	4,390	6,197 ± 1,534	$(1.7 \pm 0.47) \times 10^{-4}$	2.41 ± 0.22
EMY173	4,710	4,200	6,772	4,113	87 ± 271	$(2.7 \pm 9) \times 10^{-6}$	1.02 ± 0.07

[a] DHY150 is a homozygous diploid in which both copies of *SPC110* are tagged at the 3' end (SPC110-YFP-CFP). EMY173 is a homozygous diploid in which both copies of *SPC110* are tagged at opposite ends with YFP and CFP (YFP-SPC110-CFP). The intensity values for YFP, FRET, and CFP are the sum of the intensity values of a 5 × 5 pixel square centered on the SPB. Background is subtracted. The standard deviation is shown for the final FRET values. *YSF* was 0.232; *CSF* was 0.446.

tion facilitates a more intuitive interpretation of the data. For example, for our negative SPB control, the values for $FRET_{Youvan}$, $FRET_{Gordon}$, and $FRET_{Relative}$ were 87 ± 271, $(2.7 \pm 9) \times 10^{-6}$, and 1.02 ± 0.07, respectively (Table 7.3). For the positive control the values were 6197 ± 1534, $(1.7 \pm 0.47) \times 10^{-4}$, and 2.41 ± 0.22, respectively (Table 7.3). On the surface, the numbers suggest that FRET for the positive control is 71-fold greater than the negative control when comparing $FRET_{Youvan}$, or 62-fold greater when comparing $FRET_{Gordon}$. Of course, this interpretation is fallacious, since in essence one is dividing by 0 and only the error in the measurements masks this fact. In comparison, $FRET_{Relative}$ actually is 2.4-fold greater than the negative control since the value of $FRET_{Relative}$ for the negative control is 1. Indeed, built into the value of $FRET_{Relative}$ is the fold increase above the negative control.

In Muller et al., 2005, we show that $FRET_{Relative}$ is an effective tool to build a model of the core layer of the yeast SPB. Subsets of $FRET_{Relative}$ values could be assigned corresponding distances that were consistent within the data set, and with previous models that were based on electron microscopy and two-hybrid analysis. Thus, the $FRET_{Signal}$ between proteins tagged with CFP and YFP is a powerful tool to examine the structure of multicomponent protein complexes.

7.8 CONCLUSION

The localization of uncharacterized proteins within the landscape of the cell is an effective and powerful approach to glean an initial understanding of a protein's biological role. Several proteins labeled with the appropriate GFP color variants can be localized simultaneously to determine if they reside in functionally distinct regions of the cell or colocalize. Colocalization can delimit two or more proteins to within a substructure ~200 nm in diameter, the limit of resolution for fluorescence microscopy. Finally, the dynamic movement of proteins can be viewed in live cells as they progress through the cell cycle or respond to external cues. Through fluorescence microscopy in combination with results from high-throughput techniques such as two-hybrid analysis and mass spectrometry, a richly textured portrait of a protein emerges.

FRET is a powerful experimental tool to determine whether two proteins are positioned within 80 Å of each other and thus juxtaposed within a complex. Models built from the results of FRET measurements lead to a structure of a complex at a resolution that is between the resolution of X-ray crystallography and light microscopy. As such, it fills an important gap in the acquisition of ultrastructural information. In addition, live cell FRET measurements allow the study of structural rearrangements as they occur in vivo. As a measure of FRET, a newly defined index, $FRET_{Relative}$, allows quantification over many levels of protein expression and will facilitate the synthesis of FRET data into coherent models.

REFERENCES

Amiri H, Schultz G, Schaefer M. 2003. FRET-based analysis of TRPC subunit stoichiometry. *Cell Calcium* 33:463–470.

Berney C, Danuser G. 2003. FRET or no FRET: a quantitative comparison. *Biophys J* 84:3992–4010.

Bevis BJ, Glick BS. 2002. Rapidly maturing variants of the *Discosoma* red fluorescent protein (DsRed): *Nat Biotechnol* 20:83–87.

Breitkreutz BJ, Stark C, Tyers M. 2003. The GRID: the General Repository for Interaction Datasets. *Genome Biol* 4:R23.

Brockerhoff SE, Stevens RC, Davis TN. 1994. The unconventional myosin, Myo2p, is a calmodulin target at sites of cell growth in *Saccharomyces cerevisiae. J Cell Biol* 124:315–323.

Cyert MS. 2001. Genetic analysis of calmodulin and its targets in *Saccharomyces cerevisiae. Annu Rev Genet* 35:647–672.

Cyert MS, Kunisawa R, Kaim D, Thorner J. 1991. Yeast has homologs (CNA1 and CNA2 gene products) of mammalian calcineurin, a calmodulin-regulated phosphoprotein phosphatase. *Proc Natl Acad Sci USA* 88:7376–7380.

Damelin M, Silver PA. 2002. In situ analysis of spatial relationships between proteins of the nuclear pore complex. *Biophys J* 83:3626–3636.

Geiser JR, Sundberg HA, Chang BH, Muller EG, Davis TN. 1993. The essential mitotic target of calmodulin is the 110-kilodalton component of the spindle pole body in *Saccharomyces cerevisiae. Mol Cell Biol* 13:7913–7924.

Geli MI, Wesp A, Riezman H. 1998. Distinct functions of calmodulin are required for the uptake step of receptor-mediated endocytosis in yeast: the type I myosin Myo5p is one of the calmodulin targets. *EMBO J* 17:635–647.

Gordon GW, Berry G, Liang XH, Levine B, Herman B. 1998. Quantitative fluorescence resonance energy transfer measurements using fluorescence microscopy. *Biophys J* 74:2702–2713.

Hailey DW, Davis TN, Muller EG. 2002. Fluorescence resonance energy transfer using color variants of green fluorescent protein. *Methods Enzymol* 351:34–49.

Harpur AG, Wouters FS, Bastiaens PI. 2001. Imaging FRET between spectrally similar GFP molecules in single cells. *Nat Biotechnol* 19:167–169.

Heim R. 1999. Green fluorescent protein forms for energy transfer. *Methods Enzymol* 302:408–423.

Ho Y, Gruhler A, Heilbut A, et al. 2002. Systematic identification of protein complexes in *Saccharomyces cerevisiae* by mass spectrometry. *Nature* 415:180–183.

Idrissi FZ, Wolf BL, Geli MI. 2002. Cofilin, but not profilin, is required for myosin-I-induced actin polymerization and the endocytic uptake in yeast. *Mol Biol Cell* 13:4074–4087.

Kalab P, Weis K, Heald R. 2002. Visualization of a Ran-GTP gradient in interphase and mitotic *Xenopus* egg extracts. *Science* 295:2452–2456.

Lippincott-Schwartz J, Patterson GH. 2003. Development and use of fluorescent protein markers in living cells. *Science* 300:87–91.

Miyawaki A. 2003. Visualization of the spatial and temporal dynamics of intracellular signaling. *Dev Cell* 4:295–305.

Miyawaki A, Griesbeck O, Heim R, Tsien RY. 1999. Dynamic and quantitative Ca^{2+} measurements using improved cameleons. *Proc Natl Acad Sci USA* 96:2135–2140.

Muller EGD, Snydsman BE, Novik I, Hailey DW, Gestaut DR, Niemann CA, O'Toole ET, Giddings TH, Sundin BA, Davis TN. 2005. The organization of the core proteins of the yeast spindle pole body. *Mol Biol Cell* 16:3341–3352.

Nagai T, Ibata K, Park ES, Kubota M, Mikoshiba K, Miyawaki A. 2002. A variant of yellow fluorescent protein with fast and efficient maturation for cell-biological applications. *Nat Biotechnol* 20:87–90.

Ormo M, Cubitt AB, Kallio K, Gross LA, Tsien RY, Remington SJ. 1996. Crystal structure of the *Aequorea victoria* green fluorescent protein. *Science* 273:1392–1395.

Overton MC, Blumer KJ. 2002. The extracellular N-terminal domain and transmembrane domains 1 and 2 mediate oligomerization of a yeast G protein-coupled receptor. *J Biol Chem* 277:41463–41472.

Pausch MH, Kaim D, Kunisawa R, Admon A, Thorner J. 1991. Multiple Ca^{2+}/calmodulin-dependent protein kinase genes in a unicellular eukaryote. *EMBO J* 10:1511–1522.

Periasamy A, Elangovan M, Elliott E, Brautigan DL. 2002. Fluorescence lifetime imaging (FLIM) of green fluorescent fusion proteins in living cells. *Methods Mol Biol* 183:89–100.

Rocheleau JV, Edidin M, Piston DW. 2003. Intrasequence GFP in Class I MHC molecules, a rigid probe for fluorescence anisotropy measurements of the membrane environment. *Biophys J* 84:4078–4086.

Sekar RB, Periasamy A. 2003. Fluorescence resonance energy transfer (FRET) microscopy imaging of live cell protein localizations. *J Cell Biol* 160:629–633.

Stevens RC, Davis TN. 1998. Mlc1p is a light chain for the unconventional myosin Myo2p in *Saccharomyces cerevisiae*. *J Cell Biol* 142:711–722.

Stryer L. 1978. Fluorescence energy transfer as a spectroscopic ruler. *Annu Rev Biochem* 47:819–846.

Truong K, Ikura M. 2001. The use of FRET imaging microscopy to detect protein–protein interactions and protein conformational changes in vivo. *Curr Opin Struct Biol* 11:573–578.

Uetz P, Giot L, Cagney G, et al. 2000. A comprehensive analysis of protein–protein interactions in *Saccharomyces cerevisiae*. *Nature* 403:623–627.

van Roessel P, Brand AH. 2002. Imaging into the future: visualizing gene expression and protein interactions with fluorescent proteins. *Nat Cell Biol* 4:E15–E20.

Wouters FS, Verveer PJ, Bastiaens PI. 2001. Imaging biochemistry inside cells. *Trends Cell Biol* 11:203–211.

Yoder TJ, Pearson CG, Bloom K, Davis TN. 2003. The *Saccharomyces cerevisiae* spindle pole body is a dynamic structure. *Mol Biol Cell* 14:3494–3505.

Youvan DC, Coleman WJ, Silva CM, Peterson J, Bylina EJ, Yang MM. 1997. Fluorescence imaging micro-spectrophotometer (FIMS). *Biotechnology et alia*, http://www.et-al.com/searchable/abstracts/abstract1.html.

Zhang J, Campbell RE, Ting AY, Tsien RY. 2002. Creating new fluorescent probes for cell biology. *Nat Rev Mol Cell Biol* 3:906–918.

8

Characterization of Functional Protein Complexes

Leopold L. Ilag and Carol V. Robinson
University of Cambridge, Cambridge, United Kingdom

8.1 INTRODUCTION

It is now widely accepted that noncovalent interactions are maintained in the gas phase of the mass spectrometer with the electrostatic components being favored since the driving force for hydrophobic attraction is reduced by desolvation (Robinson et al., 1996). The advent of nanoflow electrospray (Wilm and Mann, 1994) improved the preservation of noncovalent complexes since the droplet enters the process at a much later stage than conventional electrospray and therefore requires reduced desolvation (Chung et al., 1999). Figure 8.1 shows a schematic representation of the nanoflow electrospray introduction of a solution containing a protein RNA complex. Providing the appropriate concentration of protein complexes is employed (low 1–10 μM) and the orifice in the nanoflow needle is of the order of a few micrometers, we calculate that each droplet should contain a single complex molecule. Desolvation of the droplet in which solvent evaporation takes place is aided by a flow of nitrogen gas. Maintaining noncovalent complexes requires that interactions between protein subunits are not perturbed. This is achieved by introducing the complex from neutral pH and aqueous buffer, typically 10–1000 mM ammonium acetate. These solution conditions as well as the compact nature and interactions between subunits limit the extent to which the protein molecules will

Proteomics for Biological Discovery, edited by Timothy D. Veenstra and John R. Yates.
Copyright © 2006 John Wiley & Sons, Inc.

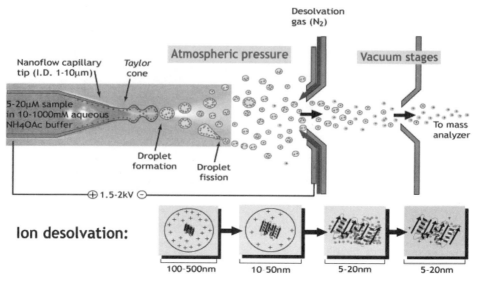

Figure 8.1. *Schematic representation of the nanoflow electrospray process used for introducing noncovalent protein RNA complexes. Complexes at concentrations typically in the range 5–20 µM are introduced from ammonium acetate buffer using a nanoflow capillary. A voltage of 1.5–2 kV is typically applied to the capillary and backing pressure is often used to initiate flow. Droplet formation takes place at atmospheric pressure, each droplet calculated to contain one complex molecule, providing the appropriate concentration and needle orifice are employed. With the aid of a counter-current flow of gas, desolvation takes place such that the droplet shrinks from a few hundred nm to ~10 nm, yielding gas-phase ions largely devoid of solvent and buffer molecules. (See color insert.)*

be protonated or charged. As a result, the m/z value of the resulting ions exceeds the working range of standard quadrupole instruments. This limitation was overcome by the coupling of electrospray methods with time-of-flight analysis with infinitely higher m/z range (Verentchikov et al., 1994). Furthermore, as a consequence of this development, collisional cooling was realized (Krutchinsky et al., 1998). Collisional cooling is a process whereby ions experience multiple collisions along their flight path, reducing their internal energy and resulting in the preservation of noncovalent interactions.

As the molecular weight of complexes that are maintained increases, a tendency to retain residual water and buffer molecules has also been observed (Nettleton et al., 1998; Rostom et al., 1998). This phenomenon leads to broader peaks and difficulty in obtaining precise mass measurements and consequently to ambiguities in the composition of noncovalent complexes. This difficulty is overcome, at least in part, by using tandem mass spectrometry or MS/MS in which ions are isolated at a particular m/z value. After collision with neutral gas molecules, admitted into the collision cell, the activated complex undergoes dissociation and the products are examined in the time-of-flight analyzer. Isolation of the ion of interest often requires the use of a quadrupole mass filter, since this analyzer is most widely coupled with electrospray. The m/z range of such devices is typically up to $4000\,m/z$. This imposes limits on the size of macromolecular complexes that can be studied and led to the implementation of a low-frequency quadrupole that has an effective range of up to $32,000\,m/z$, as shown in Figure 8.2 (Sobott et al., 2002).

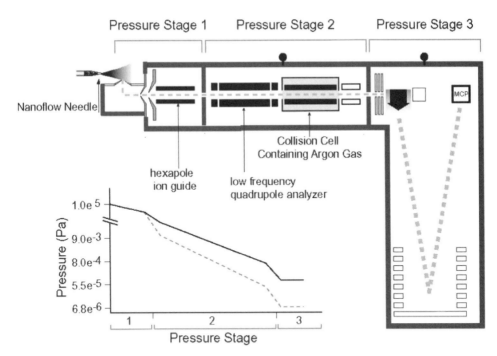

Figure 8.2. Schematic of the quadrupole time-of-flight mass spectrometer modified for the analysis and transmission of macromolecular particles. The ion beam is shown as a dotted grey line of varying width. Full details of the modifications to the instrument can be found in Sobott et al. (2002). To maintain the noncovalent interactions between components in the macromolecular complex, it is necessary to employ collisional cooling in the various vacuum stages of the mass spectrometer to reduce the translational energy and hence internal energy of the macromolecular ions. Inset: Graphical representation of the pressure gradient that is employed for recording spectra of noncovalent complexes (continuous line) that for normal operating conditions is shown with a dotted line.

Such a quadrupole enables large macromolecular assemblies to be subjected to MS/MS experiments, paving the way for the study of molecular machines. Additionally, by performing MS/MS without isolation of ions—consequently all ions are accelerated in the collision cell—conditions can be achieved whereby complexes are largely stripped of residual solvent molecules while maintaining subunit interactions. This process effectively increases the resolution of the measurements and allows the separation of overlapping complexes with differing components but with closely similar m/z values.

As a testimony to the pace of technological development in this field, just five years previously noncovalent interactions in simple tetrameric assemblies were reported with masses typically up to ~60 kDa (Nettleton et al., 1998). Rapid developments in time-of-flight technology enabled homo-oligomeric complexes in the megadalton mass range to be examined (Rostom and Robinson, 1999; Van Berkel et al., 2000) and even macromolecular assemblies such as ribosomes with masses over 2 MDa (Rostom et al., 2000). Moreover, the origin of these macromolecular complexes is no longer restricted to favorable cases involving recombinant proteins present in large amounts but rather to complexes isolated from natural sources without the benefit or enrichment strategies. As the pace of technology continues to advance, it is now becoming a reality that heterogeneous dynamic assemblies composed of protein and nucleic acid can be examined.

Among the dynamic assemblies that are crucial to the function of the cell are those involved in the synthesis, translation, and degradation of RNA. Such processes are controlled by molecular machines known as RNA polymerase, ribosomes, and the degradasome, respectively. Despite their fundamental importance, these machines are elusive targets for biophysical study. In part, because these multisubunit assemblies are not practical to overexpress, most studies relying on isolation of materials from natural sources. Furthermore, many of these macromolecular complexes tend to be asymmetric, making them less straightforward to study by techniques that profit from data averaging such as NMR and crystallography. In this chapter we discuss developments that allow us to extend applications of mass spectrometry to noncovalent molecular machines. The first example is that of the N-terminal catalytic domain of bacterial RNase E, which forms part of the *Escherichia coli* degradasome. Here the unique capabilities of mass spectrometry are exemplified in defining the precise stoichiometry of the interactions within the complex (Callaghan et al., 2003). For polynucleotide phosphorylase (PNPase) from *Streptomyces antibioticus*, also thought to be a component analogous to *E. coli* PNPase in the degradosome, mass spectrometry was able to define a trimeric structure in contrast to that predicted in databases. For *E. coli* RNA polymerase, the subunit stoichiometry is already established from the X-ray diffraction data of the analogous complex from *Thermus aquaticus* (Zhang et al., 1999). The attributes of mass spectrometry for this investigation include the ability to define on a rapid time scale the effects of challenging the particle with additional factors. For ribosomes, a mass spectrometry approach based on dissociation of proteins was developed and exploited to establish proteins whose interactions with RNA are either increased or perturbed as a result of conformational change (Hanson et al., 2003). Together these three macromolecular complexes involved in critical RNA processing events enable us to present complementary mass spectrometry approaches to the study of noncovalent complexes.

8.2 THE DEGRADOSOME

The degradosome of *E. coli* is a multienzyme complex involved in the degradation of messenger RNA. These enzymes are phosphorylases that degrade RNA from the 3′ end. RNase E and PNPase form part of the macromolecular machine (Blum et al., 1997). Previous studies of RNase E using yeast two-hybrid analyses indicated a propensity for self-association. Mass spectrometry was used to determine the stoichiometry of 1–529 N-terminal construct (Callaghan et al., 2003). The measured mass of the protein recorded from a solution of 1 M ammonium acetate revealed a molecular mass of 248,281 ± 31 Da consistent with a tetrameric association (247,537 Da), as shown in Figure 8.3. Binding of a synthetic RNA 10-mer to the tetrameric form was also assessed by nanoflow electrospray mass spectrometry. By monitoring the change in mass upon addition of the RNA oligomer, it was demonstrated that up to four molecules of RNA were capable of binding to the tetramer. These data are consistent with the binding of one oligonucleotide per subunit of the tetrameric assembly. In combination with X-ray solution scattering

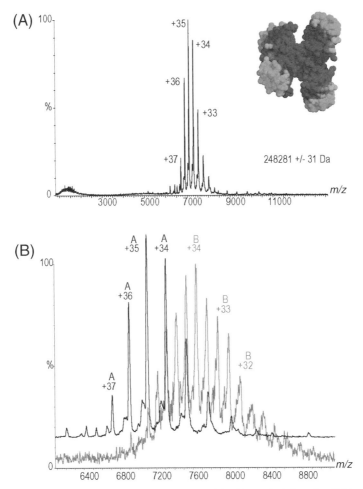

Figure 8.3. *(A) Nanoflow electrospray mass spectrum of the N-terminal domain of RNase E catalytic domain recorded under nondissociating conditions. Charge states assigned to the protein tetramer are labeled. (B) Expansion of the spectrum over the m/z range 6300–9000. The lower spectrum was recorded in the presence of a twofold stoichiometric excess of a substrate RNA analog. The additional charge states correspond in mass to a distribution of RNA molecules binding to the protein tetramer. The major series labeled with charge states corresponds to up to three molecules of RNA binding to the protein tetramer, but higher m/z species can be observed with up to four molecules on the lower charged states. Inset: A possible solution for the N-terminal domain of RNase E based on solution scattering profile and symmetry constraints. The structure was produced with Molscript (Kraulis, 1999) and reproduced with permission from Callaghan et al. (2003). (See color insert.)*

data and symmetry constraints, these mass spectrometry results contributed to a shape reconstruction of RNase E and suggest that RNA, at least in the case of the short substrates used here, can be bound simultaneously to neighboring subunits.

X-ray crystallographic analysis of PNPase from *Streptomyces antibioticus* is consistent with a hexameric structure formed by dimerization of two trimeric structures. The nanoflow electrospray mass spectrum of PNPase (Figure 8.4) clearly shows species with molecular masses of $81{,}907 \pm 18$, $163{,}564 \pm 48$, and $245{,}285 \pm 33$ Da. The calculated mass for a single subunit of PNPase from this

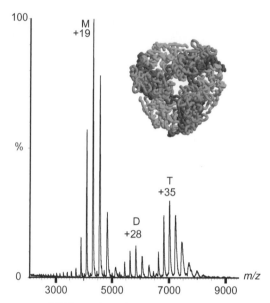

Figure 8.4. *Mass spectrum of PNPase from Streptomyces antibioticus showing the three major components assigned to monomer, dimmer, and trimer. Inset: Crystallographic structure in which subunit is represented in different shades. Adapted from Symmons et al. (2000). (See color insert.)*

source is 81,132 Da, implying that the three species correspond to monomer, dimer, and trimer, respectively. From close examination of the X-ray structure, however, it would appear that formation of a trimeric species in solution is feasible, in line with the mass spectrometry data reported here.

8.3 RNA POLYMERASE

Bacterial RNA polyermase is the basic cellular machinery for DNA transcription, which gives rise to different types of RNAs in the cell. The *E. coli* core enzyme is a 389 kDa complex with a subunit composition of $\alpha_2\beta\beta'\omega$ (Ebright, 2000). X-ray analysis of the crystal structure of the enzyme from *Thermus aquaticus* reveals a core structure akin to a crab claw with the β and β' domains making up each of the pincers (Zhang et al., 1999). The holoenzyme is composed of the core enzyme bound to σ^{70} (Murakami and Darst, 2003). There are more than six known sigma factors in *E. coli* as well as a host of other regulatory proteins that modulate transcription. It is not clearly understood how these interact with each other and how tight regulation of RNA synthesis is ensured.

A typical nanoflow electrospray spectrum of the core enzyme isolated from *E. coli* is shown in Figure 8.5 (Ilag et al., 2005). Despite the fact that the complex was isolated from natural sources and not subjected to enrichment strategies, the spectrum is well resolved and has only one major series of peaks centered around m/z 10,000, revealing that it is remarkably homogeneous. Two additional minor

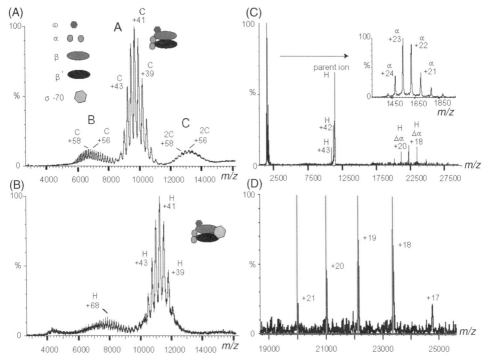

Figure 8.5. *Mass spectra of core and holo forms of E. coli RNA polymerase (A, B) and tandem mass spectrum of the +41 parent ion of the holo complex (C, D). (A) The core enzyme gives rise to a predominant series of peaks labeled A, a more highly charged series labeled B, and a low intensity series corresponding in mass to dimer labeled C. (B) Addition of recombinant s[70] leads to formation of the holo complex at higher m/z values. Inset: Schematic representation of the subunits adapted from the X-ray structure. (C) Tandem mass spectrum of the +41 charge state of the holo enzyme. (D) Expansion of the high m/z ions formed by loss of an a subunit and comparison with the theoretical isotope simulation of the holoenzyme minus the a subunit. Reproduced with permission from Ilag et al. (2005). (See color insert.)*

series of peaks can be discerned at m/z 7000 and 13,000. The major species (charge state series A, Figure 8.5A) at *m/z* 10,000 corresponds to the principal component. The measured mass (393,689 ± 46 Da) is consistent with the pentameric stoichiometry $\alpha_2\beta\beta'\omega$ (388,981). The main species is flanked at lower and higher *m/z* values by minor sets of peaks labeled B and C. The calculated mass for series B corresponds essentially to the mass of the principal complex. The fact that the complex remains intact but gives rise to a more highly charged series of ions is consistent therefore with some unfolding, presumably on the periphery of the complex, exposing more basic sites for protonation yet maintaining intersubunit contacts. The charge state series C corresponds to a mass that is twice that of the principal component and is therefore assigned to a dimer of the core enzyme. Association to form dimeric species has also been observed in solution under low salt conditions (Dyckman and Fried, 2002). This spectrum therefore demonstrates that it is possible to study the intact core enzyme isolated from natural sources, to observe minor components assigned to partial unfolding of one or more subunits within the complex and self-association to form dimeric species.

In order to examine the active form of this enzyme, σ^{70} from recombinant sources was added to core enzyme in a 1:1 ratio to reconstitute the holoenzyme. Figure 8.5B shows a spectrum recorded for this solution. By comparison with the core enzyme alone, it can be seen that for the reconstituted holoenzyme the charge state series is shifted to higher m/z values. The calculated mass for this species is 468,047 ± 45 Da and the theoretical mass for the holoenzyme is 461, 409. As this is considerably higher than the theoretical mass to confirm that the species formed is the holoenzyme, we applied a ms/ms procedure. Figure 8.5C shows the MS/MS spectrum of the +41 charge state of the holoenzyme. At m/z values below 2000, a highly charged series of ions is assigned to α subunits. At high m/z ~22,500, a lower intensity series of ions is observed. The mass of this series of ions is consistent with that calculated for the holoenzyme complex that has lost one α subunit. The subunit composition of this subcomplex is confirmed by simulation of the isotopes that would constitute the pentameric complex (Figure 8.5D). The remarkable coincidence of the theoretical simulated peaks demonstrates one of the considerable advantages of this MS/MS approach. Since residual solvent and buffer ions are lost in the collisional activation, a greater accuracy in the measured mass is achieved. These results confirm the stoichiometry of the holoenzyme and define the principal dissociation pathway as loss of an α subunit.

Recently, using a mass spectrometry approach, we have shown that the regulator of sigma D (Rsd) forms a 1:1 complex with σ^{70} (Westblade et al., 2005). It was therefore of interest to see how binding of Rsd affects σ^{70} binding to the holoenzyme. By addition of aliquots of Rsd to a solution containing the holoenzyme, additional series of peaks become apparent below m/z 5000 (Figure 8.6A). These peaks correspond in mass to a species of mass 90 kDa. This value is higher than the individual masses of any of the components present in the complex and is consistent with 1:1 complex formed between Rsd and σ^{70}. Since there is essentially no free σ^{70} in the holoenzyme preparation, the only σ^{70} that could bind to Rsd must previously have been associated with the holoenzyme. Further addition of Rsd in a twofold excess over the holoenzyme preparation gave rise to a further increase in the amount of the 90 kDa species, implying that a greater proportion of σ^{70} is now in complex with Rsd. Acceleration of the ions formed from the sample containing a twofold excess of Rsd over the holoenzyme (Figure 8.6B) gives rise to an additional series of peaks at high m/z, labeled *. The mass of this series is consistent with core enzyme binding to Rsd. The results of this experiment therefore demonstrate that Rsd displaces σ^{70} from the holoenzyme by forming a 1:1 complex with Rsd and binding to the core enzyme itself.

These mass spectrometry experiments with purified Rsd demonstrate, surprisingly, that Rsd can form a complex with core RNA polymerase. Rsd had been identified previously (Jishage and Ishihama, 1998) as an *E. coli* protein that acts as an antisigma factor by binding to the RNA polymerase σ^{70} subunit and preventing it from interacting with the core enzyme to give the transcriptionally competent holo RNA polymerase. The results from the mass spectrometry experiments show not only that Rsd displaces σ^{70} from holo RNA polymerase to form a 1:1 Rsd:σ^{70} complex, but also that Rsd interacts with the remaining core enzyme. This finding

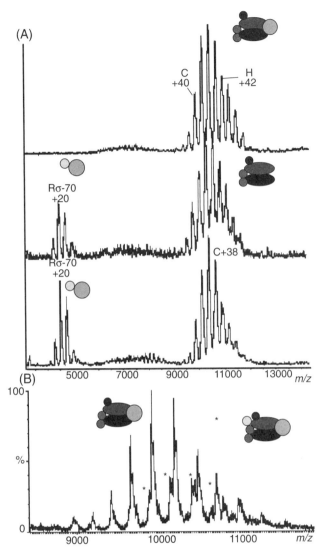

Figure 8.6. Spectra recorded after addition of Rsd to the holo RNA polymerase (A) and acceleration of the resulting high mass complex (B). Upper trace holoenzyme alone, middle trace holoenzyme with equimolar Rsd, and lower trace holoenzyme with twofold excess of Rsd. The series of ions at low m/z corresponds in mass to a 1:1 complex of Rsd with σ^{70}. The complex formed in the presence of a twofold excess of Rsd over that of holoenzyme was accelerated in the collision cell at 150 V, revealing the presence of species labeled * assigned to the core enzyme bound to Rsd. Reproduced with permission from Ilag et al. (2005). (See color insert.)

implies therefore that by binding σ^{70} Rsd effectively targets RNA polymerase to bind an alternative sigma factor. Overall, the results suggest that Rsd may have a role in channeling core RNA polymerase to particular subsets of promoters.

8.4   *ESCHERICHIA COLI* RIBOSOMES

The ribosome is among the most complex molecular machines in the cell. It is responsible for the translation of the genetic code into proteins by decoding information from messenger RNAs synthesised by RNA polymerase. The functional 70S complex is made up of two subunits, the 50S and 30S, which is composed of three large RNA molecules and over 50 different proteins occurring in single copies with the exception of L7/12 present in four copies. Mass spectra of intact *E. coli* ribosomes give rise to broad charge state distributions between 16,000 and 25,000 *m/z* assigned to the 30S, 50S, and 70S, respectively (Rostom et al., 2000). Manipulating the pressure and acceleration of intact ribosomes in the gas phase induced their dissociation to individual proteins and subcomplexes (Rostom et al., 2000). Given the unique masses in protein databases, it was possible to assign the peaks to the various ribosomal proteins. More recently, we have demonstrated that it is possible to effect dissociation of proteins and RNA by changing the solution conditions from which the sample is introduced (Hanson et al., 2003). The dissociation pattern was found to be very reproducible. At pH 7.0 only L7/L12, L11, and L10 dissociate in detectable quantities. By lowering the pH of the solution, four additional proteins dissociate as well as the 5S RNA. A total of 17 proteins are released in the gas phase in response to replacement of Mg^{2+} with Li^+ ions in solution. Clearly, these responses to changes in solution conditions reflect specific structural rearrangements within the ribosome.

In order to determine which factors were important in controlling this dissociation, we considered surface accessibility, isoelectric point of the protein, and low surface area of interaction of the protein with rRNA. The latter property was found to produce a positive correlation, such that favorable release occurred for proteins with a relatively low surface area of interaction with RNA. Therefore, we were able to use proteins that dissociate as probes of their interaction with RNA. We examined complexes of ribosomes with the elongation factor G (EF-G) inhibited by fusidic acid or thiostrepton and used protein dissociation to report on the conformational changes that take place. Mass spectra recorded for the fusidic acid-inhibited complex revealed subtle changes in peak intensities of the proteins that dissociated from the ribosome (Figure 8.7A). The thiostrepton-inhibited complex, however, was examined under the same conditions and yet markedly different spectra were obtained (Figure 8.7B). L5 and L18, two proteins that interact solely with the 5S RNA, were released. These results allowed us to propose that the ribosome elongation factor-G complex inhibited by thiostrepton, but not fusidic acid, involves destabilization of 5S RNA–protein interactions. In this case, therefore, the pattern of protein dissociation was used as an indirect indicator of conformational changes in a macromolecular machine.

8.5   FUTURE PROSPECTS

The examples presented here are by no means a comprehensive reflection of the field but have been selected to exemplify various aspects of the study of intact functional particles using mass spectrometry. The relatively straightforward determination of stoichiometry of proteins and RNA in the RNase E and PNPase

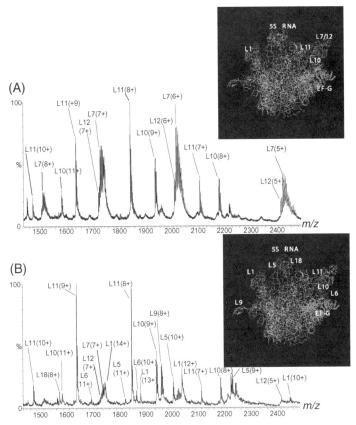

Figure 8.7. Mass spectra of the m/z region 1450–2500 of the ribosome EF-G complex in the presence of (A) fusidic acid and (B) thiostrepton. The two spectra are markedly different. The spectrum recorded in the presence of fusidic acid is similar to that observed for ribosomes in the absence of EF-G under these solution and mass spectrometry conditions. By contrast, the complex inhibited by thiostrepton demonstrates the absence of L7/L12 and the presence of additional proteins L5, L6, and L18. The structure of the 50S subunit was produced using the coordinates from Thermus thermophilus at 5.5 Å resolution PDB file 1GIY (Yusupov et al., 2001). The structure of EF-G (pdb ascension code 1FNM) was fitted according to the structure of Ban et al. (2000). The proteins shaded in the two structures represent those that are released from the two complexes. Reproduced with permission from Hanson et al. (2003). (See color insert.)

experiments benefited from the precision afforded by mass measurement in determining the stoichiometry of interacting species. Moreover, such experiments demonstrate the complementary nature of mass spectrometry and X-ray crystallography, where the largest structural unit for the PNPase was found to be a trimer using mass spectrometry and a hexamer crystallographically (Symmons et al., 2000). The challenge of complexes with protein factors, either in real time as in the case of the polymerase or after inhibition with antibiotics in the case of ribosome complexes, offers real opportunity to report, on a rapid time scale, the effects of these additional factors and the dynamics of change in such machines. This approach, we believe, offers new opportunities for defining dynamic associations in the cell, an attribute that cannot be overemphasized given that 65 sigma factors have been identified for the RNA polymerase from *Streptomyces coelicolor* (Lonetto et al., 1994).

The challenges that remain for mass spectrometry applications described in this chapter are to reduce still further the sample requirement of such experiments, to continue to improve the resolution of the resulting complexes by finding new methods of desolvation that do not perturb assemblies, and, in common with many structural biology methods, to overcome practical difficulties associated with investigating membrane-bound protein complexes. In spite of these current limitations, however, mass spectrometry has much to offer structural biologists in terms of (i) relatively small sample sizes: subpicomole quantities are often sufficient to obtain spectra of complexes; (ii) tolerance of heterogeneity or substoichiometric quantities of subunits within the complex; (iii) ability to cope with enhanced conformational dynamics of proteins within assemblies; (iv) no reliance on symmetry for averaging; and (v) speed of investigation since it is possible to determine the effect of adding a particular component within a matter of minutes. For these reasons, we anticipate that mass spectrometry will continue to stimulate research into study of macromolecular machines and their response to mimicry of the changing environment in the cell.

ACKNOWLEDGMENT

We acknowledge with thanks helpful discussions with Charlotte Hanson, Frank Sobott for Figure 8.1, and the CVR mass spectrometry group. We thank the BBSRC and Royal Society for funding.

REFERENCES

Ban N, Nissen P, Hansen J, Moore PB, Steitz TA. 2000. The complete atomic structure of the large ribosomal subunit at 2.4 Å. *Science* 289:905–920.

Blum E, Py B, Carpousis AJ, Higgins CF. 1997. Polyphosphate kinase is a component of the *Escherichia coli* RNA degradosome. *Mol Microbiol* 26:387–398.

Callaghan AJ, Grossmann JG, Redko YU, Ilag LL, Moncrieffe MC, Symmons MF, Robinson CV, McDowall KJ, Luisi BF. 2003. Quaternary structure and catalytic activity of the *Escherichia coli* ribonuclease E amino-terminal catalytic domain. *Biochemistry* 42:13848–13855.

Chung EW, Henriques DA, Renzoni D, Morton CJ, Mulhern TD, Pitkeathly MC, Ladbury JE, Robinson CV. 1999. Probing the nature of interactions in SH2 binding interfaces— evidence from electrospray ionization mass spectrometry. *Protein Sci* 8:1962–1970.

Dyckman D, Fried MG. 2002. The *Escherichia coli* cyclic AMP receptor protein forms a 2:2 complex with RNA polyermase holoenzyme *in vitro*. *J Biol Chem* 277:19064–19070.

Ebright R. 2000. RNA polymerase: structural similarities between bacterial RNA polymerase and eukaryotic RNA polymerase II. *J Mol Biol* 293:199–213.

Hanson CL, Fucini P, Nierhaus KH, Robinson CV. 2003. Dissociation of intact *Escherichia coli* ribosomes in a mass spectrometer. Evidence for conformational change in a ribosome elongation factor G complex. *J Biol Chem* 278:1259–1267.

Ilag LL, Westblade LF, Deshayes C, Kolb A, Busby SJW, Robinson CV. 2005. Mass spectrometry of *Escherichia coli* RNA Polymerase: interactions of the core enzyme with sigma 70 and Rsd protein. *Structure* 12:269–275.

Jishage M, Ishihama A. 1998. A stationary phase protein in *Eschercihia coli* with binding activity to the major sigma subunit of RNA polymerase. *Proc Natl Acad Sci USA* 95:4953–4958.

Kraulis PJ. 1999. Molscript: a program to produce both detailed and schematic plots of protein structures. *J Appl Crystallogr* 24:946–950.

Krutchinsky AN, Chernushevich IV, Spicer VL, Ens W, Standing KG. 1998. Collisional damping interface for an electrospray ionization time-of-flight mass spectrometer. *J Am Soc Mass Spectrom* 9:569–579.

Lonetto MA, Brown KL, Rudd KE, Buttner MJ. 1994. Analysis of the *Streptomyces coeli-color* sigE gene reveals the existence of a subfamily of eubacterial RNA-polymerase sigma-factors involved in the regulation of extracytoplasmic functions. *Proc Natl Acad Sci USA* 91:7573–7577.

Murakami KS, Darst SA. 2003. Bacterial RNA polymerases: the whole story. *Curr Opin Struct Biol* 13:1–9.

Nettleton EJ, Sunde M, Lai V, Kelly JW, Dobson CM, Robinson CV. 1998. Protein subunit interactions and structural integrity of amyloidogenic transthyretins—evidence from electrospray mass spectrometry. *J Mol Biol* 281:553–564.

Robinson CV, Chung EW, Kragelund BB, Knudsen J, Aplin RT, Poulsen FM, Dobson CM. 1996. Probing the nature of non-covalent interactions by mass spectrometry. A study of protein–CoA ligand binding and assembly. *J Am Chem Soc* 118:8646–8653.

Rostom AA, Robinson CV. 1999. Detection of the intact GroEL chaperonin assembly by mass spectrometry. *J Am Chem Soc* 121:4718–4719.

Rostom AA, Sunde M, Richardson SJ, Schreiber G, Jarvis S, Bateman R, Dobson CM, Robinson CV. 1998. Dissection of multi-protein complexes using mass spectrometry—subunit interactions in transthyretin and retinol-binding protein complexes. *Proteins Struct Funct Genet Suppl* 2:3–11.

Rostom AA, Fucini P, Benjamin DR, Juenemann R, Nierhaus KH, Hartl FU, Dobson CM, Robinson CV. 2000. Detection and selective dissociation of intact ribosomes in the mass spectrometer. *Proc Natl Acad Sci USA* 97:5185–5190.

Sobott F, Hernandez H, McCammon MG, Tito MA, Robinson CV. 2002. A tandem mass spectrometer for improved transmission and analysis of large macromolecular assemblies. *Anal Chem* 74:1402–1407.

Symmons MF, Jones GH, Luisi BF. 2000. A duplicated fold is the structural basis for polynucleotide phosphorylase catalytic activity, processivity, and regulation. *Structure* 8:1215–1226.

Van Berkel WJH, Van Den Heuvel RHH, Versluis C, Heck AJR. 2000. Detection of intact megadalton protein assemblies of vanillyl-alcohol oxidase by mass spectrometry. *Protein Sci* 9:435–439.

Verentchikov AN, Ens W, Standing KG. 1994. A reflecting time of flight mass spectrometer with an electrospray ion source and orthogonal extraction. *Anal Chem* 66:126–133.

Westblade LF, Ilag LL, Powell AK, Kolb A, Robinson CV, Busby SJW. 2005. Studies of *Escherichia coli Rsd-sigma*[70] complex. *J. Mol. Biol.* 335:685–692.

Wilm M, Mann M. 1994. Electrospray and Taylor-cone theory, Dole's beam of macromolecules at last? *Int J Mass Spectrom Ion Proc* 136:167–180.

Yusupov M, Yusupov G, Baucom A, Lieberman K, Earnest T, Cate J, Noller H. 2001. Crystal structure of the ribosome at 5.5 A resolution. *Science* 292:883.

Zhang G, Campbell E, Minakhin L, Richter C, Severinov K, Darst S. 1999. Crystal structure of *Thermus aquaticus* core RNA polymerase at 3.3 Å resolution. *Cell* 98:811–824.

9

Structural Proteomics by NMR

G. Marius Clore

Laboratory of Chemical Physics, National Institute of Diabetes and Digestive and Kidney Diseases, National Institutes of Health, Bethesda, Maryland

9.1 INTRODUCTION

Nuclear magnetic resonance (NMR) is a powerful spectroscopic technique that permits the detailed study at atomic resolution of the three-dimensional structure and dynamics of macromolecules and their complexes in solution. In this brief chapter, I discuss various aspects of NMR that are pertinent to structural proteomics, that is, the high-throughput study of protein–protein complexes at the atomic level. Structural work on complexes has gained increasing importance since it is evident that the structure of a complex yields far greater functional insight than the structures of its individual component proteins.

Macromolecular structure determination by NMR is intrinsically a highly specialized, labor-intensive, and time-consuming technique. In addition, for a system of any reasonable size (say, greater than about 70 residues), isotopic labeling with ^{15}N and ^{13}C is required. Numerous reviews have been written on the subject detailing the experimental and computational methodologies involved (Wuthrich, 1986; Clore and Gronenborn, 1989, 1991, 1998; Bax and Grzsiek, 1993; van de Ven, 1995; Cavanagh et al., 1996). Determining the structure of a single protein by NMR can be broken down into essentially four steps: (i) sequential resonance assignment making use of a number of experiments to identify through-bond connectivities along the backbone and side chains (usually 3D triple resonance experiments); (ii) assignment of cross-peaks in nuclear Overhauser enhancement spectra (usually 3D and 4D) to obtain short ($\leq$5–6 Å) interproton distance restraints, which provide

Proteomics for Biological Discovery, edited by Timothy D. Veenstra and John R. Yates.
Copyright © 2006 John Wiley & Sons, Inc.

the main source of geometrical information; (iii) measurement of additional NMR observables that provide useful conformational information (these may include three-bond scalar couplings that are related to torsion angles by simple empirical equations; $^{13}C\alpha/^{13}C\beta$ chemical shifts that are related empirically to backbone ϕ/ψ torsion angles; and long-range orientational restraints, such as residual dipolar couplings measured in dilute liquid crystalline media); and (iv) calculation of the three-dimensional structure from the experimental NMR restraints using simulated annealing. Generally, an iterative refinement strategy is employed (Clore and Gronenborn, 1989, 1991): calculations are initially carried out with a limited set of interproton distance restraints corresponding to NOE cross-peaks with unambiguous assignments; further interproton distance restraints from the remaining NOE cross-peaks are subsequently added in an iterative manner on the basis of a successively calculated series of structures. While improvements in spectrometer technology (e.g., the advent of cryoprobe technology that increases the signal-to-noise ratio three- to fourfold; higher field magnets that increase spectral resolution, thereby reducing spectral overlap) have reduced the measurement time to some extent, collecting all the data necessary to solve a NMR structure at high accuracy still requires several months. Similarly, improvements in spectral analysis software have permitted the introduction of some degree of automation (Montelione, 1991; Gerstein et al., 2003; Yee et al., 2003), but extensive human intervention is still necessary to fully and reliably interpret the data in all but the simplest of cases.

In this light, what contribution can NMR make to structural proteomics? There are two major methods for deriving high-resolution structural information at atomic resolution: NMR spectroscopy in solution and single crystal X-ray diffraction. In rare instances, electron microscopy is also capable of providing high-resolution information in the solid state. In addition, mass spectrometry in combination with crosslinking data is potentially capable of providing low-resolution structural information when combined with the computational techniques conventionally employed to derive structures from NMR data. If crystals can be obtained rapidly, there is little doubt that crystallography, particularly with the advent of synchrotron X-ray sources, offers the fastest route to high-resolution structure determination. However, complexes are generally more difficult to crystallize than isolated proteins, and it is usually the case that weak complexes (with K_D values in the 1–100 μM range) are extremely difficult to cocrystallize. In the case of NMR, complexes are amenable to structural investigation providing exchange is either fast (weak binding) or slow (tight binding) on the chemical shift time scale. If exchange, however, is intermediate on the chemical shift time scale, the signals are broadened out, precluding any detailed structural work.

A full structure determination of a protein–protein complex by NMR is extremely time consuming. For example, in the case of the 40 kDa EIN·HPr complex from the bacterial phosphotransferase system, the total NMR measurement time alone was ~3500 hours (or 4.8 months) (Garrett et al., 1999). Clearly, therefore, the conventional approach is not suitable for high throughput. Fortunately, new developments have significantly shortened the amount of time required by making full use of prior knowledge in the form of existing high-resolution crystal structures of the free proteins (Clore, 2000; Schwieters and Clore, 2001; Clore and Bewley, 2002). With this information in hand, it is possible to derive high-resolution structures of complexes using limited intermolecular NOE data to provide transla-tion (as well as orientational) information and residual dipolar coupling data (Prestegard et al., 2000; Bax et al., 2001) to generate very accurate orientational information. This chapter presents the underlying principles behind this approach and illustrates its application to a variety of protein–protein complexes. In addition, strategies are discussed whereby translational information from NOE data can be replaced entirely by highly ambiguous intermolecular distance restraints derived from $^{15}N/^1H_N$ chemical shift perturbation mapping (Clore and Schwieters, 2003).

9.2 INTERMOLECULAR DISTANCE RESTRAINTS

As noted earlier, the nuclear Overhauser effect (NOE) is the primary source of geometric information for NMR-based structure determination (Wüthrich, 1986; Clore and Gronenborn, 1989). The NOE (in the initial rate approximation) is pro-portional to the sixth root of the distance between two protons. Consequently, the upper limit for interproton distances that can be detected using the NOE is 5–6 Å. The key to deriving intermolecular NOE-derived interproton distance restraints lies in combining various isotope (^{15}N and ^{13}C) labeling strategies with isotope fil-tering experiments that permit one to detect NOEs on protons attached to specific isotopes of nitrogen and carbon (i.e., NMR active such as ^{15}N or ^{13}C, or NMR inac-tive such as ^{14}N and ^{12}C) (Clore and Gronenborn, 1998). For example, in a complex comprising one protein labeled uniformly with ^{13}C and the other at natural isotopic abundance (i.e., ^{12}C), one can selectively detect NOEs from protons attached to ^{13}C and protons attached to ^{12}C. Typical labeling schemes and the corresponding inter-molecular NOEs observed are illustrated in Figure 9.1, and an example of the data obtained from a 3D ^{13}C-separated/^{12}C-filtered NOE experiment is shown in Figure 9.2.

The NOE is not the only method that can be used to derive intermolecular dis-tance restraints. It is also possible to derive distance restraints using a combination of crosslinking, proteolytic digestion, and mass spectrometry (Bennett et al., 2000; Sinz and Wang, 2001; Schulz et al., 2003). In many cases, however, the data will not yield unique crosslinking partners but multiple possibilities. In addition, it is possible to use another NMR-based approach, which involves derivatizing a suit-able surface accessible cysteine (which may have to be introduced by site-directed mutagenesis) on one protein with either a spin label or a metal binding site (such as EDTA) and measuring paramagnetic relaxation enhancement effects on the other protein to yield long-range (15–35 Å) distance restraints (Voss et al., 1995; Battiste and Wagner, 2000; Mal et al., 2002; Dvoretsky et al., 2002; Iwahara et al., 2003). In general, however, such a strategy can only be applied in a rational manner if one already has a good idea of the interaction surfaces involved in complex for-mation. To some exent, such information can be derived rather easily by either $^{15}N/^1H_N$ chemical shift perturbation mapping (Walters et al., 2001; Zuiderweg, 2002) or cross-saturation experiments (Takahashi et al., 2000; Nakanishi et al., 2002). The latter experiment is far more challenging experimentally since it neces-sitates that one of the proteins is not only ^{15}N labeled but fully deuterated as well.

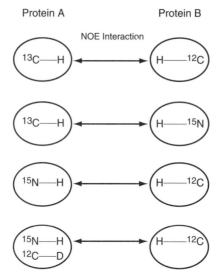

Figure 9.1. *Typical isotope labeling schemes used in the study of protein–protein complexes and corresponding intermolecular NOEs observed. If not indicated, the nitrogen or carbon isotope is ^{14}N and ^{12}C, respectively. In the fourth labeling scheme, ^{12}C attached protons in protein A are deuterated to narrow the lines of the NH resonances. This is useful for larger complexes.*

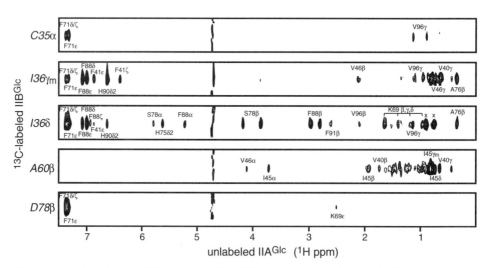

Figure 9.2. *Intermolecular NOEs in the IIA^{Glc}·IIB^{Glc} complex. Strips from a 3D ^{13}C-separated/^{12}C-filtered NOE spectrum recorded at 800 MHz on a 1 : 1 IIA^{Glc} (^{12}C/^{14}N)·IIB^{Glc} (^{13}C/^{15}N) complex, illustrating NOEs from protons attached to ^{13}C on IIB^{Glc} to protons attached to ^{12}C on IIA^{Glc}. Reproduced from Cai et al. (2003).*

9.3 ORIENTATIONAL RESTRAINTS

Long-range orientational restraints can be derived from the measurement of residual dipolar couplings (Tjandra et al., 1997b; Bax et al., 2001) and chemical shift anisotropy (Cornilescu et al., 1998; Wu et al., 2001) in liquid crystalline media and

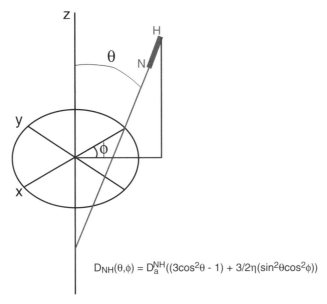

$$D_{NH}(\theta,\phi) = D_a^{NH}((3\cos^2\theta - 1) + 3/2\eta(\sin^2\theta\cos^2\phi))$$

Figure 9.3. *Schematic illustration of orientational information derived from residual dipolar coupling measurements. The observed dipolar coupling, D_{NH}, is dependent on the angle θ between the N-H interatomic vector (shown as the thick line) and the z axis of the tensor, the angle ϕ, which describes the position of the projection of the interatomic vector on the xy plane of the tensor, the magnitude (D_a^{NH}) of the principal component of the tensor, and the rhombicity (η) of the tensor. (See color insert.)*

in suitable cases from heteronuclear T_1/T_2 data (Tjandra et al., 1997a). The characteristic feature of these various parameters is that they yield direct geometric information on the orientation of an interatomic vector(s) with respect to an external axis system (e.g., the alignment tensor in liquid crystalline media, the diffusion tensor for relaxation measurements) expressed in terms of two angles: θ, the angle between the interatomic vector and the z axis of the tensor, and θ, the angle that describes the position of the projection of the interatomic vector on the xy plane of the tensor (Figure 9.3).

For most practical purposes, residual dipolar couplings provide the easiest method for deriving orientational information. In an isotropic medium, the dipolar couplings average to zero. In the solid state, the maximum value of the N–H dipolar coupling is 20.7 kHz. To effectively measure dipolar couplings in solution, therefore, it is necessary to devise means of inducing only a small (ca. 10^{-3}) degree of order such that the N–H dipolar couplings lie in the ±20 Hz range. Experimentally, this is achieved by dissolving the protein or protein complex of interest in a dilute, water-soluble, liquid crystalline medium. Examples of such media include lipid bicelles (Tjandra and Bax, 1997), filamentous phages such as fd or pf1 (Clore et al., 1998; Hansen et al., 2000), rod-shaped viruses such as tobacco mosaic virus (Clore et al., 1998), and polyethylene glycol/hexanol (Rückert and Otting, 2000).

9.4   CONJOINED RIGID BODY/TORSION ANGLE DYNAMICS

In many instances, protein complex formation involves no significant changes in backbone conformation. Thus, if the structures of the individual proteins are already known at high resolution and it can be shown that the backbone conformation remains essentially unchanged upon complex formation (e.g., by comparison of dipolar coupling data measured on the complex with the X-ray structures of the free proteins), one can then make use of conjoined rigid body/torsion angle dynamics to rapidly solve the structure of the complex on the basis of intermolecular NOE data and backbone N–H dipolar couplings (Clore, 2000; Schwieters and Clore, 2001). In this procedure, only the interfacial side chains are allowed to alter their conformation. The backbone and noninterfacial side chains of one protein are held fixed, while those of the second protein are only allowed to rotate and translate as a rigid body. This has been applied with considerable success in the case of several 30–40 kDa complexes of the bacterial phosphotransferase system (Wang et al., 2000; Cornilescu et al., 2002; Cai et al., 2003). A comparison of the structure of the EIN·HPr complex obtained using the conventional full structure determination approach (Garrett et al., 1999) with that obtained by conjoined rigid body/torsion angle dynamics is shown in Figure 9.4.

It should be emphasized that conjoined rigid body/torsion angle dynamics can readily be extended to cases where significant changes in backbone conformation are localized to specific regions of the protein, such as the binding interface. In such cases, both the interfacial side chains and the relevant portions of the protein backbone would be given torsional degrees of freedom, and the experimental data would also have to include intramolecular restraints (NOE, dipolar coupling, etc.) relating to that portion of the backbone. This, for example, is the strategy that was employed to solve the structure of the IIA^{Mtl}·HPr complex (Cornilescu et al., 2002). This was necessitated because the crystal structure of IIA^{Mtl} (van Montfort et al., 1998), which contains multiple copies of IIA^{Mtl} in the unit cell, revealed alternate conformations for four loops in relatively close proximity to the putative interaction surface with HPr.

9.5   DOCKING BASED ON ^{15}N/^{1}H$_N$ CHEMICAL SHIFT PERTURBATION AND N–H DIPOLAR COUPLINGS

Providing the complex under study can be aligned in a suitable liquid crystalline medium, the measurement of dipolar couplings is straightforward and permits one to determine the relative orientation of two proteins in a complex. Dipolar couplings, however, do not yield any translational information, which is essential for docking. Clearly, NOE-derived intermolecular interproton distance restraints provide the most useful and reliable source of translational information. However, intermolecular NOEs are not always easy to observe and their unambiguous assignment is still difficult and time consuming, particularly for larger complexes. Backbone ^{1}H$_N$ and ^{15}N chemical shifts, on the other hand, are highly sensitive to environment and have been used extensively to rapidly map interaction surfaces on proteins (Walters et al., 2001). Not surprisingly, examination of the NMR literature reveals hundreds of examples of chemical shift mapping studies; to date,

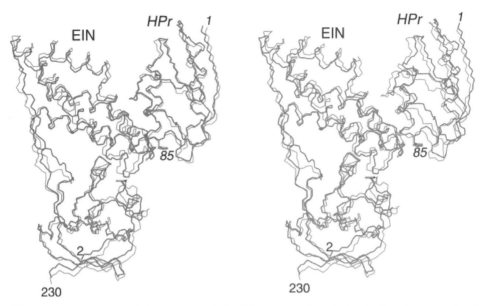

Figure 9.4. *Comparison of the structure of the EIN·HPr complex obtained using the conventional full structure determination approach (red) with that obtained by conjoined rigid body/torsion angle dynamics on the basis of 231 backbone N–H dipolar coupling data and either a full complement (blue) or partial complement (green) of NOE-derived intermolecular interproton distance restraints. The full complement of intermolecular NOEs comprises 109 interproton distance restraints; the partial complement consists of only eight intermolecular methyl proton–NH interproton distance restraints. The relative orientation of the proteins in all three calculated structures is identical. The backbone rms difference between the conventional NMR structure (red) and the structure calculated by docking the X-ray coordinates of the free proteins using the full complement of intermolecular NOEs (blue) only reflects the differences in the NMR and X-ray coordinates of the individual proteins, and these differences are within the uncertainty of the NMR coordinates. Adapted from Clore (2000). (See color insert.)*

however, only a handful of structures of macromolecular complexes have been determined by NMR.

Recently, we have shown that it is possible to convert chemical shift perturbation maps into highly ambiguous intermolecular distance restraints, which, in combination with orientational restraints from dipolar couplings, can reliably and accurately dock the partner proteins in a complex by means of rigid body/torsion angle dynamics calculations (Clore and Schwieters, 2003). Clearly, this methodology provides a powerful tool for high-throughput structural proteomics and, moreover, can greatly accelerate the determination of higher accuracy NMR structures of complexes (including the detailed placement of interfacial side chains) by providing a good starting point for the assignment of intermolecular NOE data.

The chemical shift maps are represented by a set of "r^{-6}-summed" distance restraints as follows (Clore and Schwieters, 2003). Given N_a residues on protein A and N_b residues on protein B that have been localized to the protein–protein interface by chemical shift mapping, a set of $N_a + N_b$ ambiguous distance restraints (d_{aB} and d_{bA}) is derived between all hydrogen, nitrogen, and oxygen atoms (i) of each

Figure 9.5. *Results of docking calculations for the EIN·HPr (left), IIA^{Glc}·HPr (middle) and IIA^{Mtl}·HPr (right) complexes on the basis of highly ambiguous distance restraints derived from ^{15}N/^{1}H$_N$ chemical shift perturbation maps and backbone N–H dipolar couplings. (A) Interfacial residues (blue/cyan for HPr, red/orange for the three enzymes, and purple for active site histidines) identified by ^{1}H$_N$/^{15}N chemical shift perturbation are displayed on a molecular surface representation of the proteins. (The blue and red colored interfacial residues indicate residues with an accessible surface area (ASA) in the free proteins $\geq$50% of that in an extended Gly-X-Gly peptide; the cyan and orange colored residues indicate interfacial residues in the free proteins with 5% $\leq$ ASA < 50%. (B) Plots of the dipolar coupling R-factor (R$_{dip}$) versus accuracy for the converged structures characterized by no violations >0.5 Å in the highly ambiguous intermolecular distance and R$_{dip}$ $\leq$ R$_{dip}$^{median}. In the case of the EIN·HPr (left panel) and IIA^{Mtl}·HPr complexes (right panel), the circles and diamonds indicate structures in the lower and higher energy populations, respectively, of the radius of gyration energy function (E$_{rgyr}$) distribution. (C) Histograms of the E$_{rgyr}$ distributions for the converged structures. The E$_{rgyr}$ distribution is unimodal for the IIA^{Glc}·HPr complex (middle), but bimodal for the EIN·HPr (left) and IIA^{Mtl}·HPr (right) complexes. For the bimodal distributions, the lower and higher energy E$_{rgyr}$ populations are colored red and blue, respectively. Note that in the case of the IIA^{Mtl}·HPr complex, all the structures in lower energy E$_{rgyr}$ population reside in the correct cluster 1 ensemble; all the structures in the incorrect cluster 2 ensemble reside in the higher energy E$_{rgyr}$ population. (D) Backbone (depicted as tubes) best-fit superpositions of the average coordinates (red) of the converged structures on the previously determined NMR structures (blue) solved on the basis of intermolecular NOEs and residual dipolar couplings. In the case of the IIA^{Mtl}·HPr complex, the mean coordinates are derived from the cluster 1 ensemble. The ensemble distributions of the docked structures are depicted by isosurfaces of the reweighted atomic density maps. Reproduced from Clore and Schwieters (2003). (See color insert.)*

---

residue a on protein A and all hydrogen, nitrogen, and oxygen atoms (j) of all residues b on protein B, and vice versa:

$$d_{aB} = \left(\sum_b \sum_{ij} r_{ai,bj}^{-6} \right)^{-1/6} \text{ and } d_{bA} = \left(\sum_a \sum_{ij} r_{ai,bj}^{-6} \right)^{-1/6}$$

(9.1)

where $r_{ai,bj}$ is the distance between atom i of residue a of protein A and atom j of residue b of protein B. The number of atoms per residue ranges from 5 for Gly to 18 for Arg. Each d_{aB} restraint therefore comprises a set of $r_{ai,bj}$ distances involving 5–18 atoms of residue a, depending on the nature of residue a, and anywhere between 50 and 250 atoms from protein B, depending on the number and type of selected interfacial residues b on protein B. The number of $r_{ai,bj}$ distances encompassed in a single ambiguous distance restraint may range anywhere from 400 to 3000. Each d_{aB} and d_{bA} ambiguous distance restraint is given an upper bound of 5 Å. This does not imply that any individual $r_{ai,bj}$ distance is 5 Å or less since d_{aB} is always smaller than the shortest $r_{ai,bj}$ distance. Moreover, a cutoff of 5 Å is actually quite generous owing to the nature of the ambiguous distance restraints. Thus, for example, if a given d_{aB} ambiguous distance restraint is made up of 20 individual $r_{ai,bj}$ distances, each 10 Å in length, the value of d_{aB} is 6 Å.

The results of such calculations are shown in Figure 9.5 for three complexes: EIN·HPr, IIA^{Glc}·HPr, and IIA^{Mtl}·HPr (Clore and Schwieters, 2003). In the case of the first two complexes, a unique orientation is obtained that is within 1–2 Å of the structure derived from intermolecular NOE data and dipolar couplings. For the IIA^{Mtl}·HPr complex, however, two alternative solutions are obtained, the first is ~1 Å from the correct solution and the second ~11 Å away.

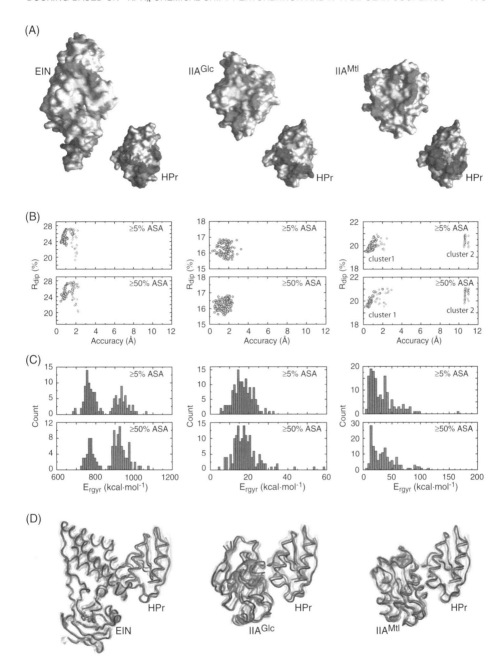

For an asymmetric alignment tensor, N–H dipolar couplings measured in a single alignment medium are consistent with four possible relative protein–protein orientations, two of which differ by a 180° rotation about the z axis of the alignment tensor, and the other two by a 180° rotation about the y axis of the alignment tensor. In most instances, exemplified by the EIN·HPr and IIA^{Glc}·HPr complexes, the ambiguous intermolecular distance restraints derived from $^1H_N/^{15}N$ chemical shift mapping resolve the fourfold degeneracy such that only a single orientation is consistent with both the ambiguous intermolecular distance restraints and the $^1D_{NH}$ dipolar couplings (Clore and Schwieters, 2003).

In unfavorable cases, such as the $IIA^{Mtl}\cdot HPr$ complex, the ambiguous intermolecular distance restraints only reduce the number of solutions to two. The twofold reduction in degeneracy is achieved because the ambiguous intermolecular distance restraints ensure that the two binding surfaces are apposed and interpenetration of the two molecules is prohibited by the van der Waals repulsion term. In the case of the $IIA^{Mtl}\cdot HPr$ complex, the persistence of twofold degeneracy arises from the fact that the x and y axes of the alignment tensor lie in the plane of the protein–protein interface, such that a 180° rotation about the z axis can occur without interpenetration of the two molecules (Figure 9.2). Fortunately, it is usually easy to distinguish the correct solution from the incorrect one using a variety of independent approaches (Clore and Schwieters, 2003). The simplest involves reexamination of the $^{15}N/^1H_N$ chemical shift perturbation maps in the light of the structures to assess whether these maps allow one to distinguish between the two alternate solution. In this instance, the incorrect solution predicts several interfacial residues that do not exhibit any $^{15}N/^1H_N$ perturbations and are therefore unlikely to be part of the binding site. Consistency with prior biochemical data can also be employed. Phosphoryl transfer occurs from His15 of HPr to His65 of IIA^{Mtl}, which places an upper limit of ~14 Å on the Cα–Cα separation between these two histidine residues. In the correct solution, this Cα–Cα distance is ~12 Å, whereas in the incorrect one it is ~17 Å. Experimentally, the twofold degeneracy can be resolved by measuring a second set of dipolar couplings in a liquid crystalline medium possessing a different alignment tensor (e.g., charged versus uncharged media). In addition, an empirical method, based on the radius of gyration, for assessing the overall packing density and surface complementarity can also be employed.

9.6 STRUCTURAL PROTEOMICS OF THE GLUCOSE ARM OF THE BACTERIAL PHOSPHOTRANSFERASE SYSTEM

In bacteria, carbohydrate transport across the membrane is mediated by the phosphoenolpyruvate:sugar phosphotransferase system (PTS), which provides tight coupling of translocation and phosphorylation (Kundig et al., 1964). The PTS is a classical example of a signal transduction pathway involving phosphoryl transfer whereby a phosphoryl group originating on phosphoenolpyruvate is transferred to the translocated carbohydrate via a series of three bimolecular protein–protein complexes (Postma et al., 1996). The first two steps of the PTS are common to all sugars: enzyme I (EI) is autophosphorylated by phosphoenolpyruvate and subsequently donates the phosphoryl group to the histidine phosphocarrier protein HPr. The proteins downstream from HPr are sugar specific, comprising four distinct families of IIA permeases. In the case of the glucose branch of the PTS, the phosphoryl group is transferred from HPr to IIA^{Glc} and thence from IIA^{Glc} to the C-terminal cytoplasmic domain (IIB^{Glc}) of the glucose transporter $IICB^{Glc}$. The complexes in this pathway are rather weak with K_D values in the 10 μM range. Although high-resolution crystal structures have been determined for three of the four proteins, namely, EIN (Liao et al., 1996), HPr (Jia et al., 1993), and IIA^{Glc} (Worthylake et al., 1991), crystallization of these protein–protein complexes has proved to be refractory, despite many years of trying. Thus, this system provides a showcase for the impact of NMR in structural proteomics.

Figure 9.6 shows the cascade of three protein–protein complexes, EIN·HPr (Garrett et al., 199), IIA^{Glc}·HPr (Wang et al., 2000), and IIA^{Glc}·IICB^{Glc} (Cai et al., 2003), involved in phosphoryl transfer in the glucose-specific arm of the PTS. These complexes shed light on understanding fundamental aspects of protein–protein recognition, mechanisms for phosphoryl transfer between proteins, and the diversity of structural elements recognized by a single protein. Specificity of the protein–protein interaction surfaces is characterized by geometric and chemical complementarity, coupled with extensive redundancy to permit the effective recognition of multiple partners. There is little or no conformational change in the protein backbone before and after association. Some interfacial side chains, however, adopt different conformations (side chain conformational plasticity) depending on the interacting partner so as to achieve optimal intermolecular interactions. A consequence of these properties is increased velocity in signal transduction by eliminating any unnecessary time delay required for significant conformational change.

The interaction surfaces for HPr on EIN and IIA^{Glc} are very similar despite the fact that their underlying structures are completely different in terms of linear sequence, secondary structure (helices for EI, β-strands for IIA^{Glc}), and topological arrangement of structural elements. HPr makes use of essentially the same surface to interact with both its upstream partner EI and its downstream partner IIA^{Glc}. Concomitantly, the binding sites for IIB^{Glc} and HPr on IIA^{Glc} overlap extensively (~85% of the binding site for IIB^{Glc} constitutes part of the binding site for HPr). One might therefore anticipate that IIB^{Glc} could also interact with EIN. However, NMR data indicate that there is absolutely no interaction between EIN and IIB^{Glc} at millimolar concentrations. From a functional perspective this is important since it ensures that the PTS cascade is not bypassed. In addition, prevention of the potential shortcut between enzyme I and IICB^{Glc} for glucose phosphorylation is necessary since these proteins also regulate the functions of proteins in other pathways (Postma et al., 1996). The structural basis for specificity and discrimination lies in the different charge distributions on the interaction surfaces of HPr and IIB^{Glc} such that binding of IIB^{Glc} to EIN is precluded by electrostatic repulsion (Cai et al., 2003).

9.7 CONCLUSION

NMR is the only solution technique capable of providing high-resolution structural information on protein–protein complexes at atomic resolution. While NMR is not a high-throughput technique, recent advances have considerably enhanced the speed of NMR structure determinations and the size and complexity of protein–protein complexes that can be studied. Indeed, NMR spectroscopy combined with prior knowledge on the structures of individual proteins from high-resolution X-ray crystallography promises to provide a very powerful approach for the efficient determination of three-dimensional structures of protein–

(A) Glucose PTS

(B) EIN·HPr complex

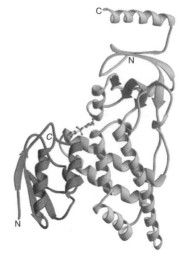

(C) IIA^Glc·HPr complex

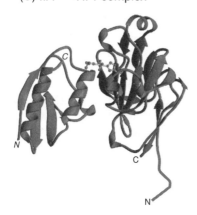

(D) IIA^Glc·IICB^Glc complex

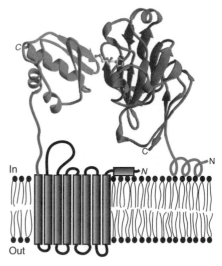

Figure 9.6. Summary of the glucose arm of the E. coli PTS. (A) Diagrammatic illustration of the PTS cascade illustrating the transfer of phosphorus originating from phosphoenolpyruvate and ending up on glucose through a series of bimolecular protein–protein complexes between phosphoryl donor and acceptor molecules. Ribbon diagrams of the (B) first (EIN·HPr), (C) second (HPr·IIA^Glc), and (D) third (IIA^Glc·IICB^Glc) complexes of the glucose PTS. The N-terminal domain of EI (EIN) is shown in gold, HPr in red, IIA^Glc in blue, and the IIB^Glc domain of IICB^Glc in green. Also shown in yellow are the active site histidine residues of EIN (His189), HPr (His15) and IIA^Glc (His90) and the active site cysteine (Cys35) of IIB^Glc, together with the pentacoordinate phosphoryl group (red atoms) in the transition states of the complexes. IIB^Glc constitutes the C-terminal cytoplasmic domain of IICB^Glc. The transmembrane IIC^Glc domain of IICB^Glc is thought to comprise eight transmembrane helices (shown diagrammatically in black). Note that the N-terminal end of IIA^Glc (residues 1–18) is disordered in free solution (C), but upon interaction with a lipid bilayer, residues 2–10 adopt a helical conformation (D), thereby further stabilizing the IIA^Glc·IIB^Glc complex, by partially anchoring IIA^Glc to the lipid membrane. Reproduced from Cai et al. (2003). (See color insert.)

protein complexes up to ~100 kDa, particularly in cases that are refractory to cocrystallization.

Focus in this chapter was specifically on three-dimensional structure determination of protein–protein complexes. However, NMR also has a role in rapid screening and providing absolute proof for the formation of protein–protein complexes. While this does require substantially more material than mass spectrometry, typically a minimum of about 50 μM solution in a volume of 250 μL, the detection of protein–protein complex formation is straightforward and extremely rapid and can be accomplished by comparing the ^{1}H–^{15}N correlation spectrum of one of the partners, uniformly labeled with ^{15}N, in the absence and presence of the unlabeled second partner. Chemical shift perturbations and line broadening provide unambiguous proof of complex formation.

REFERENCES

Battiste JL, Wagner G. 2000. Utilization of site-directed spin labeling and high resolution heteronuclear magnetic resonance for global fold determination of large proteins with limited nuclear Overhauser effect data. *Biochemistry* 39:5355–5365.

Bax A, Grzsiek S. 1993. Methodological advances in protein NMR. *Acc Chem Res* 26: 131–138.

Bax A, Kontaxis G, Tjandra N. 2001. Dipolar couplings in macromolecular structure determination. *Methods Enzymol* 339:127–174.

Bennett KL, Kussmann M, Bjork P, Godzwon M, Mikkelsen M, Sorensen P. 2000. Chemical cross-linking with thiol-cleavable reagents combined with differential mass spectrometric peptide mapping: a novel approach to assess intermolecular protein contacts. *Protein Sci* 9:1503–1518.

Cai M, Williams DC, Wang G, Lee BR, Peterkofsky A, Clore GM. 2003. Solution structure of the phosphoryl transfer complex between the signal-transducing protein IIA^{Glucose} and the cytoplasmic domain of the glucose transporter IICB^{Glucose} of the *Escherichia coli* glucose phosphotransferase system. *J Biol Chem* 278:25191–25206.

Cavanagh J, Fairbrother WJ, Palmer AG, Skelton NJ. 1996. *Protein NMR Spectroscopy: Principles and Practice*. Academic Press, New York.

Clore GM. 2000. Accurate and rapid docking of protein–protein complexes on the basis of intermolecular nuclear Overhauser enhancement data and dipolar couplings by rigid body minimization. *Proc Natl Acad Sci USA* 97:9021–9025.

Clore GM, Gronenborn AM. 1989. Determination of three-dimensional structures of proteins and nucleic acids in solution by nuclear magnetic resonance spectroscopy *CRC Crit Rev Biochem Mol Biol* 24:479–564.

Clore GM, Gronenborn AM. 1991. Structures of larger proteins in solution: three- and four-dimensional heteronuclear NMR spectroscopy. *Science* 252:1390–1399.

Clore GM, Gronenborn AM. 1998. Determining structures of larger proteins and protein complexes by NMR. *Trends Biotechnol* 16:22–34.

Clore GM, Bewley CA. 2002. Using conjoined rigid body/torsion angle simulated annealing to determine the relative orientation of covalently linked protein domains from dipolar couplings. *J Magn Reson* 143:329–335.

Clore GM, Schwieters CD. 2003. Docking of protein–protein complexes on the basis of highly ambiguous distance restraints derived from ^{1}H$_N$/^{15}N chemical shift mapping and backbone ^{15}N-^{1}H residual dipolar couplings using conjoined rigid body/torsion angle dynamics. *J Am Chem Soc* 125:2902–2912.

Clore GM, Starich MR, Gronenborn AM. 1998. Measurement of residual dipolar couplings of macromolecules aligned in the nematic phase of a colloidal suspension of rod-shaped viruses. *J Am Chem Soc* 120:10571–10572.

Cornilesci G, Marquardt JL, Ottiger M, Bax A. 1998. Validation of protein structure from anisotropic carbonyl chemical shifts in a dilute liquid crystalline phase. *J Am Chem Soc* 120:6836–5837.

Cornilescu G, Lee BR, Cornilescu CC, Wang G, Peterkofsky A, Clore GM. 2002. Solution structure of the phosphoryl transfer complex between the cytoplasmic A domain of the mannitol transporter II^{Mannitol} and HPr of the *Escherichia coli* phosphotransferase system. *J Biol Chem* 277:42289–42298.

Dvoretsky A, Gaponenko K, Rosevera PR. 2002. Derivation of structural restraints using a thiol-reactive chelator. *FEBS Lett* 528:189–192.

Garrett DS, Seok UJ, Peterkofsky A, Gronenborn AM, Clore GM. 1999. Solution structure of the 40,000 M_r phosphoryl transfer complex between the N-terminal domain of enzyme I and HPr. *Nature Stuct Biol* 6:166–173.

Gerstein M, Edwards A, Arrowsmith CH, Montelione GT. 2003. Structural genomics: current progress *Science* 299:1663.

Hansen MR, Hanson P, Pardi A. 2000. Filamentous bacteriophage for aligning RNA, DNA, and proteins for measurement of nuclear magnetic resonance dipolar coupling interactions. *Methods Enzymol* 317:220–240.

Iwahara J, Anderson DE, Murphy EC, Clore GM. 2003. EDTA-derivatized deoxythymidine as a tool for rapid determination of protein binding polarity to DNA by intermolecular paramagnetic relaxation enhancement. *J Am Chem Soc* 125:6634–6635.

Jia Z, Quail JW, Waygood EB, Delbaere LT. 1993. The 2.0 Å resolution structure of the *Escherichia coli* histidine-containing phosphocarrier protein HPr: a redetermination. *J Biol Chem* 268:22940–22501.

Kundig W, Ghosh S, Roseman S. 1964. Phosphate bound to histidine in a protein as an intermediate in a novel phosphotransferase system. *Proc Natl Acad Sci USA* 52: 1067–1074.

Liao DI, Silverton E, Seok YJ, Lee BR, Peterkofsky A, Davies DR. 1996. The first step in sugar transport: crystal structure of the amino terminal domain of enzyme I of the *E. coli* PEP:sugar phosphotransferase system and a model of the phosphotransfer complex with HPr. *Structure* 4:861–872.

Mal K, Ikura M, Kay LE. 2002. The ATCUN domain as a probe of intermolecular interactions: application to calmodulin–peptide complexes. *J Am Chem Soc* 124:14002–14003.

Montelione GT. 2001. Structural genomics: an approach to the protein folding problem. *Proc Natl Acad Sci USA* 98:13488–13489.

Nakanishi T, Miyazawa M, Sakakura M, Terasawa H, Takahashi H, Shimada I. 2002. Determination of the interface of a large protein complex by transferred cross-relaxation measurements. *J Mol Biol* 318:245–249.

Postma PW, Lengeler JW, Jacobson GR. 1996. Phospho*enol*pyruvate:carbohydrate phosphotransferase systems. In: Neidhardt FC (Ed), *Escherichia coli and Salmonella: Cellular and Molecular Biology*, ASM Press, Washington DC, pp 1149–1174.

Prestegard JH, al-Hashimi HM, Tolman JR. 2000. NMR structures of biomolecules using field-oriented media and residual dipolar couplings. *Q Rev Biophys* 33:371–424.

Rückert M, Otting G. 2000. Alignment of biological macromolecules in novel nonionic liquid crystalline media for NMR experiments. *J Am Chem Soc* 122:7793–7797.

Schulz DM, Ihling C, Clore GM, Sinz A. 2003. Mapping the topology and determination of a low-resolution three-dimensional structure of the calmodulin·melittin complex by chemical cross-linking and high resolution FTICR mass spectrometry: direct demonstration of multiple binding modes. *Biochemistry* 43:4703–4715.

Schwieters CD, Clore GM. 2001. Internal coordinates for molecular dynamics and minimization in structure determination and refinement. *J Magn Reson* 152:288–302.

Sinz A, Wang K. 2001. Mapping protein interfaces with a fluorogenic cross-linker and mass spectrometry: application to nebulin–calmodulin complexes. *Biochemistry* 40:8903–7913.

Takahashi H, Nakanishi T, Kami K, Arata Y, Shimada Y. 2000. A novel NMR method for determining the interfaces of large protein–protein complexes. *Nature Struct Biol* 7:220–223.

Tjandra N, Bax A. 1997. Direct measurement of distances and angles in biomolecules by NMR in a liquid crystalline medium. *Science* 278:1111–1114.

Tjandra N, Garrett DS, Gronenborn AM, Bax A, Clore GM. 1997a. Defining long range order in NMR structure determination from the dependence of heteronuclear relaxation times on rotational diffusion anisotropy. *Nature Struct Biol* 4:443–449.

Tjandra N, Omichinski J, Gronenborn AM, Clore GM, Bax A 1997b. Use of dipolar ^{15}N–^{1}H and ^{13}C–^{1}H couplings in the structure determination of magnetically oriented macromolecules in solution. *Nature Struct Biol* 4:732–738.

Van de Ven FJM. 1995. *Multidimensional NMR in Liquids: Basic Principles and Experimental Methods*. VCH Publishers, New York.

Van Montfort RL, Pijning T, Kalk KH, Hangyi I, Kouwijzer ML, Robillard GT, Dijkstra BW. 1998. The structure of the *Escherichia coli* phosphotransferase IIA^{Mannitol} reveals two conformations of the active site. *Structure* 6:377–388.

Voss J, Hubbell WL, Kaback HR. 1995. Distance determination in proteins using designed metal ion binding sites and site-directed labeling: application to the lactose permease of *Escherichia coli*. *Proc Natl Acad Sci USA* 92:12300–12303.

Walters KJ, Ferentz AE, Hare NJ, Hidalgo P, Jasanoff A, Matsuo H, Wagner G. 2001. Characterizing protein–protein complexes and oligomers by nuclear magnetic resonance spectroscopy. *Methods Enzymol* 339:238–258.

Wang G, Louis JM, Sondej M, Seok YJ, Peterkofski A, Clore GM. 2000. Solution structure of the phosphoryl transfer complex between the signal transducing proteins HPr and IIA^{Glucose} of the *Escherichia coli* phosphoenolpyruvate:sugar phosphotransferase system. *EMBO J* 19:5635–5649.

Worthylake D, Meadow ND, Roseman S, Liao DI, Herzberg O, Remington SJ. 1991. Three-dimensional structure of the *Escherichia coli* phosphocarrier protein III^{Glc}. *Proc Natl Acad Sci USA* 88:10382–10386.

Wu Z, Tjandra N, Bax A. 2001. ^{31}P chemical shift anisotropy as an aid in determining nucleic acid structure in liquid crystals. *J Am Chem Soc* 123:3617–3618.

Wüthrich K. 1986. *NMR of Proteins and Nucleic Acids*. Wiley, Hoboken, NJ.

Yee A, Pardee K, Christendat D, Savchenko A, Edwards AM, Arrowsmith CH. 2003. Structural proteomics: toward high-throughput structural biology as a tool in functional genomics. *Acc Chem Res* 36:183–189.

Zuiderweg ER. 2002. Mapping protein–protein interactions in solution by NMR spectroscopy. *Biochemistry* 41:1–7.

Part III

Novel Approaches in Proteomics

10

Protein Microarrays

Cassio Da Silva Baptista and David J. Munroe
*SAIC—Frederick, Inc., National Cancer Institute at Frederick, Research Technology
Program, Laboratory of Molecular Technology, Frederick, Maryland*

10.1 INTRODUCTION

The genomics revolution and the success of whole genome sequencing and genetic mapping projects have transformed biology and medicine in ways that were inconceivable one to two decades ago. The rapid development of genomic databases, bioinformatics tools, and enabling technologies such as cDNA and oligonucleotide microarrays have provided new insights and understanding into biological and disease processes through the global analysis of gene expression patterns. Although gene expression profiling at the mRNA level has proved to be a powerful and useful tool, this approach suffers from inherent limitations: (i) mRNA abundance does not typically correlate well with protein abundance (Templin et al., 2002), presumably due to variations in mRNA stability, translatability, and protein stability; (ii) protein structure, activity, and function can be altered and regulated by subcellular localization or post-translational modifications, such as phosphorylation or acylation, that cannot be predicted by mRNA abundance; and (iii) many biologically important samples, such as serum and urine, do not typically contain ample quantities of mRNA. So, although mRNA expression profiling continues to be a valuable tool, there is a growing recognition that these approaches should be complemented by profiles of the gene products or proteins themselves. The global analysis of protein expression patterns, proteomics, is a natural extension and complement to genomics and the genomics platforms that have developed over the last two decades. A major challenge during the post-genome era will be to develop protein profiling platforms and methodologies and to correlate the data sets they generate with disease and biological and environmental phenotypes.

Proteomics for Biological Discovery, edited by Timothy D. Veenstra and John R. Yates.
Copyright © 2006 John Wiley & Sons, Inc.

It has been estimated that the human genome encodes approximately 30,000 individual genes (Wang, 2004), a number that translates, after accounting for post-translational processing and modification, to as many as 100,000 to 1 million different proteins. The development of platforms for the identification and cataloging of such a large number of molecules is indeed a daunting task; albeit one that has recently attracted a lot of attention.

Currently, the most widely used approaches for proteomic analysis involve two-dimensional gel electrophoresis (2DGE) and mass spectrometry, applied either individually or in combination. Since these platforms and approaches have been reviewed and discussed in detail elsewhere in this book, this chapter focuses on protein microarrays, an emerging technological approach that complements and offers alternatives to 2DGE and mass spectrometry-based platforms.

10.2   PROTEIN MICROARRAY DEVELOPMENT

Protein microarrays typically consist of a library of target or capture reagents (peptides, proteins, tissues, affinity reagents) robotically arrayed or spotted in high density onto a solid support. Protein samples to be analyzed are labeled (usually with a fluorescent tag) and hybridized to a target or capture reagent immobilized on the array. Following a simple wash step, binding to the individual targets is measured and quantified.

Microarrays offer a number of advantages as protein profiling platforms. To begin with, microarrays allow large numbers of predetermined, specific target molecules to be assayed simultaneously with low sample consumption. Similarly, large numbers of samples can be analyzed in parallel on identical arrays at low cost. Microarrays also allow direct comparative analysis by differentially labeling two samples (dual color labeling) and cohybridizing them to the same microarray. A schematic drawing of the dual labeling approach appears in Figure 10.1.

As a direct extension from DNA microarrays, Haab et al. (2001) were the first to demonstrate the direct comparative analysis concept on a protein microarray platform. In this initial study, two complex protein samples (one re-presenting a standard for comparative quantitation and the other an experimental sample) were differentially labeled with spectrally resolvable fluorescent dyes and simultaneously hybridized to a single microarray printed with 115 monoclonal antibodies to generate a comparative protein profile. The results showed that some of the antibody/antigen pairs were successful at detecting cognate ligands at absolute concentrations below 1 ng/mL. This level of sensitivity would allow for the measurement of many clinically important proteins in actual patient blood samples.

Yet another advantage of protein microarrays is their flexibility. Advances in high-throughput peptide synthesis and the wide availability of arrayed cDNA, phage display, and single-chain antibody libraries, monoclonal antibodies, recombinant proteins, and tissue resources make possible the production of a variety of different protein microarrays with diverse content. The range of content possible with microarrays makes feasible the analysis of nucleic acid–protein, protein–protein, ligand–receptor, enzyme–substrate, antigen–antibody, or small molecule–protein interactions. This versatility and flexibility were demonstrated by MacBeath and Schreiber (2000), who showed that a single high-density microarray platform

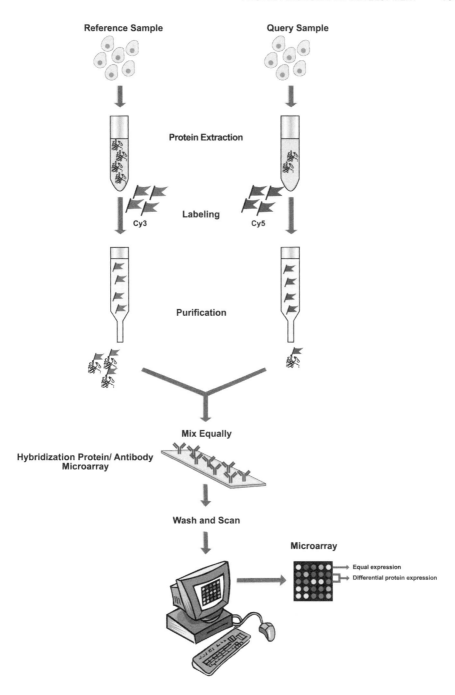

Figure 10.1. *Schematic representation of the dual color labeling approach. Proteins are extracted from reference and query samples and labeled with fluorescent dyes (typically Cy3 and Cy5). Following a "clean-up" to remove unincorporated dye, the labeled proteins are mixed and hybridized to the microarray. After washing, the microarray is scanned using a microarray scanner. Hybridization signals are extracted and differential protein levels are determined. (See color insert.)*

could be used to analyze protein–protein, protein–small molecule, and enzyme–substrate interactions.

Finally, the microarray format lends itself to increased sensitivity. As theorized by Ekins and co-workers, platforms such as microarrays, with a limited copy number of capture molecules and a low sample quantity, yield increased sensitivity and signal-to-noise, compared to systems that utilize larger amounts of material (Ekins, 1989; Ekins et al., 1990). This concept has since been directly demonstrated on many different protein microarray platforms (Haab et al., 2001; Sreekumar et al., 2001; Templin et al., 2002; Bradbury et al., 2003).

Protein microarrays have the potential to revolutionize high-throughput proteomic analysis in much the same way that DNA microarrays have revolutionized genomic analysis. However, to date, the utilization, development, and application of high-density protein, peptide, and antibody microarrays have been somewhat limited, due largely to technical challenges that are unique to protein microarrays. For example, DNA and protein are very different molecules with unique chemical and physical properties. It is these very characteristics that make protein microarrays more challenging to fabricate and apply than DNA microarrays. DNA molecules consist of four nucleotides that generate a hydrophilic and uniform structure on a negatively charged sugar backbone. Proteins, on the other hand, are composed of 20 different amino acids, the various combinations of which can result in molecules with very different physical properties (hydrophobic or hydrophilic, acidic or basic) and a wide variety of complex tertiary structures. Protein structure and function can be further modified by various post-translational modifications such as glycosylation, acetylation, deamination, or phosphorylation (Templin et al., 2002). A comparison of relevant nucleic acid and protein characteristics are summarized in Table 10.1.

Despite the technical challenges, the feasibility and utility of protein microarrays have been clearly established. Ongoing innovation and development of these platforms continue and will undoubtedly result in significant contributions to the growing field of proteomics.

TABLE 10.1. Comparison of Nucleic Acid and Protein Characteristics

Properties	Nucleic Acid	Protein
Composition	5 nucleotides	20 amino acids
Structure	Uniform double helix	Complex tertiary and quaternary structures
	Acidic	Acidic or basic
Structure stability	Stable	Marginally stable under physiological conditions
Hydrophobicity	Hydrophilic	Hydrophilic or hydrophobic
Stabilizing forces	Hydrophobic forces, base stacking, hydrogen bond, ionic interactions	Electrostatic forces, hydrophobic forces, hydrogen-bonding forces, disulfide bridges
Modification	Methylation, depurination, deamination	Glycosylation, acetylation, deamination, phosphorylation
Amplification	Polymerase chain reaction	Unavailable
Binding specificity	High	Variable
Affinity prediction	Possible	Not possible

10.3 PROTEIN MICROARRAY PLATFORMS, FORMATS, AND FABRICATION

Over the past six years, a variety of protein microarray formats and applications have been described that utilize an assortment of different spotting techniques, surface chemistries, capture/affinity reagents, and labeling/detection systems. In the following sections, these formats and the different methods used in their fabrication and operation are discussed.

10.3.1 Protein Microarray Formats

Protein microarrays can be grouped into two basic categories or application types: protein function microarrays for functional analysis and protein expression microarrays for expression profiling.

In the first application, protein arrays are used to study protein–protein interactions, protein modifications, DNA–protein interactions, RNA–protein interactions, protein–lipids interaction, small molecule–protein interactions, and post-translational modifications, that is, functional proteomics. Since these types of protein microarrays are used to demonstrate molecular interaction and/or function, they are referred to as protein function microarrays. Zhu and co-workers (2001) demonstrated the utility of this microarray format by cloning 94% of the yeast ORFs into an expression vector, purifying 5800 of the encoded proteins, and spotting them onto glass microscopic slides. These microarrays were then used to dissect protein–protein and protein–phospholipid interactions in yeast and resulted in the identification of several new calmodulin and phospholipid interacting proteins as well as common binding motif for several calmodulin binding proteins.

In the second application, protein microarrays are used for massively parallel quantitation of protein expression levels in a given sample or set of samples, such as cell extracts, tissues, or body fluid. These types of protein microarrays are referred to as protein expression arrays. They are particularly useful for protein profiling studies aimed at determining differences in protein expression patterns among defined sets of samples. These types of microarrays have been used in various applications, such as the identification of tumor-specific markers and molecular targets for therapeutic intervention in cancer (Poetz et al., 2005).

Protein expression microarrays can be further subdivided into forward-phase and reverse-phase formats, depending on whether the affinity reagent or the analyte is immobilized onto the solid surface (Liotta et al., 2003). A schematic representation of these two formats is represented in Figure 10.2. In forward-phase protein microarrays, the affinity reagent is bound to the surface of the array. In this format each feature corresponds to a unique target molecule. Each microarray is individually hybridized with a complex labeled sample or pair of samples, and a protein expression profile, defined by the library of capture reagents immobilized onto the surface of the array, is generated. This approach has been used successfully by several groups of investigators to generate differential protein profiles (Haab et al., 2001; Knezevic et al., 2001; Sreekumar et al., 2001; Anderson et al., 2003; Miller et al., 2003; Zhou et al., 2004).

In reverse-phase protein microarrays, defined sets of samples are immobilized onto the array, with each feature representing a unique sample to be analyzed. Each

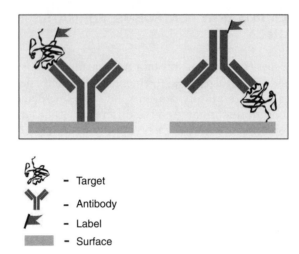

- Target
- Antibody
- Label
- Surface

Figure 10.2. *Common microarray formats. In forward-phase protein arrays, affinity reagent (antibody) is immobilized onto a solid support. In reverse-phase protein arrays, samples to be analyzed (cell extracts or tissues) are immobilized onto a solid surface. (See color insert.)*

array is hybridized with a labeled affinity reagent, and the relative expression level of an individual protein, specific to that affinity reagent, is simultaneously determined in all of the samples spotted onto the microarray (Liotta et al., 2003). The advantages of the reverse-phase approach are that complex mixtures of sample proteins need not be labeled and that only one antibody is required for each assay. The disadvantages of this format are that low-abundance proteins are often difficult to detect among the more abundant proteins and that the stability and integrity of some proteins may be adversely affected when presented on solid phase.

One of the more promising applications of the reverse microarray format is the tissue microarray. In this approach, hundreds of tissue sections can be screened for the presence or absence of specific nucleic acids or proteins. A major advantage of tissue microarrays is that histological integrity is preserved, allowing subtissue or subcellular localization, as well as relative gene expression levels, to be determined simultaneously (Jones et al., 2002; Liotta et al., 2003; Poetz et al., 2005).

10.3.2 Supports and Binding Surfaces

Protein microarrays require the immobilization of antibodies, peptides, proteins, small molecules, or cell lysates onto solid surfaces via a noncovalent absorption, covalent linkage, or affinity capture methods such as the streptavidin–biotin interaction or histidine-tag–nickel-chelate. The ideal protein microarray surface allows specific immobilization of capture reagents, while excluding or limiting the binding of nonspecific proteins, in a manner that minimizes disruption to native structure and maintains protein function.

The two most commonly used supports for protein microarrays are glass slides and membrane coated slides. While glass slides have low background fluorescence and are amendable to high-throughput robotics, they have poor protein binding capacity. To overcome this limitation, glass surfaces can be treated with a variety of different compounds, such as aminosilane, polylysine, or aldehyde surface activation. Membrane (nylon, nitrocellulose) coated slides, in contrast, have high binding capacity but are also prone to higher background and autofluorescence. Membranes commonly used in other protein analysis applications (Westerns, etc.) are commercially available adhered to glass slides, lending them amendable to high-throughput platforms. Other surfaces used in protein microarray experiments are hydrogel slides (polyacrylamide coated slides), gold surfaces, and microtiter plates (reviewed in Kusnezow and Hoheisel, 2003; Pavlickova et al., 2004).

Noncovalent protein surface interactions can be mediated via hydrophobic (nitrocellulose, polystyrene) or positively charged (polylysine, aminosilane) surfaces (Templin et al., 2002). Noncovalent binding of proteins has some advantages, such as low nonspecific binding to the glass surface, high capacity, preservation of protein, and preservation of protein structure and function. On the other hand, it presents some drawbacks, such as little control of orientation or quantitative adsorption of molecules, which can lead to reduced reproducibility and/or binding efficiency (Lee et al., 2003).

Covalent attachment of proteins onto microarray surfaces can be mediated via a range of activated surfaces, such as aldehyde, epoxy, amine, and active esters. This approach permits robust immobilization of an extensive range of proteins with excellent reproducibility. However, protein modifications resulting from binding with the activated surface may result in structural alterations or loss of protein activity (Lee et al., 2003).

Affinity capture methods such as streptavidin–biotin and histidine-tag–nickel-chelate interactions have also been extensively used for immobilization. The major problem with these methods is that biotinylation of some molecules, such as antibodies, can alter their binding specificities. On the other hand, these methods allow a stable linkage between molecules that immobilizes the capture reagent in a specific, predetermined orientation (Lee et al., 2003).

An important issue to consider regarding binding surfaces is that random orientation of immobilized proteins onto solid surfaces can impact protein activity, access to active sites, or conformation. Appropriate care with respect to these issues should be exercised to ensure that the capture reagents maintain their desired properties. Finally, immobilization of capture reagents for high-throughput applications should be a simple process. In 2003, after surveying eight different microarray surfaces, Angenendt and co-workers (2003) proposed optimal slide coatings for antibodies and protein microarrays. They concluded that, for quantitative measurements of proteins in complex biological samples, dendrimer slides that consist of a dendrimer layer with reactive epoxy groups provide an excellent signal-to-noise ratio; for quantitative measurement of antibodies, PEG slides (a PEG layer with reactive epoxy group) are the preferred surface coating; for experiments that necessitate the detection of very low-abundance proteins and antibodies, both dendrimer and amine slides are most appropriate. They also verified reports that FAST slides (a commercial nitrocellulose-based matrix) have a high signal-to-noise ratio when applied to the production of antibody microarrays (Angenendt et al., 2003).

An alternative to planar microarrays (described previously) are microsphere (bead)-based microarrays that allow multiplex analysis of a defined number of analytes. In microsphere-based systems, capture molecules are immobilized onto different fluorescent-tagged or size encoded beads. Flow cytometry systems are employed for the simultaneous discrimination of bead types. The main drawback to these platforms is that the multiplexing is limited to the number of distinguishable beads available, in contrast to planar microarrays that can analyze tens or hundreds of thousands of features (Templin et al., 2003; Poetz et al., 2005).

10.3.3 Robotic Spotting

Since the ultimate goal of protein arrays is to interrogate hundreds or thousands of protein characteristics in a single study, high-throughput production is critical. A particularly critical factor in this field is the use of precision, high-speed robot spotters for the reproducible and high-throughput deposition of affinity reagents onto the solid surface of choice.

Microarray targets can be generated using contact or noncontact printing arrayers. Contact pins commonly used for the printing of both nucleic acid and protein microarrays are shown in Figure 10.3. Contact printing arrayers are pin tool-based arrayers that deposit subnanoliter volumes of capture reagent directly onto the surface by direct contact. The amount of proteins deposited can be controlled by the size of the pin. Noncontact printing technologies, on the other hand, use commercially available ink jet printers or peizoelectronic robotic systems to deposit nanoliter/picoliter droplets of protein solutions onto the surface (Roda et al., 2000). A drawback to ink jet systems is that it can be difficult to reproducibly deposit a precise amount of a variety of different proteins. Peizoelectronic robotic systems, on the other hand, are becoming increasingly popular, mainly because of their precision deposition characteristics (Templin et al., 2002; Huang, 2003).

10.3.4 Capture Reagents

The production of capture reagents is perhaps the biggest challenge and arguably the most important step in protein microarray fabrication. Ideally, a microarray for

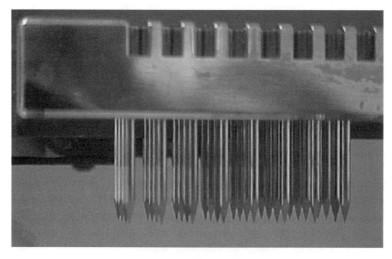

Figure 10.3. *Microarray spotting pins.*

proteomic analysis would consist of a large numbers of high-affinity and high-specificity capture reagents—at least one for each protein of interest. Several methodologies have been developed to generate large numbers of specific capture reagents (see below). In addition to high specificity and high affinity to the target molecule, the ideal capture reagent should be easy to generate and handle and have the capability to recognize a diverse set of target molecules. Presently, no capture reagent meets all of these requirements (Pavlickova et al., 2004). The most frequent capture reagents for protein microarrays are antibodies. Recombinant proteins, peptides, aptamers, and small molecules are other capture reagents utilized in protein arrays. Another protein microarray system, known as the autoantigen protein array or antigen array, enlists autoantigen capture reagents to detect auto-antibodies (Robinson et al., 2002).

One of the most common capture reagents is the antibody because of its potential high affinity and target specificity. Antibodies can be classified into three different groups based on their method of production: monoclonal antibodies, polyclonal antibodies, and phage display. While monoclonal antibodies are theoretically in unlimited supply and can often be selected for high specificity, the production of monoclonal antibodies is time consuming, labor intensive, and expensive. On the other hand, polyclonal antibodies are of relatively high affinity and are easy to produce, but are of limited supply and can be less specific than mono-clonal antibodies.

Phage display is a robust method for generating recombinant antibodies or antibody fragments (scFvs or Fabs). Recombinant antibodies and antibody fragments can also be produced using the related ribosome and yeast display methodologies (Nielsen and Geierstanger, 2004). All of these methods involve the creation of large libraries of reagents with variable binding activity, which are then selected by multiple rounds of affinity purifications. Extensive phage and yeast display libraries have been generated from naive V-gene sources, and these have the potential to provide high-affinity antibodies (Nielsen and Geierstanger, 2004). Ribosome display, on the other hand, has an advantage over the phage display in that all steps, from amplification to maturation of antibody, occur in vitro—thus negating the requirement for a bacterial or yeast host cell system. On the negative side, issues associated with the stability of the mRNA/translating polypeptide complex and the lack of a demonstrated ability to generate large numbers of different high-affinity antibodies have generated uncertainties about the application of ribosome display to the construction of diverse antibody libraries (Pavlickova et al., 2004). One general advantage of recombinant antibody technologies is that the antibody affinities may be improved by affinity maturation (Nielsen and Geierstanger, 2004). However, all of these techniques, as noted earlier, are labor intensive and costly. Recombinant antibody production has been reviewed by Pavlickova et al. (2004) and Bradbury et al. (2003).

Although antibodies are the most popular class of capture reagents, they are not the only capture agents that have been used to construct protein microarrays. Aptamers, or oligonucleotides selected for binding to proteins with high specificity and affinity, can also be used. The method used to isolate these oligonucleotides is called SELEX (systematic evolution of ligands by exponential enrichment) (Huang, 2003). Among the advantages of aptamers are their low cost, stability, and that a nonspecific protein dye can be used to label/detect the bound protein posthybridization.

Purified or recombinant proteins are another class of capture reagent. Recombinant proteins can be purified from bacterial or in vitro expression systems using fusion tags. Glutathione S-transferase (GST) and hexahistidine are the most common tags applied for fusion protein systems. Unfortunately, proteins expressed in *Escherichia coli* do not undergo post-translational modifications, which can be vital for the desired activity of the protein. Eukaryotic expression systems do exist (Predki, 2004); however, high-throughput production of recombinant protein in mammalian, insect, and yeast culture is generally expensive and does not rival the throughput of expression in *E. coli*.

Peptides are good candidates for capture reagents in that they have the advantage of lower cost and ease of preparation in addition to smaller size and increased stability relative to full-length proteins. Diverse peptide libraries can be generated either biologically or synthetically. Complex biologically generated peptide libraries can be constructed using cDNA libraries encoding random peptides that can be expressed in prokaryotic or mammalian cell culture systems and selected for binding to a specific ligand. Mammalian cell-based systems, although costly and time consuming, offer the advantage of post-translational modifications that may play an essential role in the function of the protein. Synthetic production of peptide libraries can be achieved by photolithographic synthesis or SPOT synthesis (Frank and Doring, 1988). Microarrays of peptides are a potentially valuable tool for the global analysis of cellular activities and/or for the identification of protein epitopes that define the active site (Min and Mrksich, 2004).

Yet another type of capture reagent is the small-molecule ligand. Like peptides, small-molecule ligands can be synthesized chemically and are relatively inexpensive; however, they often suffer from low affinity. Nord et al. (2000) have used small-molecule ligands as capture reagents for recombinant proteins from crude cell lysates, illustrating the application of this microarray format for the identification of small-molecule targets or interactors.

An alternative microarray fabrication scheme, known as self-assembling protein microarrays, has recently been described and demonstrated by Ramachandran et al. (2004). This approach involves the generation of recombinant protein capture reagents on the surface of the microarrays themselves. This system functions by printing cDNA expression constructs onto glass slides followed by expression of the target proteins via mammalian reticulocyte transcription/translation lysates. Epitope tags fused to the proteins allow them to be immobilized in situ. This system circumvents the need to synthesize and purify proteins and also avoids problems associated with long-term protein stability during storage. As a demonstration, the authors used this platform to map pairwise interactions among 29 proteins involved in human DNA replication initiation and recapitulated the regulation of Cdt1 binding to select replication proteins as well as mapping its DNA-binding domain.

10.3.5 Protein Microarray Labeling Systems

The majority of detection systems utilized in antibody microarray platforms are variations of sandwich assays or direct labeling. A schematic representation of the various antibody microarray detection formats is depicted in Figure 10.4.

The sandwich assay approach is simply a multiplexed version of detection systems commonly used for standard ELISA immunoassays and employs a matched pair

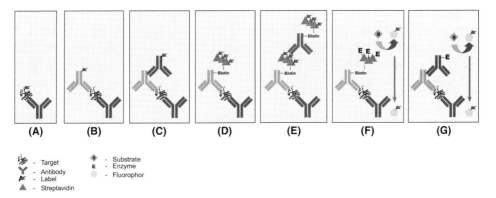

Figure 10.4. *Signal generation schemes used in antibody microarray applications. (A) Direct labeling, (B) and (C) indirect labeling—sandwich assays, (D) and (E) indirect labeling—biotin mediated; (F) and (G) Indirect labeling—enzyme mediated. (See color insert.)*

of antibodies specific for each protein target. Basically, these approaches rely on an immobilized antibody for capturing the protein of interest, with a second or a third labeled antibody used for detection. The sandwich assay is the most accurate method of detection because each analyte must be recognized by at least two antibodies to provide a detectable signal. A major drawback of this approach, however, is that two different epitope-specific antibodies are required for each analyte (Haab, 2003; Nielsen and Geierstanger, 2004). This type of detection system has been used for the parallel measurements of low-abundance cytokines in culture supernatants and body fluids (Huang et al., 2001; Lin et al., 2002).

Direct labeling methods offer a convenient alternative to sandwich assays for protein profiling of complex mixtures. Direct labeling methods rely on the direct labeling of all proteins in a complex mixture, followed by incubation of the labeled samples with a library of capture reagents immobilized onto a microarray. An obvious advantage of this approach is that only one antibody per target is necessary. A second advantage of direct labeling is that two or more samples can be differentially labeled with spectrally resolvable fluorescent dyes, allowing direct comparison of two or more samples on a single microarray. Several groups have utilized this dual color labeling approach as a differential protein profiling strategy with samples from a variety of sources such as serum (Haab et al., 2001; Miller et al., 2003; Zhou et al., 2004), cell culture (Sreekumar et al., 2001), and tissue lysates (Knezevic et al., 2001). As with DNA microarrays, the dual color/direct labeling strategy is especially well suited for comparative expression profiling of disease versus normal samples as has been demonstrated for spinal muscular atrophy (Anderson et al., 2003) and cancer (Knezevic et al., 2001; Sreekumar et al., 2001).

The major drawbacks of direct labeling are: (i) not all proteins are labeled uniformly; (ii) labeling may interfere with antibody recognition/binding sites; (iii) only relative, not absolute, protein quantification is possible; (iv) proteins frequently assemble in complexes; as such, signal associated with a given capture reagent may emanate not only from the target protein but also from other proteins associated with the target in a complex, making data interpretation more difficult; and (v) higher background due to nonspecific binding can result. The background problems associated with direct labeling experiments, however, can be reduced by development of new printing surfaces and/or improved blocking and washing conditions (Haab, 2003; Espina et al., 2004; Zhou et al., 2004; Poetz et al., 2005).

10.3.6 Protein Microarray Detection Systems

Fluorescence Detection   A fluorescence detection system for protein microarrays was first described by Ekins co-workers in 1990 and still remains the most common approach. A microarray can either be directly probed with a fluorescently labeled protein or be indirectly probed using a sandwich assay (described previously, see Figure 10.4) or with a biotin labeled probe followed by incubation with a fluorescently labeled affinity reagent (e.g., streptavidin). Fluorescently labeled molecules such as Cy3 and Cy5 are compatible with a standard DNA microarray confocal laser or CCD camera scanners (Huang, 2003; Espina et al., 2004; Sakanyan, 2005). They are also simple, safe, and possess high resolution. A further advantage of fluorescence labeling/detection systems are that multiple spectrally resolvable dyes may be used simultaneously, permitting direct comparative analysis (dual color labeling) and/or multiplexing. Drawbacks to fluorescence labeling/ detection systems are their sensitivity, as compared to chemiluminescence, and the labeling process may disrupt the protein's structure and hence its native interactions with other biomolecules. This disadvantage has been partially addressed through the use of nonlabeling methods such as surface enhanced laser desorption/ionization mass spectrometry (SELDI-MS). Unfortunately, while SELDI is a good approach for the detection of small proteins and peptides, its sensitivity is inferior compared to fluorescence-based methods (Templin et al., 2003; Poetz et al., 2005).

Chromogenic and Chemiluminescence Detection   In addition to fluorescence, chromogen and chemiluminescent substrate-based signaling systems have been used for protein microarray detection. These systems utilize a conjugated enzyme such as horseradish peroxidase or alkaline phosphatase to catalyze the enzymatic cleavage of a substrate to produce a colored product or emit light. The resulting signal can then be recorded with a CCD camera, phosphor imager, or X-ray film. While such systems offer good sensitivity, they are limited by signal instability, low resolution, and single-channel detection (Huang, 2003; Espina et al., 2004; Sakanyan, 2005).

Radioisotopic Detection   Isotopic labeling has also been used as a detection system in protein function microarrays designed for detecting and identifying protein–protein interactions, protein–nucleic acid interactions, protein–ligand interactions, and protein modifications (Huang, 2003). The advantages of radioisotopic labeling and detection include sensitivity and preservation of protein structure and function. However, health and safety issues and long detection times seriously limit its use in high-throughput protein microarray platforms (Huang, 2003; Espina et al., 2004).

10.4 PROTEIN MICROARRAY DATA ANALYSIS AND NORMALIZATION

Primary microarray data analysis includes the collection of raw data (i.e., "scanning"), application of a grid to localize array features, the linkage of array features to a feature "print list," background subtraction, elimination of "bad spots," and, in the case of platforms employing dual color labeling, normalization of signal intensity. Well-established protocols for microarray scanning, grid application, and linkage to print lists have been developed for DNA microarrays and may be applied directly to protein microarrays (Angulo and Serra, 2003). Likewise, a host of algorithms for background subtraction and the elimination of "bad spots" are also available (Brown et al., 2001). Finally, for platforms using dual or multiple fluorescent label formats, normalization of signal intensities must also be performed to correct for differential dye intensity, stability, and incorporation effects. Simple normalization of overall dye signal strength is generally regarded as sufficient to account for differences in dye intensities. Several additional protocols have also been developed to compensate for differential stability and incorporation effects. One of these, dye swapping, is a well-established approach commonly used in dual color DNA microarray analysis. Dye swapping involves the hybridization of two different microarrays. In the first array, sample A (reference) is labeled with dye #1, and sample B (query) is labeled with dye #2. In the second or reverse-labeled array, sample A is labeled with dye #2, and sample B is labeled with dye #1. The dye#1/dye#2 ratio is calculated for each feature represented in the array and a normalized ratio (NR) is calculated, whereby

$$NR = \sqrt{Ratio1 / Ratio2}$$

where ratios 1 and 2 correspond to microarray 1 and 2.

In the above equation, Ratio 1 = Sample A reference-dye#1 relative fluorescent units (rfu)/samples B query-dye#2 rfu and Ratio 2 = Sample B query-dye#1 rfu/ Sample A reference-dye#2 rfu. NR is the value that represents the abundance of protein A in sample A relative to protein A in sample B. A NR value >2.0 or <0.5 represents a significant differential protein expression (Anderson et al., 2003).

Another normalization approach for dual color microarray platforms has been proposed by Haab et al. (2001). Here, the normalization factor is established by comparing the dye#1/dye#2 ratios of three to four well-established array features or internal standards (Haab et al., 2001). The normalization factor is then applied to the dye ratios determined for the remaining proteins to determine if a significant change in the abundance of a protein between two experimental samples is observed. Other alternative or less well-established approaches to dual color microarray normalization have been developed by Sreekumar et al. (2001), Miller et al. (2003), and Zhou et al. (2004).

10.5 CONCLUSION

In this chapter we have reviewed the various aspects of protein microarray platform development and application. A multitude of protein microarray platforms are currently available and continue to be developed. The best system or approach is dependent on the experimental question being asked and the constraints dictated by the individual investigator's resources. Despite the large variety of readily accessible protein microarray platforms, the development of this technology is still in its early stages with three key areas in particular need of further enhancement: (i) the development of a wider variety of affinity reagents (i.e., monoclonal and single-chain antibody libraries, aptamers, peptide libraries, recombinant proteins), (ii) improved surface chemistries for immobilization of capture or affinity reagents, and (iii) further development of so-called self-assembling protein microarrays (Ramachandran et al., 2004) to circumvent the need for large-scale protein synthesis and capture reagent immobilization.

In summary, protein microarrays have great potential to complement current proteomic technologies and bridge the transition between genomics and proteomics. With further development, protein microarrays will undoubtedly become a useful and important component of the proteomic researcher's toolbox alongside more widely utilized technological platforms such as 2DGE and mass spectrometry.

REFERENCES

Anderson K, Potter A, Baban D, Davies KE. 2003. Protein expression changes in spinal muscular atrophy revealed with a novel antibody array technology. *Brain* 126:2052–2064.

Angenendt P, Glokler J, Sobek J, Lehrach H, Cahill DJ. 2003. Next generation of protein microarray support materials: evaluation for protein and antibody microarray applications. *J Chromatogr A* 1009:97–104.

Angulo J, Serra J. 2003. Automatic analysis of DNA microarray images using mathematical morphology. *Bioinformatics* 19:553–562.

Bradbury A, Velappan N, Verzillo V, Ovecka M, Chasteen L, Sblattero D, Marzari R, Lou J, Siegel R, Pavlik P. 2003. Antibodies in proteomics II: screening, high-throughput characterization and downstream applications. *Trends Biotechnol* 21:312–317.

Brown CS, Goodwin PC, Sorger PK. 2001. Image metrics in the statistical analysis of DNA microarray data. *Proc Natl Acad Sci USA* 98:8944–8949.

Ekins RP. 1989. Multi-analyte immunoassay. *J Pharm Biomed Anal* 7:155–168.

Ekins R, Chu F, Biggart E. 1990. Multispot, multianalyte, immunoassay. *Ann Biol Clin (Paris)* 48:655–666.

Espina V, Woodhouse EC, Wulfkuhle J, Asmussen HD, Petricoin EF 3rd, Liotta LA. 2004. Protein microarray detection strategies: focus on direct detection technologies. *J Immunol Methods* 290:121–133.

Frank R, Doring R. 1988. Simultaneous multiple peptide-synthesis under continuous-flow conditions on cellulose paper disks as segmental solid supports. *Tetrahedron* 44:6031–6040.

Haab BB. 2003. Methods and applications of antibody microarrays in cancer research *Proteomics* 3:2116–2122.

Haab BB, Dunham MJ, Brown PO. 2001. Protein microarrays for highly parallel detection and quantitation of specific proteins and antibodies in complex solutions. *Genome Biol* 2:RESEARCH0004.

Huang RP. 2003. Protein arrays, an excellent tool in biomedical research. *Front Biosci* 8:559–576.

Huang RP, Huang R, Fan Y, Lin Y. 2001. Simultaneous detection of multiple cytokines from conditioned media and patient's sera by an antibody-based protein array system. *Anal Biochem* 294:55–62.

Jones MB, Krutzsch H, Shu H, Zhao Y, Liotta LA, Kohn EC, Petricoin EF 3rd. 2002. Proteomic analysis and identification of new biomarkers and therapeutic targets for invasive ovarian cancer. *Proteomics* 2:76–84.

Knezevic V, Leethanakul C, Bichsel VE, Worth JM, Prabhu VV, Gutkind JS, Liotta LA, Munson PJ, Petricoin EF 3rd, Krizman DB. 2001. Proteomic profiling of the cancer microenvironment by antibody arrays. *Proteomics* 10:1271–1278.

Kusnezow W, Hoheisel JD. 2003. Solid supports for microarray immunoassays. *J Mol Recognit* 16:165–176.

Lee Y, Lee EK, Cho YW, Matsui T, Kang IC, Kim TS, Han MH. 2003. ProteoChip: a highly sensitive protein microarray prepared by a novel method of protein immobilization for application of protein–protein interaction studies. *Proteomics* 3:2289–2304.

Lin Y, Huang R, Santanam N, Liu YG, Parthasarathy S, Huang RP. 2002. Profiling of human cytokines in healthy individuals with vitamin E supplementation by antibody array. *Cancer Lett* 187:17–24.

Liotta LA, Espina V, Mehta AI, Calvert V, Rosenblatt K, Geho D, Munson PJ, Young L, Wulfkuhle J, Petricoin EF 3rd. 2003. Protein microarrays: meeting analytical challenges for clinical applications. *Cancer Cell* 4:317–325.

MacBeath G, Schreiber SL. 2000. Printing proteins as microarrays for high-throughput function determination. *Science* 289:1760–1763.

Miller JC, Zhou H, Kwekel J, Cavallo R, Burke J, Butler EB, Teh BS, Haab BB. 2003. Antibody microarray profiling of human prostate cancer sera: antibody screening and identification of potential biomarkers. *Proteomics* 3:56–63.

Min DH, Mrksich M. 2004. Peptide arrays: towards routine implementation. *Curr Opin Chem Biol* 8:554–558.

Nielsen UB, Geierstanger BH. 2004. Multiplexed sandwich assays in microarray format. *J Immunol Methods* 290:107–120.

Nord K, Gunneriusson E, Uhlen M, Nygren PA. 2000. Ligands selected from combinatorial libraries of protein A for use in affinity capture of apolipoprotein A-1M and taq DNA polymerase. *J Biotechnol* 80:45–54.

Pavlickova P, Schneider EM, Hug H. 2004. Advances in recombinant antibody microarrays. *Clin Chim Acta* 343:17–35.

Poetz O, Schwenk JM, Kramer S, Stoll D, Templin FM, Joos TO. 2005. Protein microarrays: catching the proteome. *Mech Ageing Dev* 126:161–170.

Predki PF. 2004. Functional protein microarrays: ripe for discovery. *Curr Opin Chem Biol* 8:8–13.

Ramachandran N, Hainsworth E, Bhullar B, Eisenstein S, Rosen B, Lau AY, Walter JC, LaBaer J. (2004) Self-assembling protein microarrays. *Science* 305:86–90.

Robinson WH, DiGennaro C, Hueber W, Haab BB, Kamachi M, Dean EJ, Fournel S, Fong D, Genovese MC, de Vegvar HE, Skriner K, Hirschberg DL, Morris RI, Muller S, Pruijn GJ, van Venrooij WJ, Smolen JS, Brown PO, Steinman L, Utz PJ. 2002. Autoantigen microarrays for multiplex characterization of autoantibody responses. *Nature Med* 8:295–301.

Roda A, Guardigli M, Russo C, Pasini P, Baraldini M. 2000. Protein microdeposition using a conventional ink-jet printer. *Biotechniques* 28:492–496.

Sakanyan V. 2005. High-throughput and multiplexed protein array technology: protein–DNA and protein–protein interactions. *J Chromatogr B Analyt Technol Biomed Life Sci* 815:77–95.

Sreekumar A, Nyati MK, Varambally S, Barrette T, Ghosh D, Lawrence T, Chinnaiyan A. 2001. Profiling of cancer cells using protein microarrays: discovery of novel radiation-regulated proteins. *Cancer Res* 61:7585–7593.

Templin MF, Stoll D, Schrenk M, Traub PC, Vohringer CF, Joos TO. 2002. Protein microarray technology. *Drug Discov Today* 7:815–822.

Templin MF, Stoll D, Schwenk JM, Potz O, Kramer S, Joos TO. 2003. Protein microarrays: promising tools for proteomic research. *Proteomics* 3:2155–2166.

Wang Y. 2004. Immunostaining with dissociable antibody microarrays. *Proteomics* 4:20–26.

Zhou H, Bouwman K, Schotanus M, Verweij C, Marrero JA, Dillon D, Costa J, Lizardi P, Haab BB. 2004. Two-color, rolling-circle amplification on antibody microarrays for sensitive, multiplexed serum-protein measurements. *Genome Biol* 5:R28.

Zhu H, Bilgin M, Bangham R, Hall D, Casamayor A, Bertone P, Lan N, Jansen R, Bidlingmaier S, Houfek T, Mitchell T, Miller P, Dean RA, Gerstein M, Snyder M. 2001. Global analysis of protein activities using proteome chips. *Science* 293:2101–2105.

11

Microfluidics-Based Proteome Analysis

Yan Li

University of Maryland, College Park, Maryland

Don L. DeVoe

University of Maryland, College Park, Maryland

Cheng S. Lee*

University of Maryland, College Park, Maryland

11.1 INTRODUCTION

Since the concept of Micro-Total Analysis Systems (μTAS) was first proposed (Manz et al., 1990), the field has advanced rapidly, with ongoing developments promising to profoundly revolutionize modern bioanalytical methodologies. Whether termed μTAS, lab-on-a-chip, or microfluidics, the collection of technologies that define the field are proving to be an important innovation capable of transforming the ways in which bioanalytical techniques are performed. In particular, miniaturized bioanalytical devices based on microfluidics technology provide various important advantages over benchtop instruments.

The very act of miniaturization provides significant benefits for many microfluidic instruments. Reduced size and power requirements lead to improved portability, with higher levels of integration possible, for example, on-chip micropumps for sample manipulation or liquid chromatography separations. Microfabrication methods lend themselves to the formation of complex microfluidic systems, opening

* To whom correspondence should be addressed.

Proteomics for Biological Discovery, edited by Timothy D. Veenstra and John R. Yates.
Copyright © 2006 John Wiley & Sons, Inc.

the way to highly parallel analytical tools, while realizing low per-unit cost for disposable applications.

Furthermore, the low volume fluid control enabled by microfluidics allows smaller dead volume and reduced sample consumption, while the low Reynolds numbers that characterize most microfluidic systems lead to highly laminar flow, eliminating the need for considering turbulent effects during instrument design. Many efficient pumping methods, including capillary action and electroosmotic flow, scale favorably in these systems, enabling valveless flow control at the microscale. Similarly, thermal time constants tend to be extremely small due to the large surface area to volume ratio, reducing the onset of significant Joule heating during electrokinetic separations and thus allowing higher separation voltages for shorter analysis times and equivalent or better separation resolution for complex mixtures in an integrated format.

Choices of material and fabrication procedure are critical aspects of microfluidic devices. Currently, the majority of commercial devices are made from glass or silicon. In some cases, these materials are chosen for their inherent properties, for example, the use of established derivatization methods for silica surfaces. However, these materials are often chosen in order to readily leverage established fabrication procedures from the semiconductor and microelectromechanical systems industries. For a wide range of applications, glass and silicon microfluidic devices suffer from high fabrication costs and poor mechanical robustness due to their brittle nature, leading to a demand for alternative substrate materials for commercial applications. To this end, plastics offer a very promising solution, enjoying advantages including lower cost, a wide range of surface properties, and ease of manufacture. One salient advantage is that plastic microfluidic systems are readily fabricated using replication techniques such as casting, embossing, or injection molding (Soper et al., 2000).

For example, the use of bulk-micromachined silicon templates containing raised three-dimensional patterns for imprinting microchannels into plastic substrates by hot embossing has been widely demonstrated (Martynova et al., 1997). By taking advantage of anisotropic silicon etching, in which the family of crystal planes dictates the final geometry of silicon template features, well-defined microchannel molds may be realized with sidewalls sloped at 54.7° from vertical. Alternately, deep reactive-ion etching of silicon may be used to form templates with nearly vertical sidewalls. This imprinting technique, however, requires heating to soften the plastic near its glass transition temperature, enabling pattern transfer from the silicon template during imprinting. While effective, the disadvantage is that the silicon templates tend to break upon cooling due to mismatched thermal expansion coefficients between the silicon template and plastic substrate. In practice, a typical 100 mm diameter silicon template can only be used to imprint on the order of 10 devices before template failure occurs. More recently, a room temperature imprinting method for the fabrication of plastic microfluidic devices has been reported (Xu et al., 2000). Since no external heating is involved, a single silicon template can be used to imprint scores of microdevices without concern for thermal mismatch failures. Furthermore, overall fabrication time per device is shortened from about 30 min to 5 min.

Templates based on electroplated metals, rather than silicon, are capable of producing embossing templates with features as small as several microns. Fabricated by electroplating thick metal films around lithographically defined polymer features (photoresist in the case of photolithographic patterning, or poly(methyl methacrylate) (PMMA) in the case of X-ray lithography), these templates are more difficult to manufacture than silicon devices but offer several key advantages. By eliminating silicon as the mold material, the electroplating process results in templates with significantly longer lifetimes, even when using higher temperature embossing conditions. In addition, the templates are produced with vertical microchannel walls, rather than sloped $54.7°$ sidewalls produced by bulk etching of single crystal silicon. This allows both greater precision and flexibility in channel geometry and reduces variations in electric field distribution across the channel cross section for improved uniformity during on-chip electrokinetic separations.

A wide range of rigid plastics have been used for microfluidic systems, including PMMA, poly(carbonate) (PC), poly(ethylene terephthalate) (PETP), poly(ethyl ethylketone) (PEEK), and poly(vinylidene difluoride) (PVDF), to name only a few. Each of these materials offers different benefits in terms of chemical, mechanical, thermal, and optical properties. PC, for example, offers particular advantages in terms of excellent thermal and mechanical stability, high UV transmission for optical detection, and repeatable fabrication by injection molding or hot embossing, with dimensions of imprinted microchannels typically varying by less than 2% in height and width. Irreversible thermal sealing of polycarbonate microchannels has been demonstrated, with bond energies equal to the native surface energy of bulk PC (Rosenberger et al., 2002) achieved in devices such as the sealed PC micro fluidic chip shown in Figure 11.1.

Another popular method for microfluidic fabrication is based on curing a silicone elastomeric material, such as poly(dimethylsiloxane) (PDMS), on a rigid mold to transfer the mold features into the elastomer. This method has been shown to successfully reproduce features with 50 nm resolution (Jackman et al., 1995) and has been extensively used for prototyping microfluidic systems. Permanent bonding of PDMS to a variety of rigid substrates to seal the microchannels is often achieved by pretreating PDMS with oxygen plasma. The plasma introduces silanol groups

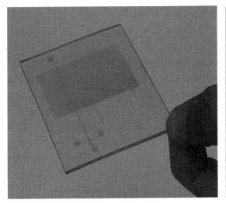 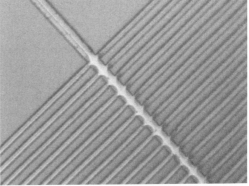

Figure 11.1. *(Left) Fabricated polycarbonate two-dimensional microfluidic chip. (Right) Micrograph showing detail of intersecting microchannels. (Image courtesy of Calibrant Biosystems.) (See color insert.)*

(Si—OH) at the expense of methyl groups (Si—CH₃) (Duffy et al., 1998; McDonald et al., 2000), which condense with appropriate groups (OH, COOH, ketone) on another surface when the two layers are brought into conformal contact. Oxidized PDMS also seals irreversibly to other materials, including glass, silicon, silicon dioxide, quartz, silicon nitride, poly(ethylene), poly(styrene), and glassy carbon. Oxidation of the PDMS has the additional advantage that it renders the surface hydrophilic because of the presence of silanol groups. These negatively charged channels have greater resistance to adsorption of hydrophobic and negatively charged analytes and are capable of creating strong bond energies over twice the native PDMS surface energy, and comparable to the surface energy of rigid polymers such as PC. PDMS surface chemistry is well characterized, and various derivatization methods have been developed to modify PDMS surfaces for hydrophobic/hydrophilic control, charge state modification, and reduction of unwanted molecular adsorption. It exhibits excellent long-term mechanical, electrical, optical, and chemical stability, is biocompatible (Tang et al., 1999), and requires minimal investment for prototype fabrication.

Despite the benefits of PDMS fabrication methods, the material suffers from several key drawbacks. PDMS exhibits a high permeability for both gasses and, in some cases, analytes. It suffers from poor mechanical properties including low elastic modulus and low tensile strength, which can lead to poor mechanical robustness and low geometric stability. Its low stiffness can also lead to difficulties with structural collapse when forming large open channels, chambers, and reservoirs. The ionic nature of the siloxane backbone in PDMS makes it susceptible to hydrolysis, and long-term exposure to water is capable of breaking the siloxane bond, especially at a pH lower than 2.5 or higher than 11. It is well known that PDMS suffers from limited life span for devices requiring stable surface charge for electroosmotic flow and, in general, possesses low solvent stability. Thus, while PDMS offers a simple and attractive approach for microfluidic fabrication, it may not be suitable for all situations, particularly when a wide range of pH levels or long-term exposure to solution is required as in many proteomics applications.

11.2 APPLICATION OF MICROFABRICATED DEVICES FOR PROTEOME SAMPLE TREATMENTS

The field of microfluidics for genomic and proteomic analysis on small integrated platforms is quickly demonstrating its potential for providing novel bioanalytical, clinical, and research tools. While a large portion of microfluidics research has historically been directed toward the development of genomic analytical instrumentation, such as microscale PCR systems and electrophoresis-based DNA sequencing platforms, an increasingly broad range of microfluidics technologies has emerged in recent years for the study of cells and their constituent proteins. These technologies are aimed at miniaturizing and addressing shortcomings in traditional instrumentation for cell manipulation/concentration, cell lysis, and protein fractionation/digestion steps using on-chip methods. Recent advances toward the implementation of proteome sample treatments in a microfluidic format are therefore summarized and discussed.

11.2.1 Cell Manipulation, Concentration, and Lysis

Several methods, including optical tweezers, fluorescence or magnetic activated cell sorting, centrifugation, filtration, and electric field-based approaches, are currently employed in biological laboratories for manipulation, concentration, and separation of bioparticles. Among these methods, the electric field-based approaches are well suited for miniaturization due to relative ease of electrode integration and electric field generation (Cheng et al., 1998; Wang et al., 2000a). By utilizing ac and dc electric fields on microfabricated semiconductor chips, Huang and co-workers (2001) have achieved dielectrophoretic enrichment of target bioparticles such as *Escherichia coli* bacteria, peripheral blood mononuclear cells, and white blood cells with a concentration factor ranging from 20- to 30-fold. Besides offering efficient electric concentration of bioparticles, these electrically driven microchips provide the advantages of speed, flexibility, controllability, and ease of automation.

Several research groups have developed microfluidics-based cell lysis devices based on chemical, electrical, thermal, or sonic poration methods. For example, Waters and co-workers (1998) have applied electrokinetic fluid actuation and thermal cycling to achieve the lysis of *E. coli* and PCR amplification of DNA. Electrokinetically induced mixing of erythrocytes with sodium dodecyl sulfate (SDS) as the lysing agent was illustrated in a microfluidic system fabricated on a glass chip (Li and Harrison, 1997). Schilling et al. (2002) have accomplished the continuous lysis of bacterial cells by exploiting transverse diffusion across two parallel laminar flow streams containing lytic agent and cells. Utilization of the minisonicator within microfluidics greatly improved the speed and the effectiveness of spore disruption and contributed to the decrease in the detection limit for PCR-based DNA analysis (Belgrader et al., 1999; Taylor et al., 2001). The combination of both electrical and chemical lysis in a microfluidic system has been shown to result in highly efficient cell membrane disruption by taking advantage of rapid electroporation in concert with long-term chemical membrane solubilization (McClain et al., 2001). In a recent work, viable *E. coli* cells were trapped in hydrogel micropatches photopolymerized within microfluidic systems (Heo et al., 2003). Cell entrapments within hydrogels followed by the introduction of lytic agent could provide a convenient means for the preparation of cell lysates needed in the proteome analysis.

11.2.2 Microdialysis Sample Cleanup and Fractionation

Due to the porous structures of membrane media, polymeric membranes exhibit a large surface to volume ratio, which serves to facilitate rapid solution exchange and analyte filtration based on their differences in size. Rapid cleanup of DNA and protein samples for electrospray ionization mass spectrometry (ESI-MS) analysis has been demonstrated by sandwiching a microdialysis membrane between two PC substrates containing serpentine channels (Xu et al., 1998). Signal-to-noise ratios were significantly enhanced compared to direct infusion of the original non-dialyzed samples.

In addition to desalting, a microfabricated dual-microdialysis device in an integrated platform (see Figure 11.2) was constructed for the rapid fractionation and

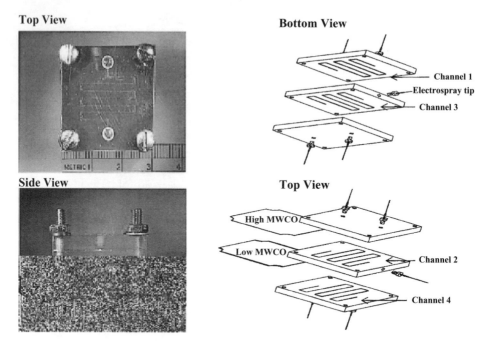

Figure 11.2. (Left) Photographs of a microfabricated dual-microdialysis device. (Right) Exploded views showing construction of this device. (From Xiang et al., 1999.)

cleanup of complex biological samples (Xiang et al., 1999). The dual-microdialysis system was composed of three PC plates containing four serpentine channels. Two pieces of polymeric membranes with high and low molecular weight cutoffs (MWCOs) were alternatively sandwiched between these plastic plates for sample fractionation and cleanup, respectively.

11.2.3 On-Chip Proteolytic Digestion

An integrated microfluidic platform was constructed for rapid and sensitive protein identification by on-line protein digestion and analysis of digested proteins using ESI-MS (Gao et al., 2001). A miniaturized membrane reactor, similar to the one shown in Figure 11.3, was assembled by fabricating microfluidic channels on two PDMS substrates and coupling the microfluidics to a PVDF porous membrane containing adsorbed trypsin. An extremely high local concentration of adsorbed trypsin inside the PVDF membrane provided ultrafast catalytic turnover. This microfluidic system enabled rapid identification of proteins in minutes instead of hours, consumed very little sample (nanogram or less), and provided on-line interface with upstream protein separation schemes for the analysis of complex protein mixtures such as cell lysates.

The advantages of miniaturizing proteolytic digestion include the increased reaction kinetics in low sample volumes and the potential for development of high-throughput sample-handling procedures. For example, a microchip trypsin reactor was constructed by covalent attachment of trypsin onto the porous silicon micro-

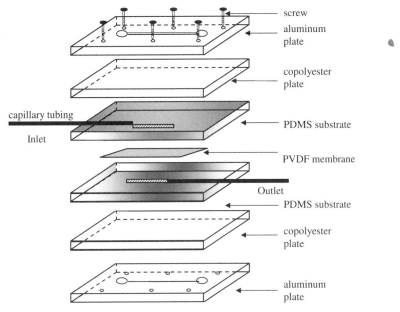

Figure 11.3. *Schematic of a membrane reactor assembly. (From Gao et al., 2001.)*

channels (Ekstrom et al., 2000). The reactor in combination with the microdispenser and shallow nanovials enabled rapid and automated sample preparation for
protein identification using matrix assisted laser desorption/ionization mass spectrometry (MALDI-MS). Wang and co-workers (2000b) reported the use of trypsin-
immobilized beads, either packed within the sample inlet reservoir or in a packed
bed, for on-chip protein digestion. Instead of using trypsin-immobilized beads,
trypsin-encapsulated sol-gel has been fabricated in situ within a sample reservoir
(Sakai-Kato et al., 2003) or microchannels (Peterson et al., 2002). The optimized
porous properties of the monoliths resulted in very low backpressure, enabling the
use of simple mechanical pumping to perform on-chip proteolytic digestion.

11.3 MICROFLUIDICS-BASED PROTEIN/PEPTIDE SEPARATIONS

Different microfluidic systems have been applied to achieve protein/peptide separations as the essential aspects of the proteomics process. This section describes the
promising proof-of-principle microfabricated devices that have been employed to
perform chromatographic- or electrokinetic-based separations. Furthermore,
recent developments of multidimensional protein/peptide separations in microfluidic networks are critically needed, particularly for the analysis of complex protein
mixtures such as cell lysates.

11.3.1 Shrinking Liquid Chromatography Landscape

The trend toward increasingly smaller columns and stationary phase structures
in liquid chromatography can be accommodated by miniaturizing liquid phase

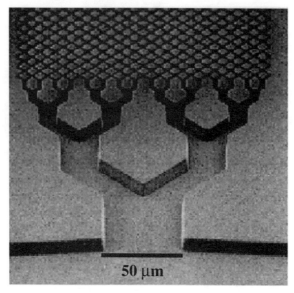

Figure 11.4. SEM image of the microchannel with a microscale monolith array fabricated by dry reactive-ion etching on a quartz substrate. (From He et al., 1998.)

separations in microfluidic networks. Even though microchannels can be micromachined and packed with particles (Ericson et al., 2000; Oleschuk et al., 2000), substantial concerns, including the fabrication of frits, the nonuniformity of packing at the walls and corners of the channels, and the difficulty of packing columns through the tortuous channel network, still remain in on-chip liquid chromatography.

Open tubular columns with the stationary phase supported on the channel walls provide potential solution to circumvent the use of particles and the accompanying packing problems. Thus, microscale monolith arrays (see Figure 11.4) were fabricated on quartz substrates using dry reactive-ion etching (He et al., 1998). In order to prepare the array for performing chromatography in reverse-phase mode, the surface of the array was coated with poly(styrene sulfonate) following surface derivatization with (γ-aminopropyl)trimethoxysilane. The monolith array was designed to act as flow splitters for enhanced interaction between the stationary phase and the chemical species in the mobile phase.

Additionally, photolithographically patterned rigid polymer monoliths have been fabricated within the microchannels by Throckmorton and co-workers (2002) and employed to achieve reverse-phase separation of amino acids and peptides. Still, the fabrication of high-efficiency stationary phase remains the toughest part in chip-based liquid chromatography systems (Harris, 2003). Miniaturized chromatography separations have to be competitive with most of lab-on-a-chip technologies that are electrophoretically based.

11.3.2 Electrokinetic-Based Separations

The use of chromatography far outweighs that of electrophoresis in analytical and microscale separations. However, the opposite is true for chip-based separations at the current stage. Besides difficulties involved in the fabrication and the packing of stationary phase for miniaturized chromatography, it is easier to apply a high voltage to a chip than pressure.

Zone Electrophoresis   Owing to its simplicity of implementation, zone electrophoresis remains the most popular method among miniaturized electrokinetic-based techniques (Figeys and Pinto, 2001). Under the influence of electric potential applied across the buffer-filled microchannels, the charged surface on the channel wall offers an electrically driven pump, the electroosmotic flow. The presence of electroosmotic pumping, the bulk flow of liquid, facilitates the zone resolution of neutral, cationic, or anionic analytes and mobilizes all the analytes toward the same direction in the channel for detection.

Efforts including the manipulation of buffer conditions (Liu et al., 2000a) and the use of surface modifications (Henry et al., 2000), however, are needed to eliminate or significantly reduce any surface–analyte interactions. The adsorption of protein/peptide onto the channel wall not only degrades the separation efficiency but also can cause irreversible loss of analytes. Furthermore, sample loading is already a critical issue for capillary electrophoresis and even more serious for on-chip zone electrophoresis with limited injection volume. Thus, various sample preconcentration strategies, including sample stacking (Li et al., 1999; Lichtenberg et al., 2001), isotachophoresis (Kaniansky et al., 2000; Wainright et al., 2002), and solid-phase extraction (Figeys and Aebersold, 1998), have been reported to increase the amount of sample injected and analyzed.

Isoelectric Focusing   Due to the use of an entire channel as the injection volume, the sample loading capacity of isoelectric focusing (IEF) is inherently greater than most chip-based electrokinetic separation techniques. Furthermore, IEF not only contributes to a high-resolution protein/peptide separation based on their differences in isoelectric point (pI), but also potentially allows the analysis of low-abundance proteins with a typical concentration factor of 50–100 times. To perform IEF, the entire channel is initially filled with a solution mixture containing proteins/peptides and carrier ampholytes for the creation of a pH gradient. Several research groups have recently reported the use of glass, quartz, or plastic-based microfluidics for demonstrating on-chip IEF separations (Hofmann et al., 1999; Mao and Pawliszyn, 1999; Rossier et al., 1999; Wen et al., 2000; Xu et al., 2000; Chen et al., 2002; Herr et al., 2003). Instead of using carrier ampholytes, Macounova et al. (2000, 2001) have generated pH gradients as a result of the electrochemical decomposition of water for achieving protein separations in a microchannel.

However, the sample loading for on-chip IEF separations is still dependent on the initial analyte concentrations and microchannel dimensions, particularly the channel length. To significantly enhance sample loading and therefore the concentrations of focused analytes, dynamic introduction of proteins/peptides using electrokinetic injection from the solution reservoir has been demonstrated by Li and co-workers (2003). The proteins/peptides continuously migrated into the plastic microchannel and encountered a pH gradient established by carrier ampholytes originally present in the channel for focusing and separation. Dynamic sample introduction and analyte focusing were directly controlled by various electrokinetic conditions, including the injection time and the applied electric field strength. The sample loading was enhanced by approximately 10–100-fold in comparison with conventional IEF.

Field Gradient Focusing Techniques   In contrast to IEF, field gradient focusing methods (O'Farrell, 1985; Koegler and Ivory, 1996) offer a broader field of application, particularly for protein/peptide separation and concentration. For example, proteins/peptides with extreme pI values may be outside the working pH range of IEF due to limited availability of commercial ampholytes. Furthermore, the focused proteins/peptides in IEF are at their pI values where they have an increased tendency to precipitate. To achieve field gradient focusing, proteins/peptides are focused by balancing the electrophoretic velocity of an analyte against the bulk velocity of the buffer containing the analyte. If there is an appropriate gradient in the electric field, the total velocity of the analyte as the sum of the bulk and electrophoretic velocities can be set to zero at a unique point along a channel. Instead of using a combination of electrodes and membranes for the generation of electric field gradient (Koegler and Ivory, 1996), Ross and Locascio (2002) have demonstrated the application of a temperature gradient together with an appropriate buffer for creating a corresponding gradient in the electrophoretic velocity of the analyte.

Size Separation Using SDS Gel Electrophoresis   The current standard method for protein sizing, SDS–polyacrylamide gel electrophoresis (PAGE) is a labor-intensive technique. In comparison to SDS-PAGE, the significant advantages of performing SDS–protein separations in a capillary electrophoresis or microfluidic format (Yao et al., 1999; Liu et al., 2000b; Bousse et al., 2001) include: (i) the ability to use higher electric fields resulting in rapid and ultrahigh resolution separations, (ii) improved reproducibility resulting from the use of replaceable polymeric sieving solution, and (iii) automated operation.

It has been reported that the formation of SDS–protein complexes is the critical step in determining separation resolution of capillary gel electrophoresis (Benedek and Guttman, 1994). Once the SDS–protein complexes are properly formed, they remain relatively stable and the presence of SDS in the separation buffer is no longer needed for further stabilization. This is particularly true for rapid protein separation in capillary gel electrophoresis and microfluidics-based devices.

For laser-induced fluorescence detection of resolved proteins, however, SDS concentration has a significant effect on noncovalent protein labeling using the popular SYPRO dyes (Molecular Probes, Eugene, OR), which offer the benefit of low variations in fluorescent intensity between different proteins. The SYPRO probes, such as SYPRO Orange and SYPRO Red, are nonfluorescent in water but highly fluorescent in detergent, in which they take advantage of SDS binding to proteins to build a fluorescence-promoting environment. When the SDS concentration is above its critical micellar concentration (CMC), 8.3 mM (~0.24% in water and somewhat less in buffer solutions), the major portion of the staining dye in the electrophoresis buffer becomes attached to the SDS micelles instead of the SDS–protein complexes. In order to sensitively detect the resolved SDS–protein complexes, an on-chip SDS dilution step was incorporated between the separation channel and fluorescence detection (Bousse et al., 2001). The dilution step reduced the SDS concentration in the electrophoresis buffer from 0.25% (above CMC) to

~0.025% (below CMC). Such dilution disrupted the SDS micelles, thereby allowing more dye molecules to bind to the SDS–protein complexes, prior to the detection.

11.3.3 Multidimensional Separation Techniques

The vast number of proteins present in the proteome of a typical organism requires that separations be performed on the mixture prior to introduction into the mass spectrometer for identification. For example, proteolytic cleavage of all yeast proteins with trypsin presents a highly complex peptide mixture of at least 300,000 peptides. The large variation of protein relative abundances having potential biological significance in mammalian systems (>6 orders of magnitude) also presents a major analytical challenge for proteomics. Thus, total peak capacity improvements in multidimensional separation platforms are critically needed to enhance the dynamic range and detection sensitivity of MS.

Assuming the separation techniques used in the two dimensions are orthogonal—that is, the two separation techniques are based on different physicochemical properties of analytes—the peak capacity of two-dimensional (2D) separations is the product of the peak capacities of individual one-dimensional methods (Giddings, 1991). Several systems have been reported, which combine two individual orthogonal separations in coupled microchannels. Rocklin and co-workers (2000) have demonstrated 2D separation of peptide mixtures in a microfluidic device using micellar electrokinetic chromatography and zone electrophoresis as the first and second dimensions, respectively. Gottschlich et al. (2001) have also fabricated a spiral shaped glass channel coated with a C-18 stationary phase for performing reverse-phase chromatographic separation of trypsin-digested peptides. By employing a cross-interface, the eluted peptides from the micellar electrokinetic chromatography (Rocklin et al., 2000) or reverse-phase chromatography channel (Gottschlich et al., 2001) were sampled by a rapid zone electrophoresis separation in a short glass microchannel. Additionally, Herr and co-workers (2003) have coupled IEF with zone electrophoresis for 2D separations of model proteins using plastic microfluidics. Slentz et al. (2003), on the other hand, have combined immobilized metal affinity chromatography for the selection of histidine-containing peptides with reverse-phase chromatography on a chip.

In each of these examples, the multiple separation dimensions are performed serially, without the ability to simultaneously sample all proteins or peptides separated in the first dimension for parallel analysis in the second dimension. As a step toward this goal, a microfabricated quartz device has been proposed by Becker and co-workers (1998) with a single channel for the first dimension and an array of 500 parallel channels with submicron dimensions as the second dimension positioned orthogonally to the first dimension channel. However, the conceptual system is difficult to implement as described by Becker and co-workers, and since its introduction in 1998 no published use of this device in performing 2D bioseparations has been demonstrated. In a further step, Chen et al. (2002) recently described a 2D capillary electrophoresis system based on a six-layer PDMS microfluidic platform. The system consisted of a 25 mm long microchannel for performing IEF, with an intersecting array of parallel 60 mm long microchannels for achieving SDS-PAGE. The system combines six individual layers of flexible PDMS, requiring the alignment, bonding, removal, realignment, and rebonding of various combinations of the six layers to perform a full 2D protein separation. This innovative system is somewhat cumbersome to implement, subject to band broadening due to diffusion during the postseparation assembly process, fairly complex due to the need for a reconfigurable three-dimensional microfluidic arrangement, and potentially subject to cross-contamination; in addition, potential arises for imperfect sealing of the various layers containing solutions of protein sample, buffers, and SDS during these manual manipulations. Nevertheless, instead of continuously sampling protein analytes eluted from the first separation dimension as in previously reported systems (Rocklin et al., 2000; Gottschlich et al., 2001; Herr et al., 2003), this PDMS implementation represents a successful demonstration of parallel 2D separations of multiple proteins (carbonic anhydrase, bovine serum albumin, and ovalbumin) in a microfluidic format.

More recently, a single-layer 2D microfluidic network was demonstrated for rapidly separating protein analytes with ultrahigh resolution based on their differences in pI and molecular weight (Calibrant Biosystems, Rockville, MD). A schematic of the system, which is fabricated in a low-cost PC substrate, is shown conceptually in Figure 11.5. Briefly, the single microchannel extending from left to right and connecting reservoirs A and B is employed for performing a nonnative IEF separation. Once the focusing is complete, the focused proteins are simultaneously transferred using an electrokinetic method into the microchannel array connecting reservoir C with the focusing microchannel for performing a parallel and high-throughput size-dependent separation. The transferred proteins were first incubated and complexed with SDS and noncovalent, environment-sensitive, fluorescent probes of SYPRO Orange. The protein complexes were then separated in the microchannel array containing a replaceable polymer sieving matrix and measured using a fluorescence microscope.

As an initial demonstration of the system, four model proteins consisting of parvalbumin (pI 4.10, MW 12.8 kDa), trypsin inhibitor (pI 4.55, MW 21.5 kDa), bovine serum albumin (pI 4.60, MW 66 kDa), and actin (pI 5.20, MW 43.0 kDa)

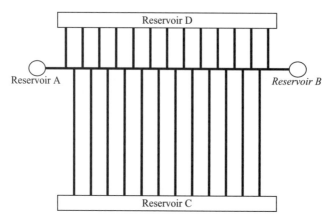

Figure 11.5. Simplified schematic of 2D protein separation platform using plastic microfluidics. (Image courtesy of Calibrant Biosystems.)

were separated in a test chip with 80 µm wide microchannels. As shown in Figure 11.6A, the first-dimension focusing was unable to resolve trypsin inhibitor and albumin with a p*I* difference of only 0.05 pH unit. However, following transfer to the second dimension and size-based separation shown in Figures 11.6B and C, all four model proteins were successfully resolved.

The separation of more complex samples has also been demonstrated using this multidimensional system. Yeast cell lysate has been successfully separated based on the same approach. A fluorescent image of a small region of a test chip following the 2D separation is shown in Figure 11.7. The 80 µm wide microchannel used for IEF was located at the rightmost edge of this image. The separated proteins shown

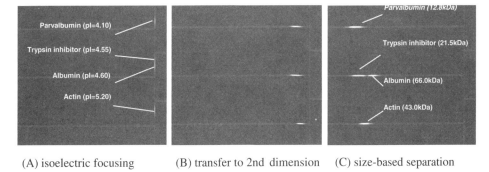

(A) isoelectric focusing   (B) transfer to 2nd dimension   (C) size-based separation

Figure 11.6. Initial evaluation of 2D protein separation in plastic microfluidic network using four model proteins. (Image courtesy of Calibrant Biosystems.)

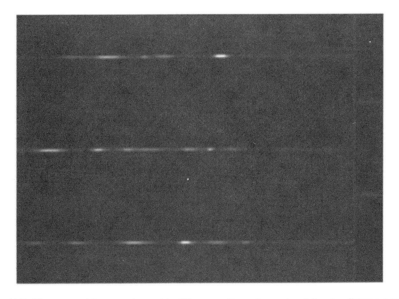

Figure 11.7. Fluorescent image of on-chip 2D separation of yeast cell lysate. (Image courtesy of Calibrant Biosystems.)

in this image were transferred from a narrow 0.5 pH region of the first dimension and represented only a small fraction of total proteins resolved on the chip. Although this simple study involved a limited number of second-dimension microchannels, and has not been optimized in terms of IEF conditions, first to second dimension transfer, or size-based separation conditions, the ability for a microfluidic platform to perform parallel 2D separations of complex protein samples has been demonstrated.

11.4 COUPLING MICROFLUIDIC DEVICES WITH ESI-MS OR MALDI-MS

Direct generation of electrospray from the planar edge of an on-chip microchannel is rather difficult because the fluid in the microchannel tends to spread over the surface before the onset of electrospray, even for relatively hydrophobic materials such as glass (Ramsey and Ramsey, 1997). Two approaches to solve this problem have been explored, including the modification of the microchannel exit surface to decrease the wettability (Ramsey and Ramsey, 1997; Xue et al., 1997a,b) and the attachment of an external nanospray tip at the microchannel exit (Bings et al., 1999; Lazar et al., 1999). Although promising spray results were obtained, extensive work was required to fabricate and manufacture the chip. Furthermore, while microfluidic devices equipped with external electrospray emitters are effective for infusion analysis, the large dead volume associated with these tips contributes to band broadening and places limitations on coupling microfluidics-based separations with ESI-MS for proteome analysis (Figeys et al., 1997; Figeys and Aebersold, 1998; Li et al., 1999; Wen et al., 2000; Zhang et al., 1999, 2000).

Alternately, shaped electrospray tips integrated directly into microfluidic systems have also been investigated. Researchers have employed various approaches, including mechanical cutting (Yuan and Shiea, 2001), laser etching (Rohner et al., 2001; Tang et al., 2001), and polymer molding (Kim and Knapp, 2001) for the fabrication of on-chip multiple electrospray emitters in plastic microfluidics. Kameoka and co-workers (2002), on the other hand, have integrated an array of tips made from a shaped polymer film into an array of microfluidic channels, demonstrating the ability to selectively electrospray from individual tips with no crosstalk between adjacent tips spaced only 80 µm apart. For silicon-based microfluidic devices, Licklider and co-workers (2000) developed a process to fabricate a parylene nozzle on the edge of a silicon microchip. Furthermore, Schultz et al. (2000) have demonstrated the use of deep reactive-ion etching technology to fabricate high aspect ratio tips on the planar surface of a silicon wafer (see Figure 11.8).

Microfluidic devices, with their increased surface-to-volume ratio, offer the potential to accelerate protein digestion through surface immobilized proteolytic enzymes. To this end, Ekstrom and co-workers (2000) described a microfluidic system that integrated a porous silicon-based enzyme reactor with a sample pretreatment robot and a microfabricated microdispenser to transfer protein digests to a MALDI target for MS analysis. Furthermore, Gyros AB (Upsalla, Sweden) has developed a microfluidic system in the same form as a compact disc for protein sample preparation including digestion, concentration, and desalting using an on-chip reverse-phase column integrated into the polymer disc (Holmquist et al., 2003). The system used centrifugal force to drive fluids on the disc, with hydropho-

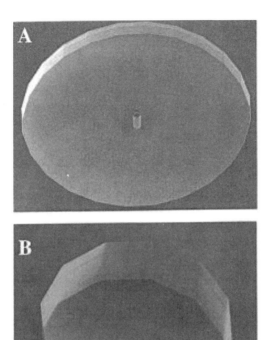

Figure 11.8. SEM images of a microfabricated silicon electrospray device show (A) an electrospray nozzle tilted at a 30° angle and (B) the reservoir etched 100 μm into the backside of the substrate. The 10 μm channel extending from the tip of the nozzle to the reservoir is visible as a dark circle in the middle of the reservoir. (From Schultz et al., 2000.)

bic patches to control fluid gating. The concentrated peptides were crystallized on prepared targets along the edge of the CD and analyzed by MALDI-MS through direct ionization from the on-chip targets.

11.5   CONCLUSION

The field of microfabrication of bioanalytical devices has grown significantly over the last decade from academic research to several commercially available systems. This chapter focuses on the enclosed fluidic network instead of the array devices that display microscopic arrays of immobilized DNA/RNA and proteins on their surfaces. Different microfluidic systems have been constructed and employed to specific aspects of proteomic analysis, including cell manipulation/concentration, protein purification, protein digestion, protein/peptide separation, and protein identification by MS. Microfabricated systems have the potential to automate multi-plexing sample processing steps and combine high-throughput multidimensional protein separations in a microfluidic network that are readily available in the macroscopic world.

Although the applications of microfluidic devices in proteomics have demonstrated great promise, the assessment of the performances of the microfluidic systems in terms of reproducibility, resolution, sensitivity, and speed can only be achieved by applying these technologies to complex biological samples. It is also crucial to combine various microfluidic components, which enable all required proteome technologies in an integrated platform for a true lab-on-a-chip.

REFERENCES

Becker H, Lowack K, Manz A. 1998. Planar quartz chips with submicron channels for two-dimensional capillary electrophoresis applications. *J Micromech Microeng* 8:24–28.

Belgrader P, Hansford D, Kovacs, GTA, Venkateswaran K, Mariella R, Milanovich F, Nasarabadi S, Okuzumi M, Pourahmadi F, Northrup MA. 1999. A minisonicator to rapidly disrupt bacterial spores for DNA analysis. *Anal Chem* 71:4232–4236.

Benedek K, Guttman A. 1994. Ultra-fast high-performance capillary sodium dodecyl sulfate gel electrophoresis of proteins. *J Chromatogr A* 680:375–381.

Bings NH, Wang C, Skinner CD, Colyer CL, Thibault P, Harrison DJ. 1999. Microfluidic devices connected to fused-silica capillaries with minimal dead volume. *Anal Chem* 71:3292–3296.

Bousse L, Mouradian S, Minalla A, Yee H, Williams K, Dubrow R. 2001. Protein sizing on a microchip. *Anal Chem* 73:1207–1212.

Chen X, Wu H, Mao C, Whitesides GM. 2002. A prototype two-dimensional capillary electrophoresis system fabricated in poly(dimethylsiloxane). *Anal Chem* 74:1772–1778.

Cheng J, Sheldon EL, Wu L, Uribe A, Gerrue LO, Carrino J, Heller MJ, O'Connell JP. 1998. Preparation and hybridization analysis of DNA/RNA from *E. coli* on microfabricated bioelectronic chips. *Nat Biotechnol* 16:541–546.

Duffy DC, McDonald JC, Schueller JA, Whitesides GM. 1998. Rapid prototyping of microfluidic systems in poly(dimethylsiloxane). *Anal Chem* 70:4974–4984.

Ekstrom S, Onnerfjord P, Nilsson J, Bengtsson M, Laurell T, Marko-Varga G. 2000. Integrated microanalytical technology enabling rapid and automated protein identification. *Anal Chem* 72:286–293.

Ericson C, Holm J, Ericson T, Hjerten, S. 2000. Electroosmosis- and pressure-driven chromatography in chips using continuous beds. *Anal Chem* 72:81–87.

Figeys D, Aebersold R. 1998. Nanoflow solvent gradient delivery from a microfabricated device for protein identifications by electrospray ionization mass spectrometry. *Anal Chem* 70:3721–3727.

Figeys D, Pinto D. 2001. Proteomics on a chip: promising developments. *Electrophoresis* 22:208–216.

Figeys D, Ning Y, Aebersold R. 1997. A microfabricated device for rapid protein identification by microelectrospray ion trap mass spectrometry. *Anal Chem* 69:3153–3160.

Gao J, Xu J, Locascio LE, Lee CS. 2001. Integrated microfluidic system enabling rapid protein digestion, peptide separation, and protein identification. *Anal Chem* 73:2648–2655.

Giddings JC. 1991. *United Separation Science*, Wiley, Hoboken, NJ.

Gottschlich N, Jacobson SC, Culbertson CT, Ramsey JM. 2001. Two-dimensional electrochromatography/capillary electrophoresis on a microchip. *Anal Chem* 73:2669–2674.

Harris CM. 2003. Shrinking the LC landscape. *Anal Chem* 75:65A–69A.

He B, Tait N, Regnier F. 1998. Fabrication of nanocolumns for liquid chromatography. *Anal Chem* 70:3790–3797.

Henry AC, Tutt TJ, Galloway M, Davidson YY, McWhorter CS, Soper SA, McCarley RL. 2000. Surface modification of poly(methyl methacrylate) used in the fabrication of microanalytical devices. *Anal Chem* 72:5331–5337.

Heo J, Thomas KJ, Seong GH, Grooks RM. 2003. A microfluidic bioreactor based on hydrogel-entrapped *E. coli*: cell viability, lysis, and intracellular enzyme reactions. *Anal Chem* 75:22–26.

Herr AE, Molho JI, Drouvalakis KA, Mikkelsen JC, Utz PJ, Santiago JG, Kenny TW. 2003. On-chip coupling of isoelectric focusing and free solution electrophoresis for multi-dimensional separations. *Anal Chem* 75:1180–1187.

Hofmann O, Che D, Cruickshank KA, Müller UR. 1999. Adaptation of capillary isoelectric focusing to microchannels on a glass chip. *Anal Chem* 71:678–686.

Holmquist M, Engstrom J, Selditz U, Andersson P, Wallenborg S. 2003. High-speed protein digestion and sample preparation for MALDI-MS on a microfluidic CD. HPCE 2003, January 17–22, San Diego, California.

Huang Y, Ewalt KL, Tirado M, Haigis R, Forster A, Ackley D, Heller MJ, O'Connell JP, Krihak M. 2001. Electric manipulation of bioparticles and macromolecules on micro-fabricated electrodes. *Anal Chem* 73:1549–1559.

Jackman RJ, Wilbur JL, Whitesides GM. 1995. Fabrication of submicrometer features on curved substrates by microcontact printing. *Science* 269:664–666.

Kameoka J, Orth R, Ilic B, Czaplewski D, Wachs T, Craighead HG. 2002. An electrospray ionization source for integration with microfluidics. *Anal Chem* 74:5897–5901.

Kaniansky D, Masar M, Bielcikova J, Ivanyi F, Eisenbeiss F, Stanislawski B, Grass B, Neyer A, Johnck M. 2000. Capillary electrophoresis separations on a planar chip with the column-coupling configuration of the separation channels. *Anal Chem* 72:3596–3604.

Kim JS, Knapp DR. 2001. Microfabricated PDMS multichannel emitter for electrospray ionization mass spectrometry. *J Am Soc Mass Spectrom* 12:463–469.

Koegler WS, Ivory CF. 1996. Focusing proteins in an electric field gradient. *J Chromatogr A* 726:229–236.

Lazar IM, Ramsey RS, Sundberg S, Ramsey JM. 1999. Subattomole-sensitivity microchip nanoelectrospray source with time-of-flight mass spectrometry detection. *Anal Chem* 71:3627–3631.

Li PCH, Harrison DJ. 1997. Transport, manipulation, and reaction of biological cells on-chip using electrokinetic effects. *Anal Chem* 69:1564–1568.

Li J, Thibault P, Bings NH, Skinner CD, Wang C, Coyler C, Harrison J. 1999. Integration of microfabricated devices to capillary electrophoresis–electrospray ionization mass spectrometry using a low dead volume connection: application to rapid analysis of pro-teolytic digests. *Anal Chem* 71:3036–3045.

Li Y, DeVoe DL, Lee CS. 2003. Dynamic analyte introduction and focusing in plastic microfluidic devices for proteome analysis. *Electrophoresis* 24:193–199.

Lichtenberg J, Verpoorte E, de Rooij NF. 2001. Sample preconcentration by field amplifica-tion stacking for microchip-based capillary electrophoresis. *Electrophoresis* 22:258–271.

Licklider L, Wang X, Desai A, Tai Y, Lee TD. 2000. A micromachined chip-based electro-spray source for mass spectrometry. *Anal Chem* 72:367–375.

Liu YJ, Foote RS, Culbertson CT, Jacobson SC, Ramsey RS, Ramsey JM. 2000a. Electro-phoretic separation of proteins on microchips. *J Microcol Sep* 12:407–411.

Liu YJ, Foote RS, Jacobson SC, Ramsey RS, Ramsey JM. 2000b. Electrophoretic separation of proteins on a microchip with noncovalent, postcolumn labeling. *Anal Chem* 72:4608–4613.

Macounova K, Carbrera CR, Holl MR, Yager P. 2000. Generation of natural pH gradients in microfluidic channels for use in isoelectric focusing. *Anal Chem* 72:3745–3751.

Macounova K, Carbrera CR, Yager P. 2001. Concentration and separation of proteins in microfluidic channels on the basis of transverse IEF. *Anal Chem* 73:1627–1633.

Manz A, Graber N, Widmer HM. 1990. Miniaturized total chemical analysis systems: a novel concept for chemical sensing. *Sensors Actuators* B1:244–248.

Mao Q, Pawliszyn J. 1999. Demonstration of isoelectric focusing on an etched quartz chip with UV absorption imaging detection. *Analyst* 124:637–641.

Martynova L, Locascio LE, Gaitan M, Kramer GW, Christensen RG, MacCrehan WA. 1997. Fabrication of plastic microfluidic channels by imprinting methods. *Anal Chem* 69:4783–4789.

McClain MA, Culbertson CT, Jacobson SC, Ramsey JM. 2001. Single cell lysis on microfluidic devices. In: *Proceedings of the Micro Total Analysis Systems*, Kluwer Academic Publishers, Boston, p. 301.

McDonald JC, Duffy DC, Anderson JR, Chiu DT, Wu H, Schueller OJA, Whitesides GM. 2000. Fabrication of microfluidic systems in poly(dimethylsiloxane). *Electrophoresis* 21:27–40.

O'Farrell PH. 1985. Separation techniques based on the opposition of two counteracting forces to produce a dynamic equilibrium. *Science* 227:1586–1589.

Oleschuk RD, Shultz-Lockyear LL, Ning Y, Harrison DJ. 2000. Trapping of bead-based reagents within microfluidic systems: on-chip solid-phase extraction and electrochromatography. *Anal Chem* 72:585–590.

Peterson DS, Rohr T, Svec F, Frechet JMJ. 2002. Enzymatic microreactor-on-a-chip: protein mapping using trypsin immobilized on porous polymer monoliths molded in channels of microfluidic devices. *Anal Chem* 74:4081–4088.

Ramsey RS, Ramsey JM. 1997. Generating electrospray from microchip devices using electroosmotic pumping. *Anal Chem* 69:1174–1178.

Rocklin RD, Ramsey RS, Ramsey JM. 2000. A microfabricated fluidic device for performing two-dimensional liquid-phase separations. *Anal Chem* 72:5244–5249.

Rohner TC, Rossier JS, Girault HH. 2001. Polymer microspray with an integrated thick-film microelectrode. *Anal Chem* 73:5353–5357.

Rosenberger F, Jones E, Lee CS, DeVoe DL. 2002. High pressure thermal bonding for sealing of plastic microstructures. In: *Proceedings of the Micro Total Analysis Systems*, Kluwer Academic Publishers, Boston, p. 404.

Ross D, Locascio LE. 2002. Microfluidic temperature gradient focusing. *Anal Chem* 74:2556–2564.

Rossier JS, Schwarz A, Reymond F, Ferrigno R, Bianchi F, Girault HH. 1999. Microchannel networks for electrophoretic separations. *Electrophoresis* 20:727–731.

Sakai-Kato K, Kato M, Toyo'oka T. 2003. Creation of an on-chip enzyme reactor by encapsulating trypsin in sol-gel on a plastic microchip. *Anal Chem* 75:388–393.

Schilling EA, Kamholz AE, Yager P. 2002. Cell lysis and protein extraction in a microfluidic device with detection by a fluorogenic enzyme assay. *Anal Chem* 74:1798–1804.

Schultz GA, Corso TN, Prosser SJ, Zhang S. 2000. A fully integrated monolithic microchip electrospray device for mass spectrometry. *Anal Chem* 72:4058–4063.

Slentz BE, Penner NA, Regnier FE. 2003. Protein proteolysis and the multi-dimensional electrochromatographic separation of histidine-containing peptides on a chip. *J Chromatogr A* 984:97–107.

Soper SA, Ford SM, Qi S, McCarley RL, Kelly K, Murphy MC. 2000. Polymeric micro-electromechanical systems. *Anal Chem* 72:643A–651A.

Tang L, Sheu MS, Chu T, Huang YH. 1999. Anti-inflammatory properties of triblock siloxane copolymer-blended materials. *Biomaterials* 20:1365–1370.

Tang K, Lin Y, Matson DW, Kim T, Smith RD. 2001. Generation of multiple electrosprays using microfabricated emitter arrays for improved mass spectrometric sensitivity. *Anal Chem* 73:1658–1663.

Taylor MT, Belgrader P, Furman BJ, Pourahmadi F, Kovacs GTA, Northrup MA. 2001. Lysing bacterial spores by sonication through a flexible interface in a microfluidic system. *Anal Chem* 73:492–496.

Throckmorton DJ, Shepodd TJ, Singh AK. 2002. Electrochromatography in microchips: reversed-phase separation of peptides and amino acids using photopatterned rigid polymer monoliths. *Anal Chem* 74:784–789.

Wainright A, Williams SJ, Ciambrone G, Xue Q, Wei J, Harris D. 2002. Sample preconcentration by isotachophoresis in microfluidic devices. *J Chromatogr A* 979:69–80.

Wang XB, Yang J, Huang Y, Vykoukal J, Becker FF, Gascoyne PRC. 2000a. Cell separation by dielectrophoretic field-flow-fractionation. *Anal Chem* 72:832–839.

Wang C, Oleschuk R, Ouchen F, Li JJ, Thibault P, Harrison DJ. 2000b. Integration of immobilized trypsin beds for protein digestion within a microfluidic chip incorporating capillary electrophoresis separations and an electrospray mass spectrometry interface. *Rapid Commun Mass Spectrom* 14:1377–1383.

Waters LC, Jacobson SC, Kroutchinina N, Khandurina J, Foote RS, Ramsey JM. 1998. Microchip device for cell lysis, multiplex PCR amplification, and electrophoretic sizing. *Anal Chem* 70:158–162.

Wen J, Lin Y, Xiang F, Matson DW, Udseth HR, Smith RD. 2000. Microfabricated isoelectric focusing device for direct electrospray ionization–mass spectrometry. *Electrophoresis* 21:191–197.

Xiang F, Lin, Y, Wen J, Matson DW, Smith RD. 1999. An integrated microfabricated device for dual microdialysis and on-line ESI-ion trap mass spectrometry for analysis of complex biological samples. *Anal Chem* 71:1485–1490.

Xu N, Lin Y, Hofstadler SA, Matson D, Call CJ, Smith RD. 1998. A microfabricated dialysis device for sample cleanup in electrospray ionization mass spectrometry. *Anal Chem* 70:3553–3556.

Xu J, Locascio LE, Gaitan M, Lee CS. 2000. Room-temperature imprinting method for plastic microchannel fabrication. *Anal Chem* 72:1930–1933.

Xue Q, Dunayevskiy YM, Foret F, Karger BL. 1997a. Integrated multichannel microchip electrospray ionization mass spectrometry: analysis of peptides from on-chip tryptic digestion of melittin. *Rapid Commun Mass Spectrom* 11:1253–1256.

Xue Q, Foret F, Dunayevskiy YM, Zavracky PM, McGruer NE, Karger BL. 1997b. Multichannel microchip electrospray mass spectrometry. *Anal Chem* 69:426–430.

Yao S, Anex DS, Caldwell WB, Arnold DW, Smith KB, Schultz PG. 1999. SDS capillary gel electrophoresis of proteins in microfabricated channels. *Proc Natl Acad Sci USA* 96:5372–5377.

Yuan CH, Shiea J. 2001. Sequential electrospray analysis using sharp-tip channels fabricated on a plastic chip. *Anal Chem* 73:1080.

Zhang B, Liu H, Karger BL, Foret F. 1999. Microfabricated devices for capillary electrophoresis–electrospray ionization mass spectrometry. *Anal Chem* 71:3258–3264.

Zhang B, Foret F, Karger BL. 2000. A microdevice with integrated liquid junction for facile peptide and protein analysis by capillary electrophoresis/electrospray mass spectrometry. *Anal Chem* 72:1015–1022.

12

Single Cell Proteomics

**Norman J. Dovichi, Shen Hu, David Michels, Danqian Mao,
and Amy Dambrowitz**

University of Washington, Seattle, Washington

12.1   INTRODUCTION

Any attempt to explain the function of a protein must account for its spatial and temporal location. Not all proteins are present in all cells, and many proteins have fleeting existence as synthesis, modification, and degradation occur while the cell responds to its environment, passes through the cell cycle, and recycles its contents. Each cell will behave differently, and ultimately proteomics must descend to the level of the single cell to provide an accurate and complete description of protein function and expression. Ideally, that study should be correlated with other characteristics of the cell, such as phase of the cell in the cell cycle for proliferating cells.

We cite three examples where single cell proteomics will be valuable. The first is in development. The cells of an embryo undergo profound changes in protein expression as the organism grows from a zygote to a fully developed individual. Similar changes in protein expression occur as pluripotent stem cells progress first into precursor cells and then into terminally differentiated progeny cells. A fundamental understanding of the steps involved in development will be aided greatly by monitoring the proteome on a cell-by-cell basis.

The second example is in oncology. We have hypothesized that the cell-to-cell heterogeneity in protein expression of a tumor is correlated with the prognosis of that cancer (Hu et al., 2003a). Just as aneuploidy is correlated with poor prognosis in some cancers, it may prove that the cell-to-cell heterogeneity in protein expression is correlated with prognosis. If that hypothesis is correct, then large-scale studies of the protein content of single cells will be of great value in the clinic.

Proteomics for Biological Discovery, edited by Timothy D. Veenstra and John R. Yates.
Copyright © 2006 John Wiley & Sons, Inc.

The third example is in neuroscience, where individual neurons in the central nervous system are extremely heterogeneous. That heterogeneity reflects differences in the content and architecture of the cells. The creation of a vocabulary to describe cells and cell types will benefit greatly from the analysis of the protein content of the cell.

Single cell proteomics is of obvious value in basic and clinical studies. Nevertheless, single cell proteomics presents significant challenges because of the minute amount of protein present in the cell. This chapter describes those challenges and several approaches to the characterization of proteins in single eukaryotic cells.

12.2   THE CHALLENGE

Most proteins are present at minuscule levels within a cell as shown in Table 12.1. We consider four examples. A 100 μm diameter giant neuron contains perhaps 50 ng of protein, assuming that the cell is 10% protein by weight. The average molecular weight for a protein is about 30,000 g/mol, so that a giant neuron contains perhaps 2 pmol, or 10^{12} copies, of protein.

A typical mammalian cell has a diameter of roughly 10 μm, with a volume of 0.5 pL and a total protein content of 50 pg. Again, assuming an average molecular weight of 30 kDa, the cell contains about 2 fmol total protein, or about one billion copies of protein molecules.

A 5 μm diameter yeast cell contains about 5 pg or 0.2 fmol of protein. This primitive eucaryote functions with about 100 million copies of protein.

A 1 μm diameter bacterium contains only 50 fg or 2 amol of protein. Those organisms survive and replicate using only one million copies of protein.

The number of proteins expressed in a single cell is unknown. An animal might express 10,000 different proteins per cell. If we use this quite arbitrary number, the average protein is expected to be present at the 200 zmol level in a 10 μm diameter mammalian cell. Yeast expresses fewer proteins, perhaps 2000 per cell; the typical protein would be present at the 100 amol level.

However, the concept of the average protein is inappropriate because the distribution of protein expression in a cell is expected to be highly heterogeneous. As an analogy, if Bill Gates were added to the list of authors for this chapter, then the average wealth of the group would be about 10 billion dollars. Clearly, this average does not provide insight into the lifestyle of the authors! Similarly, the average

TABLE 12.1. Protein Content of a Single Cell

	Giant Neuron	Mammalian Cell	Yeast Cell	Bacterium
Volume	0.5 nL	0.5 pL	50 fL	0.5 fL
Total mass	0.5 μg	0.5 ng	50 pg	0.5 pg
Mass protein[a]	50 ng	50 pg	5 pg	50 fg
Moles protein[b]	2 pmol	2 fmol	0.2 fmol	2 amol
Copies protein	1×10^{12}	1×10^9	1×10^8	1×10^6

[a] Assumes the cell is 10% protein by mass.
[b] Assumes an average molecular weight for proteins of 20 kDa.
Prefixes: n = nano = 10^{-9}; p = pico = 10^{-12}; f = femto = 10^{-15}; a = atto = 10^{-18}; z = zepto = 10^{-21}; y = yocto = 10^{-24}.

protein level of a cell does not provide insight into the challenges of the study of a single cell's proteome.

There are limited data available on protein expression levels. In one of the best studies, Gygi et al. (1999a) treated *Saccharomyces cerevisiae* with a radiotracer followed by two-dimensional gel electrophoretic separation of the proteins from the yeast homogenate. Spots were excised and quantitated using scintillation counting. The proteins were identified by mass spectrometry.

The authors obtained expression data for 156 spots corresponding to 124 different gene products. Of these, eleven proteins had expression levels that were estimated to be greater than 100,000 copies (0.2 amol) per cell. This set of 124 gene products accounts for a total of ~8 million copies of protein per yeast cell, which is ~10% of the total protein content of a mid-log yeast cell. It appears that the vast majority of the protein content of the cell was either lost during sample preparation or is found in the many proteins expressed at low levels.

If we extrapolate this data from yeast to a mammalian system, we expect the total protein concentration to be an order of magnitude higher. The number of genes in mammals is also roughly an order of magnitude larger than the number of genes in yeast, so that the average protein level per cell is likely to be similar. If there are ~10 proteins present in a single yeast cell that are expressed at greater than 100,000 copies (0.2 amol or ~5 fg) per cell, then it is reasonable to expect at least 100 proteins would be expressed at that level in a single mammalian cell.

12.3 SINGLE CELL PROTEOMICS: MASS SPECTROMETRY

Mass spectrometry is undoubtedly the tool of choice in protein analysis. Technologies such as multidimensional protein identification technology (MudPIT) and isotope-coded affinity tag (ICAT) provide powerful means of identifying proteins and determining changes in their expression (Gygi et al., 1999b; Washburn et al., 2001). Unfortunately, these methods require relatively large amounts of proteins, certainly much more than contained in a single cell. Smith (2002) has estimated detection limits of 100 zmol for tryptic peptides by use of Fourier transform–ion cyclotron resonance (FT-ICR) instrumentation. However, that detection limit is for the amount of peptide introduced into the spectrometer. Significant losses accompany protein isolation, digestion, peptide extraction, and chromatographic separation, and it is not clear that analysis of tryptic digests from single cells will be practical.

Instead, it appears necessary to perform mass spectrometry on intact proteins to realize single cell analysis. Matrix assisted laser-desorption and ionization (MALDI)-based methods have been used with some success in detecting peptides and low molecular weight proteins from single cells and tissue slices. Whittal et al. (1998) reported the detection of hemoglobin from a single erythrocyte by MALDI time-of-flight (TOF) mass spectrometry. In their approach, the cell was lysed, mixed with protease, and then microspotted on a MALDI plate. Each cell contains about 450 amol of hemoglobin, and only extremely highly expressed proteins can be detected and identified in a single cell.

Xu et al. (2002) has reported the use of MALDI-TOF to analyze tissue samples. As few as 10 cells have been used for analysis. That sample generated roughly 10 peaks that exceeded the noise in the baseline. This TOF mass spectrometer has relatively poor mass resolution and does not employ MS/MS analysis, which prevents direct identification of proteins.

Page et al. (2002) have reported the use of MALDI-TOF to analyze neuropeptides in single giant neurons. This method is not applicable to peptides greater than a few kilodaltons in molecular mass.

Pasa-Tolic et al. (2002) considered the use of accurate mass tags for protein analysis with high resolution and accuracy FTICR instruments. That group reported the use of an 11 tesla FTICR allowed determination of a number of proteins obtained from a 5 pg protein sample from a *D. radiodurans* protein homogenate that was injected onto a 15 μm inner diameter capillary liquid chromatography column. The proteome of this prokaryote is certainly much simpler than that of eukaryotes, and the introduction of the protein lysate from a single cell onto the chromatography column is not demonstrated; nevertheless these results are encouraging that the protein content of a single mammalian cell will likely be performed using mass spectrometry within the decade.

12.4 SINGLE CELL SEPARATIONS

The use of separation methods to characterize the composition of a single cell has a fifty year history. The first study in 1953 considered ribosomal ribonucleic acid (rRNA) analysis in single cells based on electrophoresis on a silk fiber (Edstrom, 1953). The earliest single cell protein analysis was a study of hemoglobin in single erythrocytes by electrophoresis through an acrylamide fiber, published in 1965 (Matioli and Niewisch, 1965). Repin et al. (1975) reported the characterization of lactate dehydrogenase isoenzymes in single mammalian oocytes by electrophoresis in 1975. Ruchel (1976) reported the first protein analysis from a single giant neuron from a sea snail the following year.

Kennedy et al. (1987, 1989) inaugurated the modern era of single cell analysis by using open tubular capillary chromatography for the amino acid analysis of a single giant neuron from a snail. At the same time, Wallingford and Ewing (1988) reported the use of a capillary to sample the internal contents of a single giant neuron; they used the same capillary for electrophoresis of biogenic amines. In 1992, Hogan and Yeung (1992) reported the use of a specific label to derivatize thiols in individual erythrocytes; once the derivatization reaction was completed, a single cell was injected into a capillary and lysed, and the contents were separated by capillary electrophoresis (CE). In 1995, Gilman and Ewing (1995) reported the on-column labeling of amines from a cell that had been injected into and lysed within a capillary; the capillary was used for electrophoretic separation of amines. Meredith et al. (2000) reported the use of a pulsed laser to lyse a cell before analysis of kinase activities by CE. In 2002, Han and Lillard (2002) reported the use of cell synchronization based on the shake-off method for the characterization of RNA synthesis in single cells as a function of cell cycle.

Our group's activity in single cell analysis began in collaboration with Monica Palcic and Ole Hindsgaul at the University of Alberta, in a study of carbohydrate metabolism in single yeast and cancer cells (Krylov et al., 1999; Le et al., 1999). In that project, cells were treated with a fluorescent enzymatic substrate. Cells took up this substrate and metabolized it to create biosynthetic and biodegradation products. To study this behavior in single cells, we developed techniques to manipulate and inject single cells into a fused silica capillary, to lyse the cells within the capillary, to separate the cellular lysate by capillary electrophoresis, and to detect the separated lysate by laser-induced fluorescence (Krylov et al., 1999, 2000).

In our instrument (Figure 12.1) an inverted microscope is equipped with a capillary manipulator manufactured from Lexan (GE Plastics, Pittsfield, MA) and held by a set of micromanipulators. The capillary is filled with separation buffer that contains a surfactant, such as sodium dodecyl sulfate (SDS). The capillary tip is then centered over the cell of interest; to inject the cell, a brief pulse of vacuum is applied to the distal end of the capillary through a computer-controlled valve, aspirating the cell about 200 μm into the capillary. Mammalian cells are lysed within 30 seconds by the action of SDS and the osmotic shock from the buffer (Krylov et al., 2000).

Electrophoresis is effected by application of high voltage to the injection end of the capillary, which is held within the interlock-equipped Lexan manipulation block. The distal end of the capillary is placed in a sheath-flow cuvette, and fluorescence is excited by an argon ion laser beam and collected with a microscope objective, which images fluorescence onto a pinhole and through a spectral filter to block scattered laser light. The fluorescence is finally detected by a photomultiplier tube and recorded by a computer.

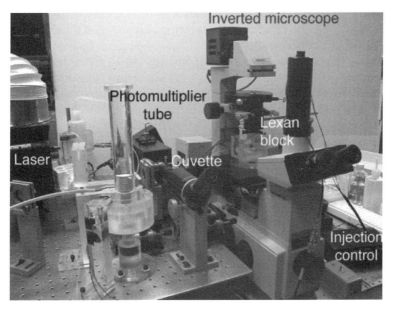

Figure 12.1. *Photograph of single cell electrophoresis instrument.*

12.5 ULTRASENSITIVE PROTEIN ANALYSIS: CAPILLARY ELECTROPHORESIS WITH LASER-INDUCED FLUORESCENCE DETECTION

The single cell methods have been developed for injection, lysis, and analysis of single cells by capillary electrophoresis. CE provides exquisite separation of complex mixtures. Laser-induced fluorescence (LIF) provides exquisite sensitivity for highly fluorescent molecules. The combination of the two tools provides an extremely powerful method for the analysis of biological molecules. To illustrate the analysis of a complex mixture, capillary array instruments have been the workhorse of large-scale genomic DNA sequencing efforts (Dovichi and Zhang, 2000). As an example of ultrasensitive analysis, CE-LIF has been used to detect and count single molecules of *b*-phycoerythrin, resolving isoforms of this protein (Chen and Dovichi, 1996).

The study of *b*-phycoerythrin was based on that molecule's native fluorescence. That molecule has unusual spectroscopic properties. Most proteins exhibit native fluorescence only when excited in the ultraviolet portion of the spectrum. The molar absorptivity and fluorescent quantum yields of the aromatic amino acids that generate native fluorescence are modest, and the background signal from fluorescent impurities tends to be high in this portion of the spectrum, leading to relatively poor detection performance. More practically, lasers that operate in the ultraviolet tend to be expensive and temperamental, which discourages their widespread use.

Instead, our approach to ultrasensitive protein analysis relies on the use of derivatization technology to introduce a fluorescent tag that is excited with more robust lasers that operate in the visible portion of the spectrum. These tags have relatively high molar absorptivity and fluorescent quantum yields, which assists sensitive detection. However, tagging chemistry is not without its challenges. We highlight several issues here.

First, it is necessary to perform the labeling reaction on dilute proteins, which may be present at a concentration in the picomole per liter range or less. The reaction is usually governed by second-order kinetics, so that it is necessary to keep the concentration of the derivatizing reagent in the millimolar range so that the reaction proceeds at a reasonable rate. The vast excess of derivatizing reagent can result in a huge background signal from unreacted reagent. While the reagent itself can be separated from proteins during the electrophoretic step, the sea of fluorescent impurities that accompany the reagent at the part-per-million level will swamp the signal from the labeled proteins. Fortunately, this fluorescent background issue can be eliminated by use of fluorogenic reagents; these reagents are nonfluorescent until they react with the protein, creating a fluorescent product (Liu et al., 1991).

Second, the use of fluorogenic reagents inevitably results in the production of a complex mixture of fluorescent products. All fluorogenic reagents of which we are aware react with the ε-amine of lysine residues, which is one of the most common amino acids. We have analyzed the yeast genome to estimate the relative abundance of lysine residues. The average open reading frame codes for ~26 lysine residues, so that most proteins will incorporate more than one fluorescent label (Pinto et al., 2003). Unfortunately, extreme denaturation conditions are required to drive the labeling reaction to completion (Liu et al., 2001). These conditions appear impractical when studying the proteome of single cells. If the reaction does not go to completion, then a complex reaction mixture will inevitably be produced. The number of possible fluorescent reaction products is 2^N-1, where N is the number of lysine residues (and other primary amines) present in the molecule (Zhao et al., 1992). The average protein in yeast, with 26 lysine residues, can produce 67,108,863 different reaction products. Unfortunately, these fluorescent products from a single protein can have different electrophoretic mobility, and multiple-labeling results in extremely complex electropherograms that are essentially worthless for protein analysis.

Fortunately, we have discovered one solution to the multiple-labeling problem. The reaction of proteins with the fluorogenic reagent 5-furoylquinoline-3-carboxaldehyde (FQ), followed by use of appropriate buffers, results in remarkably efficient electrophoretic separations (Pinto et al., 1997; Lee et al., 1998). FQ, unlike other fluorogenic reagents, produces a neutral reaction product; cationic lysine residues are converted to neutral products. In the absence of buffer additives, the heterogeneity in the number of lysine residues that are labeled results in multiple electrophoretic peaks from a single protein. However, we discovered that the addition of an anionic surfactant, such as SDS, to the buffer results in the collapse of that complex envelope into a single, sharp peak for each protein. We believe that the surfactant ion-pairs with unreacted lysine residues, producing a neutral complex with the same mobility as the FQ-labeled molecule.

The interactions between SDS and proteins are complicated. At low surfactant concentration, SDS binds specifically to high-energy sites of the protein through electrostatic interactions; the anionic surfactant ion-pairs with easily accessible lysine and arginine residues (Oakes, 1974; Turro and Lei, 1995; Santos et al., 2003). At intermediate concentrations, SDS binds through hydrophobic interactions with the surface of the protein. At higher surfactant concentration, there is a massive increase in the binding of SDS as the protein unfolds and its interior becomes accessible. The surfactant is thought to form a random set of micelles along the protein's backbone. Under saturation conditions, 1 g of many proteins will bind ~1.4 g of SDS (Oakes, 1974). This constant binding ratio is important in SDS-PAGE. At high SDS concentration, the complex of protein and SDS is assumed to generate a constant size-to-charge ratio. Like DNA electrophoresis in polymeric medium, SDS–protein complexes are separated based on size during polyacrylamide gel electrophoresis (PAGE).

12.6 CAPILLARY SIEVING ELECTROPHORESIS OF PROTEINS FROM A SINGLE CANCER CELL

We have developed a number of forms of electrophoresis to analyze the protein content of single cells. In the analysis of the protein content of a single cell, we sandwich the proteins from the single cell between two plugs of the FQ derivatizing reagent. A plug of reagent, followed by the single cell, and then another plug of reagent are sequentially injected into the capillary. The reagent solution contains SDS to lyse the cell, and the capillary tip is heated to ~90 °C to speed the labeling reaction, which typically is performed for 4 minutes.

We have developed methods to perform the capillary equivalent of SDS-PAGE on single mammalian cells (Hu et al., 2001a, 2002, 2003a). We find that acrylamide is a difficult polymer to work with; its free radical polymerization is difficult to control in the laboratory. Instead, we have investigated a number of commercially available polymers for the separation, including polyethylene oxide and the polysaccharide pullulan, for the separation of proteins from a single cell. These polymers are not crosslinked and have relatively low viscosity, so that a cell can easily be aspirated into the capillary. Separation with buffers containing polymers is called sieving electrophoresis to reflect the size-based separation mechanism.

The capillary sieving separation of the proteins from a single HT-29 human adenocarcinoma cell is shown in Figure 12.2. Perhaps 25 components are partially resolved from this cell. The earliest migrating peaks correspond to low molecular weight proteins, peptides, and biogenic amines. The peak migrating at 24 minutes has a molecular weight of about 100 kDa. The number of components resolved from this single cell is similar to the number of bands resolved on a 10 cm long SDS-PAGE separation of the proteins extracted from a few million cells, but less than that separated by a high-resolution gel.

The dynamic range of the separation is exquisite. The noise in the baseline is typically a part-per-thousand or less of the maximum signal. As discussed later, we are modifying our instrument to incorporate higher sensitivity fluorescence detectors based on avalanche photodiodes. This improved instrument has a fivefold higher sensitivity, which will further extend the dynamic range of the instrument.

We worried about contamination of the single cell electropherogram by residual culture medium. To minimize the amount of culture medium present, we wash the cells and resuspend them in phosphate-buffered saline. Injection of the cellular supernatant generates a low amplitude, featureless background signal that does not interfere with the single cell data.

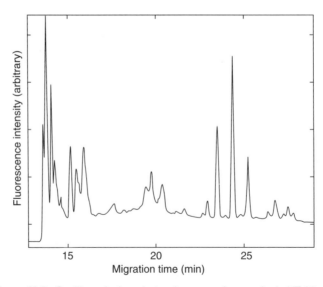

Figure 12.2. Capillary sieving electropherogram from a single HT-29 cell.

12.7 CELL CYCLE DEPENDENT SINGLE CELL CAPILLARY SIEVING ELECTROPHORESIS

We generated a number of capillary sieving electropherograms from single HT-29 cells. Those data showed reasonable reproducibility in the migration time for individual peaks (Hu et al., 2003a). However, there was a large cell-to-cell variation in the peak amplitude, averaging 40% in relative standard deviation. Some components varied in amplitude by more than an order of magnitude.

We were confident that the cell-to-cell variation in peak amplitude was not due to variation in the performance of the electrophoresis or detection system because replicate injections of the same cellular homogenate generated quite reproducible peaks. Instead, it became clear that the variation in amplitude was associated either with the cell lysis or was inherent in the cells themselves.

These cells proliferate, and a likely cause of cell-to-cell variation in protein expression is the phase of the cell cycle. The fluorescence microscope of Figure 12.1 can be used to monitor the presence of classic cytometry stains used to treat the cells before analysis. We treated the cells with Hoechst 33342, which is a vital nuclear stain. This stain is taken up by living cells and transported to the nucleus, where it intercalates within double stranded oligonucleotides. The intercalated dye becomes highly fluorescent, and the resulting fluorescence intensity can be used to estimate the DNA content of the cell. Recently divided cells are diploid, with two pairs of chromosomes, while cells about to undergo mitosis are tetraploid, with four pairs of chromosomes. We use a photomultiplier tube to monitor the fluorescence signal generated by the Hoechst stain before injection into the capillary for electrophoretic analysis, and we classify cells into G1 and G2/M phase based on that signal.

We discovered that the cell-to-cell variation in peak amplitude was indeed dominated by the phase of the cell in the cell cycle (Hu et al., 2003a). On average, cells in the G2/M phase of the cell cycle generated electrophoresis peaks that were twice the amplitude of cells in the G1 phase; this result is expected because two G1 phase daughter cells are produced by division of one G2/M parent cell. Cells in the G1 phase of the cell cycle generated electropherograms whose peaks had a 27% relative standard deviation while cells in the G2/M phase of the cell cycle generated electropherograms with a 20% relative standard deviation. The majority of the cell-to-cell variation in protein expression is associated with differences in the phase of the cell in the cell cycle. Residual cell-to-cell variation in protein electropherograms likely reflects differences of the cells *within* the phase of the cell cycle.

We also performed electrophoretic analysis of an anomalous cell that appeared to contain six pairs of chromosomes. This hexaploid cell was identified based on its Hoechst staining characteristics. This cell line has been karyotyped, and some cells in this cell line reveal extra copies of chromosomes. The protein electropherogram of this single unusual cell was more intense than that of other cells, and contained one 45 kDa component that was dramatically upregulated compared to G1 and G2/M phase cells. Analysis of this unusual cell reveals a strength of single cell protein analysis. Cells with peculiar properties can be plucked for analysis, whereas preparation of a cellular homogenate for classic analysis will dilute the signal from unusual cells to an undetectable extent.

12.8   TENTATIVE IDENTIFICATION OF PROTEINS IN SINGLE CELL ELECTROPHEROGRAMS

We normalized the single cell electropherograms to their DNA content. The resulting electropherograms did not differ to a large extent with cell cycle. Only one component differed at the 99% confidence limit between the G1 and G2/M phase cells (Hu et al., 2003a). That component migrated with an apparent molecular weight of 45 kDa, which was the same molecular weight as the highly expressed protein in the hexaploid cell.

We worked in collaboration with Ruedi Aebersold and Rick Newitt of the Institute for Systems Biology to tentatively identify this protein (Hu et al., 2003b). First, we prepared a large-scale homogenate of this cell line. Capillary sieving electrophoresis analysis of the homogenate was similar to the average single cell electropherogram but much different from electropherograms generated from the cytosolic, membrane/organelle, nuclear, and cytoskeletal/nuclear matrix fractions of this cell line. The single cell analysis appears to sample a homogeneous portion of the single cell's content.

We next used classic SDS-PAGE to separate the cellular homogenate. Fortunately, the banding pattern was similar to the capillary electrophoresis peaks, and we were able to isolate a 45 kDa band from the gel. We took a portion of the purified proteins contained within this band and spiked the cellular homogenate before capillary electrophoresis analysis. The proteins isolated from the 45 kDa SDS-PAGE gel comigrated with the target 45 kDa peak in the capillary sieving electropherogram. Whichever cellular component demonstrated the cell cycle dependent change in expression was present in the 45 kDa band isolated from the gel and in the 45 kDa peak observed in the single cell electropherograms.

In-gel digestion, liquid chromatographic separation of the extracted peptides, MS/MS analysis of the peptides, and database searching were performed to identify the proteins present within the 45 kDa band. Five proteins were identified in the band. However, only one protein was identified from more than one peptide. That protein, cytokeratin 18, is the product of one of the most highly expressed genes in this cell line. It is known to undergo a massive change in phosphorylation state in the G2/M phase of the cell cycle. Careful inspection of the single cell electropherograms revealed that the peak corresponding to this component underwent a mobility shift to faster migration time in the G2/M phase of the cell cycle, consistent with addition of negatively charged phosphate groups to the protein. It was this mobility shift that caused the apparent change in amplitude of the 45 kDa component.

This general approach provides a tedious, but ultimately successful, method for the tentative identification of proteins present in single cells. In the general case, a library of proteins is prepared using classic one- (1D) or two-dimensional (2D) electrophoretic separation of proteins prepared from a large-scale cellular homogenate. The library is split into two parts. One part is used to identify the components using classic mass spectrometric methods and the other is archived in a freezer and used to spike CE samples, where comigration is taken as evidence for the identity of the component from the single cell. While tedious, this procedure needs to be performed only once for each cell type. Migration patterns appear to be sufficiently robust in CE that a single comigration study should suffice for most applications.

The similarity of the capillary sieving electropherograms and the SDS-PAGE electropherograms is important. The identification of an interesting peak in the single cell data can be simplified by analyzing the corresponding band in the SDS-PAGE electropherogram.

However, this study points out a limitation of one-dimensional electrophoresis for the analysis of complex proteomes: only a limited number of components are resolved. A capillary analog of 2D electrophoresis is required to characterize more fully the proteome of a single cell.

12.9 CAPILLARY MICELLAR AND SUBMICELLAR SEPARATION OF PROTEINS FROM A SINGLE CELL

In the ideal case, the second dimension for protein characterization would be isoelectric focusing, to provide the capillary equivalent of classic 2D gels. Unfortunately, the multiple labeling issue discussed earlier is devastating in isoelectric focusing, leading to complex and essentially worthless electrophoresis data from fluorescently labeled proteins (Richards et al., 1999). We are not aware of a method to correct for this effect.

To generate an alternative form of electrophoresis of single cells, we turned in desperation to a micellar and submicellar electrophoresis (Wallingford and Ewing, 1988; Hogan and Yeung, 1992). As noted earlier, the addition of SDS to labeled proteins results in the collapse of the multiple labeling envelope to a single sharp peak. However, this phenomenon is only of value if the peaks from different proteins can be resolved. We were concerned that this separation would not be possible. As reported by Oakes (1974) nearly 30 years ago, proteins take up a saturating amount of 1.4g of SDS per gram of protein. If that ratio were accurate, then the charge-to-size ratio of the SDS-treated protein should be constant. Electrophoresis in the absence of a sieving medium tends to separate proteins based on their size-to-charge ratio, so that we expected poor resolution of proteins in the presence of SDS and the absence of polymer.

Fortunately, this fear was unfounded, and we applied the technology to the analysis of the proteins from single cancer cells and from a single cell embryo (Zhang et al., 2000; Hu et al., 2001b). The CE separation of proteins from a single *Caenorhabditis elegans* zygote is shown in Figure 12.3. Sample handling was a challenge in this experiment. It was necessary to dissect a single worm to isolate eggs. The egg shell was removed with treatment with chitinase and chymotrypsin before the cell was injected into the capillary and lysed. Roughly 25 components were partially resolved over a quite wide separation window.

This wide separation window is consistently observed with different cell types. We observe that the separation efficiency tends to maximize with surfactant near, but below, its critical micelle concentration. Although the separation in Figure 12.3 was performed with a simple sodium phosphate buffer, we have more recently observed improved resolution with the addition of alcohol to the separation buffer. We believe that the alcohol decreases the partitioning of hydrophobic proteins from the surfactant, leading to improved resolution.

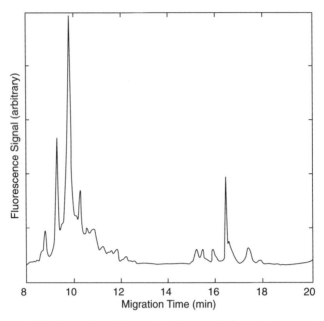

Figure 12.3. *Separation of the proteins from a single* C. elegans *embryo.*

12.10   TWO-DIMENSIONAL CAPILLARY ELECTROPHORESIS OF THE PROTEINS IN A SINGLE CELL

Giddings (1991) realized that multidimensional separations provide resolution that can be as high as the product of the resolution of the individual dimensions. For example, if 32 components can be resolved in one-dimensional separation and 32 components resolved in another type of separation, the combination of those two methods would be able, in principle, to resolve $32 \times 32 = 1024$ components.

Use of conventional two-dimensional gel electrophoresis technology is inappropriate for single cell analysis. The minute amount of protein would be severely diluted. Instead, we have developed an automated 2D CE instrument, as shown in Figure 12.4 (Michels et al., 2002). In this instrument, our injection system is coupled to a capillary sieving electrophoresis column. The cell is injected and lysed, and proteins are labeled and separated, as in the single capillary system. However, in this instrument, a second capillary is coupled to the exit of the first. When analyte reaches the exit of the first capillary, power supply 1 is turned off, and power supply 2 is turned on. Analyte present in the interface is injected onto the second capillary, separated by micellar electrophoresis, and detected with our ultrasensitive LIF detector. Once analyte has been separated in the second capillary, power is briefly applied to the first capillary, migrating another fraction into the interface for subsequent injection and separation. This process is repeated for 100–200 iterations, performing a comprehensive separation of the proteins in a single cell.

This instrument is based on the coupled liquid chromatography/capillary electrophoresis systems developed by Jorgenson's group in the 1990s (Bushey and

Jorgenson, 1990). We employ a simple coaxial transfer interface (Figure 12.5), where the tips of the two capillaries are separated by about 50 μm in a buffer-filled chamber. Electrical connection is provided to this chamber so that an electric field can be applied across either capillary as needed. In the figure, a plug of fluorescein is being transferred from the first to the second capillary.

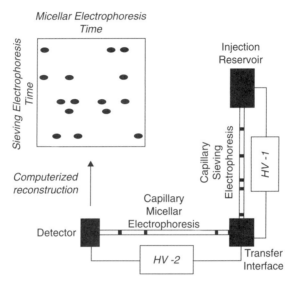

Figure 12.4. *Two-dimensional capillary electrophoresis instrument. HV-1 and HV-2 are high-voltage power supplies.*

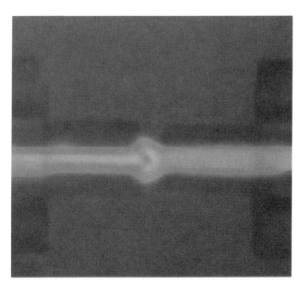

Figure 12.5. *Coaxial sample interface. A fluorescent microscope is used to image the transfer of a plug of fluorescein between two capillaries. (See color insert.)*

We have analyzed a number of samples and cell types using this instrument. A 2D electropherogram of the separation of the proteins from a single MC3T3 osteo-precursor cell is presented in Figure 12.6. These cells are associated with bone growth and repair, and they are intermediate between primitive stem cells and fully differentiated bone.

The data exist in the computer and may be presented in several different formats. This figure shows the data as an intensity plot, where the density is proportional to the logarithm of the fluorescence intensity. The image has been overexposed to highlight some of the lower intensity components present in the sample. However, this presentation format does not reproduce well the dynamic range of the fluorescence signal, which extends for nearly four orders of magnitude.

There are two streaks that are present in each sieving electrophoresis fraction, one at 10 seconds and the other at 85 seconds in the micellar electrophoresis dimension. These streaks are system peaks generated by the difference in buffer composition in the first- and second-dimension capillaries. We believe that the first component is generated by sieving material that is transferred to the second capillary with each sample. This material has a different refractive index than the sheath buffer. The laser beam can be scattered by this refractive index inhomogeneity, which results in the observed signal. The second streak is of much lower amplitude and represents a minute amount of scattered light, likely due to the presence of small ions used to prepare the buffers.

The intensity plot of Figure 12.6 does a poor job of displaying relative abundances. We instead prefer to use a landscape format to present the data as shown in Figure 12.7. The streaks that extend across the image are resolved into a set of ridges. The highest amplitude component that migrates at low molecular weight generated a fluorescence signal that saturated the detector. Unfortunately, the low

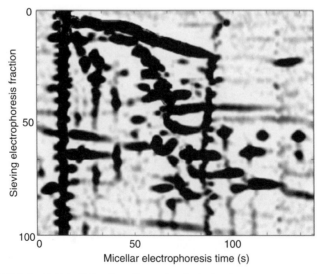

Figure 12.6. *Intensity image of the two-dimensional electropherogram generated from a single MC3T3 cell.*

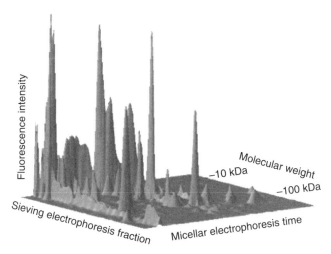

Figure 12.7. *Landscape view of the data presented in Figure 12.6.*

molecular weight components are poorly resolved by this micellar electrophoresis buffer system.

We are investigating different micellar and submicellar buffer systems to better resolve the complex protein content of a single cell. Once a satisfactory buffer system is available, we will begin the process of tentatively identifying components in the electropherogram. We will perform classic 2D gel electrophoresis on a cellular homogenate, and we will excise spots for mass spectrometric identification. As in our 1D capillary sieving electrophoresis experiment, we will spike the sample before 2D electrophoresis to perform comigration analysis. We will employ a combinatorial spiking protocol to speed the identification of peaks.

12.11 SINGLE COPY DETECTION OF SPECIFIC PROTEINS IN SINGLE CELLS

The multidimensional CE system has detection limits of a few thousand copies of many proteins. We are modifying the instrument by replacing the photomultiplier tube with a single-photon counting avalanche photodiode. This photodetector has a fivefold higher quantum yield, which will result in a modest improvement in detection limit. This improved instrument will certainly have many applications in both basic and clinical studies.

However, many researchers are interested in the study of proteins that are expressed at very low levels. The chemical labeling approaches that are currently available do not provide that capability. We are developing genetic engineering and improved instrumentation to detect and count single copies of a specific protein in a single cell, and to monitor the post-translational modifications of that protein. We believe that this technology will have wide application, particularly in the basic sciences.

One serious challenge in the study of a single, specific protein at low copy levels is the signal generated by all of the other proteins present in the cell. Clearly, the isolation of a specific protein at very low levels from the complex mixture produced by chemical labeling is unacceptable. In principle, it may prove possible to use a highly fluorescent antibody to tag the protein of interest. Instead, we rely on genetic engineering to fuse green fluorescent protein (GFP) to the protein of interest. This genetic engineering step has the advantage of being highly specific to the target protein, albeit with the disadvantage of requiring biochemical steps not available in all laboratories.

An electropherogram of the Gal4/GFP fusion expressed in yeast is presented in Figure 12.8. The bottom trace was generated from a wild-type yeast lysate, while the top was generated from the genetically modified organism, which expresses the Gal4/GFP fusion. The wild-type yeast shows several autofluorescent components that migrate around 4 minutes; these components are likely endogenous flavins. A part-per-trillion contamination of the protease inhibitor generates a peak at 2.5 minutes. The Gal4/GFP fusion creates a set of three peaks that migrate between 3 and 3.5 minutes after injection. The first peak comigrates with GFP itself and is likely the proteolytic product, where the Gal4 protein has been fully digested, leaving only the tag. The second and third peaks are different phosphorylation states, presumably of the intact fusion protein. Treatment with alkaline phosphatase increases the amplitude of the last peak, which is consistent with removal of a negative charge from the protein.

We have recently begun to investigate detection of GFP in single yeast cells. The walls of these cells are quite robust and resist lysis with simple SDS treatment. We employ a series of enzymatic steps to degrade the carbohydrate and proteinaceous components of the wall before lysis. Figure 12.9 presents an electropherogram

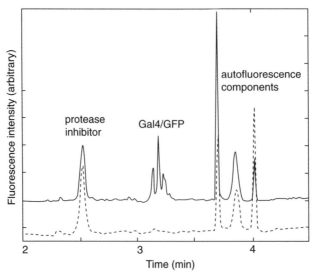

Figure 12.8. Capillary electrophoretic separation of the Gal4/GFP fusion protein from a yeast homogenate.

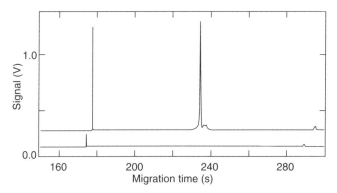

Figure 12.9. *Separation of GFP expressed under the control of the Gal1 promoter in a single yeast cell.*

generated from a single control yeast cell (bottom) and a single engineered cell (top). In this case, the cell had been genetically engineered to express GFP under control of the Gal1 promoter. The cell was grown in the presence of galactose, which resulted in the expression GFP in the single cell.

Single molecule detection has been reported for GFP that is immobilized in a thin gel film and illuminated for long periods under a microscope (Dickson et al., 1997). Unfortunately, simple microscopic detection provides no information on the post-translational modifications of the molecule; in the worst case, the fusion protein may be proteolytically digested, leaving only the GFP tag.

The photophysics of GFP is rather complicated, which makes its detection more challenging in a flowing system as in CE. We use a number of conventional modifications to the instrument to optimize detection of GFP. We minimize the detection volume by tightly focusing the laser beam and by use of a small diameter separation capillary. We optimize the laser power, which is near the photobleaching level for the molecule. We use a high-sensitivity photodiode for detection. We sample the photodetector output at a rate that matches the transit time of a molecule through the laser beam. The result is detection of single molecules of GFP with a signal-to-noise ratio of about 3 as shown in Figure 12.10.

The molecule generates a burst of fluorescence as it passes through the laser beam. This burst is a millisecond in duration and is detected above the fluctuations in the background signal. We are in the process of modifying our instrument to improve the signal-to-noise ratio for single molecule detection, and we are investigating improvements to the cell lysis conditions. This combination should make routine the characterization of GFP-fusion proteins expressed at extremely low levels.

12.12 CONCLUSION

Analysis of the protein content of a single mammalian cell is a formidable challenge. Most proteins are present at the zeptomole level, and their separation and identification push analytical capabilities to the limit. We have developed instruments that allow us to generate 2D electropherograms of the protein content from

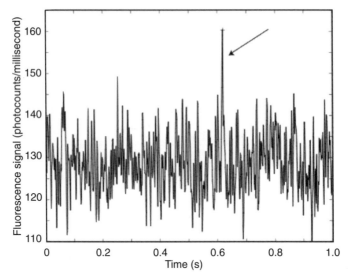

Figure 12.10. *The photon burst generated by a single molecule of GFP migrating from a capillary electrophoresis column through an ultrasensitive laser-induced fluorescence detector.*

single mammalian cells, and to detect and characterize extremely low levels of specific proteins in yeast and other systems that may be genetically engineered.

These systems are in their infancy, and many years of work will be required to optimize their performance and evaluate their utility. We certainly need to improve the resolution of the electrophoretic separation. We need to improve the sensitivity of the 2D electrophoresis instrument to detect proteins expressed at lower levels. We need to demonstrate the tentative identification of a significant fraction of the components in our 2D single cell electropherograms.

Nevertheless, these preliminary results hint at the power of single cell proteomics. We envision applications where we monitor the evolution of protein expression at the single cell level during development and differentiation and in response to specific stresses and hormones. We have begun a number of collaborations to monitor the changes in protein expression associated with cancer progression and viral infection. We are interested in the behavior of extremely rare regulatory proteins associated with regulation of the cell cycle.

As should be obvious from the success of the Human Genome Project, new science flows from the development of new technology. Unfortunately, a number of federal funding agencies, particularly those programs with a mandate to support innovative molecular analysis technologies, have been unable to provide continued support for this project. New technology can be brought to maturity only with sustained support.

ACKNOWLEDGMENTS

This group is currently supported by the National Human Genome Research Institute, the National Institute of Drug Abuse, the Department of Energy Genomes to Life Program, the Department of Defense Congressionally Directed Medical Research Program, and MDS-Sciex. DQM gratefully acknowledges a fellowship from the University of Washington Nanotechnology. We are grateful that these visionary agencies are able to support the development of innovative bioanalytical technology.

REFERENCES

Bushey MM, Jorgenson JW. 1990. Automated instrumentation for comprehensive two-dimensional high-performance liquid chromatography of proteins. *Anal Chem* 62:161–167.

Chen DY, Dovichi NJ. 1996. Single-molecule detection in capillary electrophoresis: molecular shot noise as a fundamental limit to chemical analysis. *Anal Chem* 68:690–696.

Dickson RM, Cubitt AB, Tsien RY, Moerner WE. 1997. On/off blinking and switching behaviour of single molecules of green fluorescent protein. *Nature* 388:355–358.

Dovichi NJ, Zhang J. 2000. How capillary electrophoresis sequenced the human genome. *Angew Chem Int Ed Engl* 39:4463–4468.

Edstrom JE. 1953. Nucleotide analysis on the cyto-scale. *Nature* 172:908.

Giddings JC. 1991. *Unified Separation Science*, Wiley, Hoboken, NJ.

Gilman SD, Ewing, AG. 1995. Analysis of single cells by capillary electrophoresis with on-column derivatization and laser-induced fluorescence detection. *Anal Chem* 67:58–60.

Gygi SP, Rochon Y, Franza BR, Aebersold R. 1999a. Correlation between Protein and mRNA Abundance in Yeast. *Mol Cell Biol* 19:1720–1230.

Gygi SP, Rist B, Gerber SA, Turecek F, Gelb MH, Aebersold R. 1999b. Quantitative analysis of complex protein mixtures using isotope-coded affinity tags. *Nat Biotechnol* 17:994–999.

Han FT, Lillard SJ. 2002. Monitoring differential synthesis of RNA in individual cells by capillary electrophoresis. *Anal Biochem* 302:136–143.

Hogan BL, Yeung ES. 1992. Determination of intracellular species at the level of a single erythrocyte via capillary electrophoresis with direct and indirect fluorescence detection. *Anal Chem* 64:2841–2845.

Hu S, Zhang L, Cook LM, Dovichi NJ. 2001a. Capillary sodium dodecyl sulfate–DALT electrophoresis of proteins in a single human cancer cell. *Electrophoresis* 22:3677–3682.

Hu S, Lee R, Zhang Z, Krylov SN, Dovichi NJ. 2001b. Protein analysis of an individual *Caenorhabditis elegans* single-cell embryo by capillary electrophoresis. *J Chromatogr B* 752:307–310.

Hu S, Jiang J, Cook LM, Richards DP, Horlick L, Wong B, Dovichi NJ. 2002. Capillary sodium dodecyl sulfate–DALT electrophoresis with laser-induced fluorescence detection for size-based analysis of proteins in human colon cancer cells. *Electrophoresis* 23:3136–3142.

Hu S, Zhang L, Krylov SN, Dovichi NJ. 2003a. Cell-cycle dependent protein fingerprint from a single cancer cell: image cytometry coupled with single-cell capillary sieving electrophoresis. *Anal Chem* 75:3495–3501.

Hu S, Zhang L, Newitt R, Aebersold R, Kraly JR, Jones M, Dovichi NJ. 2003b. Identification of proteins in single-cell capillary electrophoresis fingerprints based on co-migration with standard proteins. *Anal Chem* 75:3502–3505.

Kennedy RT, St. Claire RL, White JG, Jorgenson JW. 1987. Chemical analysis of single neurons by open tubular liquid chromatography. *Mikrochim Acta* II:37–45.

Kennedy RT, Oates MD, Cooper BR, Nickerson B, Jorgenson JW. 1989. Microcolumn separations and the analysis of single cells. *Science* 246:57–63.

Krylov SN, Zhang Z, Chan NW, Arriaga E, Palcic MM, Dovichi NJ. 1999. Correlating cell cycle with metabolism in single cells: combination of image and metabolic cytometry. *Cytometry* 37:14–20.

Krylov SN, Starke DA, Arriaga EA, Zhang Z, Chan NW, Palcic MM, Dovichi NJ. 2000. Instrumentation for chemical cytometry. *Anal Chem* 72:872–877.

Le XC, Tan W, Scaman CH, Szpacenko A, Arriaga E, Zhang Y, Dovichi NJ, Hindsgaul O, Palcic MM. 1999. Single cell studies of enzymatic hydrolysis of a tetramethylrhodamine labeled triglucoside in yeast. *Glycobiology* 9:219–225.

Lee IH, Pinto D, Arriaga EA, Zhang Z, Dovichi NJ. 1998. Picomolar analysis of proteins using electrophoretically mediated microanalysis and capillary electrophoresis with laser-induced fluorescence detection. *Anal Chem* 70:4546–4548.

Liu JP, Hsieh YZ, Wiesler D, Novotny M. 1991. Design of 3-(4-carboxybenzoyl)-2-quinolinecarboxaldehyde as a reagent for ultrasensitive determination of primary amines by capillary electrophoresis using laser fluorescence detection. *Anal Chem* 63: 408–412.

Liu H, Cho BY, Krull IS, Cohen SA. 2001. Homogeneous fluorescent derivatization of large proteins. *J Chromatogr A* 927:77–89.

Matioli GT, Niewisch HB. 1965. Electrophoresis of hemoglobin in single erythrocytes. *Science* 150:1824–1826.

Meredith GD, Sims CE, Soughayer JS, Allbritton NL. 2000. Measurement of kinase activation in single mammalian cells. *Nat Biotechnol* 18:309–312.

Michels D, Hu S, Schoenherr R, Eggertson MJ, Dovichi NJ. 2002. Fully automated two-dimensional capillary electrophoresis for high-sensitivity protein analysis. *Mol Cell Proteomics* 1:69–74.

Oakes, J. 1974. Protein–surfactant interactions. Nuclear magnetic resonance and binding isotherm studies of interactions between bovine serum albumin and sodium dodecyl sulfate. *J Chem Soc Farad T 1* 70:2200–2209.

Page JS, Rubakhin SS, Sweedler JV. 2002. Single-neuron analysis using CE combined with MALDI MS and radionuclide detection. *Anal Chem* 74:497–503.

Pasa-Tolic L, Lipton MS, Masselon CD, Anderson GA, Shen Y, Tolic N, Smith RD. 2002. Gene expression profiling using advanced mass spectrometric approaches. *Mass Spectrom* 37:1185–1198.

Pinto DM, Arriaga EA, Craig D, Angelova J, Sharma N, Ahmadzadeh H, Dovichi NJ, Boulet CA. 1997. Picomolar assay of native proteins by capillary electrophoresis—precolumn labeling, sub-micellar separation and laser induced fluorescence detection. *Anal Chem* 69:3015–3021.

Pinto D, Arriaga EA, Schoenherr RM, Chou SS, Dovichi NJ. 2003. Kinetics and apparent activation energy of the reaction of the fluorogenic reagent 5-furoylquinoline-3-carboxaldehyde with ovalbumin. *J Chromatogr B* 793:107–114.

Repin VS, Akimova IM, Terovskii VB. 1975. Detection of lactate dehydrogenase isoenzymes in single mammalian oocytes during cleavage by a micromodification of disc electrophoresis. *Bull Exp Biol Med* 77:767–769.

Richards DP, Stathakis C, Polakowski R, Ahmadzadeh H, Dovichi NJ. 1999. Labeling effects on the isoelectric point of green fluorescent protein. *J Chromatogr A* 853:21–25.

Ruchel R. 1976. Sequential protein analysis from single identified neurons of *Aplysia californica*. A microelectrophoretic technique involving polyacrylamide gradient gels and isoelectric focusing. *J Histochem Cytochem* 24:773–791.

Santos SF, Zanette D, Fischer H, Itri R. 2003. A systematic study of bovine serum albumin (BSA) and sodium dodecyl sulfate (SDS) interactions by surface tension and small angle X-ray scattering. *J Colloid Interf Sci* 262:400–408.

Smith RD. 2002. Trends in mass spectrometry instrumentation for proteomics. *Trends Biotechnol* 20(12 Suppl):S3–S7.

Turro NJ, Lei X-G. 1995. Spectroscopic probe analysis of protein–surfactant interactions: the BSA/SDS system. *Langmuir* 11:2525–2533.

Wallingford RA, Ewing AG. 1988. Capillary zone electrophoresis with electrochemical detection in 12.7 microns diameter columns. *Anal Chem* 60:1972–1975.

Washburn MP, Wolters D, Yates JR. 2001. Large-scale analysis of the yeast proteome by multidimensional protein identification technology. *Nat Biotechnol* 19:242–247.

Whittal RM, Keller BO, Li L. 1998. Nanoliter chemistry combined with mass spectrometry for peptide mapping of proteins from single mammalian cell lysates. *Anal Chem* 70: 5344–5347.

Xu BJ, Caprioli RM, Sanders ME, Jensen RA. 2002. Direct analysis of laser capture microdissected cells by MALDI mass spectrometry. *J Am Soc Mass Spectrom* 13:1292–1297.

Zhang Z, Krylov S, Arriaga EA, Polakowski R, Dovichi NJ. 2000. One-dimensional protein analysis of an HT29 human colon adenocarcinoma cell. *Anal Chem* 72:318–322.

Zhao JY, Waldron KC, Miller J, Zhang JZ, Harke H, Dovichi NJ. 1992. Attachment of a single fluorescent label to peptides for determination by capillary zone electrophoresis. *J Chromatogr* 608:239–242.

13

Diagnostic Proteomics

DaRue A. Prieto and Haleem J. Issaq*

Laboratory of Proteomics and Analytical Technologies, SAIC-Frederick, Inc., National Cancer Institute at Frederick, Frederick, Maryland

13.1 INTRODUCTION

The past decade has witnessed a surge in technological developments in various biomedical-related disciplines, namely, genomics, transcriptomics, and proteomics (National Cancer Institute News Center, 2004). These advances have changed the way in which experiments are designed. The core technologies that have been responsible for this growing trend include high-speed DNA sequencing (i.e., genomics), microarray technologies for the study of mRNA (i.e., transcriptomics), and mass spectrometry (MS) for studying protein function and expression (i.e., proteomics). These three major scientific areas have in many ways developed independently of one another; however, each field is biologically and computationally connected to one another. For example, global proteomics, where the goal is to characterize as many proteins as possible within a given system, is made possible only through the generation of large genomic databases. Also, a complete understanding of cell function is only possible by integrating data acquired at the genomic, transcriptomic, proteomic, as well as metabonomic levels. Consequently, proteomics-based approaches complement the genome initiatives and may be the next step in attempts to understand the biology of cancer (Yanagisawa et al., 2003).

Whereas genetic analyses will identify individuals with a predisposition to certain disease, and therefore long-term risk, proteomic analyses provide the opportunity to detect diseases as they occur (Bock et al., 2004). Alterations in

---

* To whom correspondence should be addressed.

Proteomics for Biological Discovery, edited by Timothy D. Veenstra and John R. Yates.
Copyright © 2006 John Wiley & Sons, Inc.

protein abundance, structure, and function act as useful indicators of pathological abnormalities prior to developing clinical symptoms and as such are often useful diagnostic and prognostic biomarkers. The underlying mechanism of diseases such as cancer are, however, quite complicated in that often multiple dysregulated proteins are involved. It is for this reason that recent hypotheses suggest that detection of panels of biomarkers may provide higher sensitivities and specificities for disease diagnosis than is afforded with single markers.

The discovery, identification, and validation of disease-associated proteins from biological samples (serum, plasma, urine, tissue) is a difficult and labor-intensive task, which often requires the analysis of hundreds, if not thousands, of samples under strict and reproducible experimental conditions. It has been reported that approximately 20,000 proteins are present in human serum with an overall protein concentration of about 60–80 mg/mL (Anderson and Anderson, 2002). Analysis of a serum proteome will allow the detection and identification of hundreds and thousands of proteins; however, a complete characterization of the human serum proteome (i.e., identifying all the protein constituents) is not possible using current technology and requires new protein separation, detection, quantitation, and identification approaches as well as instrumentation. Even if the methodology is optimized, it is still impossible to know if every possible protein within a proteome has been identified, as there is no possible means to precisely know the entire proteomic makeup of a sample a priori.

13.2 BIOMARKERS

A biomarker is a substance found in the blood, urine, cerebrospinal fluid, or tissues and is often detected in higher-than-normal amounts in patients with a certain disease. A biomarker can include patterns of single nucleotide polymorphisms (SNPs), DNA methylation, or changes in mRNA, protein, or metabolite abundances, providing that these patterns can be shown to correlate with the characteristics of the disease (MacNeil, 2004). It has been demonstrated, however, that there is often no predictive correlation between mRNA abundances and the quantity of the corresponding functional protein present within a cell. Hence, since proteins represent the preponderance of biologically active molecules responsible for most cellular functions, it is believed that the direct measurement of protein expression can more accurately indicate cellular dysfunction underlying the development of disease (MacNeil, 2004). Cancer tumor markers are produced either by the tumor itself or by the body in response to the presence of cancer or certain benign conditions. Measurement of tumor marker levels can be useful in detecting and diagnosing certain types of cancer. However, measurements of tumor marker levels alone are not sufficient to diagnose cancer for the following reasons:

- Tumor marker levels can be elevated in people with benign conditions.
- Tumor marker levels are not elevated in every person with cancer, especially in the early stages of the disease.
- Many tumor markers are not specific to a particular type of cancer; the level of a tumor marker can be raised by more than one type of cancer.

The utility of single markers in diagnosis and monitoring of disease (e.g., prostate specific antigen (PSA) for prostate cancer, cancer antigen-125 (CA125) for ovarian cancer, cancer antigen 15-3 (CA15-3) for breast cancer, and carcinoembryonic antigen (CEA) for ovarian, lung, breast, pancreas, and gastrointestinal tract cancer) is limited by the poor association of any single protein with a specific disease or stage of disease. The value of any single marker is significantly enhanced by combination with other markers (Bock et al., 2004).

In addition to their role in cancer diagnosis, some tumor marker levels are measured before treatment to help physicians plan appropriate therapy. In certain types of cancer, tumor marker levels reflect the extent (stage) of the disease and can be useful in predicting how well the disease will respond to treatment. Tumor marker levels may also be measured during treatment to monitor the patient's response to treatment. A decrease or normalization in the level of a tumor marker may indicate that the cancer has responded favorably to therapy. However, if the tumor marker level rises, it may indicate that the cancer is progressing. Finally, measurements of tumor marker levels may be used after treatment has ended as a part of follow-up care to monitor for recurrence. Currently, the main application of tumor markers is for assessing response to treatment and to monitor recurrence. Scientists continue to study these uses of tumor markers as well as their potential role in the early detection and diagnosis of cancer. (http://www.nci.nih.gov).

13.2.1   Characteristics of Clinically Useful Biomarkers

A biomarker, to have the greatest impact, should be present in an easily obtainable sample, such as urine or blood. In addition, the assay to clinically measure and validate the overall positive predictive value (PPV) of the biomarker must be capable of screening a large number of samples in a high-throughput manner. Validation of a biomarker requires the analysis of thousands of samples to ensure that the potential biomarker is indeed related to a disease state and is not simply a function of the variability within the blood (serum, plasma) of patients due to differences in diet, genetic background, lifestyle, and so on. While at first glance the biomarker with the highest sensitivity and specificity would appear to be the most useful, if the parameters for its measurement are too stringent to achieve reproducibility it may not be the best choice. Oftentimes a biomarker that can be measured more reliably in different laboratories will be the best choice even if its overall PPV is not as high as another candidate marker.

13.2.2   Weaknesses of Current Cancer Biomarkers

Although biomarker discovery is a major focus of proteomics, there are a surprisingly small number of clinically relevant markers in use. These include CA125, PSA, cytokeratin 7 (CK7), cytokeratin 20 (CK20), CEA, and cancer antigen 19-9 (CA19-9). The following examples of their usage clearly demonstrate the need for more accurate biomarkers that can be used to detect early stage cancer.

CA125 is the earliest and most widely used diagnostic biomarker for ovarian cancer. It is present on the surface of ovarian tumor cells and is detected by the monoclonal antibody OC125. Although CA125 is elevated in 80% of epithelial ovarian cancer cases, it is detected in only 25–50% of early stage patients (Jones et al., 2002). Another sobering fact is the apparent elevation of CA125 in patients with benign and nongynecologic conditions, resulting in significant false-positive rates for ovarian cancer (Jacobs et al., 1999).

The gold standard for early prostate cancer detection is testing for elevated PSA levels in conjunction with manual rectal examination. PSA has been hailed as the most effective tumor marker for prostate cancer. In reality, the PSA screening test generates a substantial number of false positives and has a 20% probability of incorrectly diagnosing a prostate cancer patient as normal (Imai and Yamanaka, 1998). These inaccurate results often lead to unnecessary invasive prostate biopsies.

Seven percent of ovarian tumors are metastases originating from primary lesions in the gastrointestinal tract (Wauters et al., 1995). This makes primary ovarian and metastatic colon carcinomas difficult to diagnose. Because ovarian and colon cancer chemotherapies differ dramatically, misdiagnosis can result in inappropriate drug therapy and thereby delay effective treatment of both the primary and metastatic tumor (Lagendijk et al., 1998). Epithelial tissues differ in their pattern of keratin expression and this pattern is often maintained in carcinomas and their metastases; therefore, keratin subtypes have been used to determine the site of origin of the metastatic tumor. Two such keratins are currently in use for this purpose. CK7 is ubiquitously present on ovarian carcinomas, but not on colon carcinomas. CK20 shows the opposite pattern of expression. These biomarkers appear promising at first glance, but in fact predict the primary site in metastatic carcinomas only 60–80% of the time (Nishizuka et al., 2003). This low level of predictive capability is also reflected by CEA, a common marker for evaluating effect and monitoring reoccurrence of advanced stage colon cancer. This marker not only results in a relatively low PPV (40–60%) but also is ineffective for diagnosing early stage disease (Shiwa et al., 2003).

Pancreatic cancer is the fourth leading cause of cancer-related death in both men and women and has the lowest survival rate of any solid cancer. The five-year survival rate for individuals with surgically resectable pancreatic tumors is 15–40% (Yeo et al., 1995). Sadly, patients typically show symptoms late in the course of the disease, so only 10–15% of tumors are resectable at the time of diagnosis (DiMagno et al., 1999). CA19-9 is the most widely used biomarker for pancreatic cancer. Its inaccuracy limits its efficacy as a useful screening tool; therefore, its main utility is simply monitoring the effects of treatment in patients known to have pancreatic cancer (Koopman et al., 2004). All of these examples underscore the need for biomarkers with high sensitivity and specificity in the diagnosis of debilitating conditions such as cancer.

13.3 PROTEOMIC PATTERNS: A NEW CONCEPT FOR DISEASE DIAGNOSIS

Proteomic patterns as a clinical test for disease diagnosis is a revolutionary concept that was first introduced in a seminal paper describing its application in the diagnosis of women suffering from ovarian cancer (Petricoin et al., 2002a). Proteomic pattern analysis does not rely on the identification of any of the species (proteins/peptides) observed during the analysis, instead relying solely on the changes in relative abundance of a number of different points (mass-to-charge ratio (m/z) and intensity) within the mass spectrum. The use of proteomic patterns as a diagnostic test represents a contrast to how the discovery of biomarkers has been carried out in MS-based proteomic approaches. These approaches are designed to separate and compare species across different complex samples with the goal of identifying the specific entity and its changes between the sample sets.

13.3.1 Surface Enhanced Laser Desorption Ionization Time-of-Flight Mass Spectrometry

Surface enhanced laser desorption ionization TOF-MS is a novel MS-based approach for analyzing complex protein mixtures introduced in 1993 by Hutchens and Yip (1993), which was further developed and marketed by Ciphergen Biosystems, Inc. This technique combines the principles of retentate chromatography with MS. When we think of classical chromatography, we naturally envision binding a biomolecule (i.e., DNA, peptide, or protein) to column packing material. The molecule is then eluted from the column using a mobile phase. Retentate chromatography differs in that biological samples such as serum, plasma, urine, cell lysate, or purified peptide/proteins are selectively bound to proteinchip array surfaces. Nonspecifically bound proteins, salts, and other interfering compounds are washed away. A large excess of a ultraviolet-absorbing material (matrix) is then applied to the proteins retained on the array surface and the sample is analyzed directly by TOF-MS. After acquiring the proteomic patterns, bioinformatic tools are used to identify key features within the proteomic patterns that enable source of the samples (i.e., from control or disease-affected patients) to be discerned. Once these discriminating features or peaks are identified, they can then be applied to a set of unknown samples to provide a diagnosis.

Recent reports employing a SELDI-TOF-MS-based approach have demonstrated that protein profiles of serum are useful for the early detection of ovarian (Petricoin et al., 2002a; Rai et al., 2002), prostate (Adam et al., 2002; Petricoin et al., 2002b; Banez et al., 2003; Li et al., 2004), cervical (Wong et al., 2004), breast (Wulfkuhle et al., 2001; Li et al., 2002), lung (Xiao et al., 2003), bladder (Vlahou et al., 2001), colon (Lawrie et al., 2001), head and neck (Wadsworth et al., 2004), and pancreatic (Koopmann et al., 2004) cancers. In addition to cancer, SELDI-TOF-MS has also been applied to study proteins related to other maladies such as Alzheimer's disease (Carrette et al., 2003; Lewczuk et al., 2004), rheumatoid arthritis (Uchida et al., 2002), and HIV/AIDS (Zhang et al., 2002).

The complete proteinchip/SELDI-TOF-MS system is comprised of three components, the proteinchip arrays, the mass analyzer, and data analysis software. The proteinchip arrays distinguish this technique from other MS-based systems. The arrays are available in 8 or 16 spot format and are designed to capture proteins from complex biological mixtures. A number of proteinchip array types are available. Chemically modified surfaces retain whole classes of proteins and are available in a variety of chromatographic surfaces (anionic, cationic, hydrophobic, hydrophilic, and metal affinity capture). These general purpose arrays provide high-capacity binding and are used primarily for protein profiling and peptide mapping. The combination of array chemistry and correct buffer selection is key when working with chemically modified arrays. Reproducible data is only gained by thorough knowledge of these parameters and it is this aspect of the technique that is often trivialized and underestimated. For instance, weak cationic exchange surfaces will preferentially bind serum proteins at a 50–100 mM sodium acetate buffer pH of 4.0–4.5. As the pH of the buffer increases (i.e., its stringency), the chip surface becomes more selective and fewer proteins will bind. Biochemically modified reactive surface arrays are designed to target specific proteins. An antibody, DNA, RNA, receptor, or other "bait" molecule is covalently coupled to the array surface. This immobilized molecule, in turn, captures proteins from biological samples via specific, noncovalent interactions. SEND proteinchip arrays are manufactured with matrix incorporated directly into the chemistry of the array. This process reduces the chemical noise typically associated with the manual application of matrix to the spot. The advantage is the discrimination of very low molecular mass peptides from the overlapping matrix signals.

Ciphergen's first generation mass analyzer was the PBSII. This relatively simple instrument is a time-of-flight mass spectrometer equipped with a pulsed UV nitrogen laser. The principle of operation is similar to that of MALDI-TOF-MS. A laser beam is focused on the sample, causing the proteins of interest to become desorbed from the proteinchip array surface and then ionized. Positively charged ions are accelerated under vacuum and "fly" through a TOF tube toward an ion detector. The m/z of each species is recorded based on the time required to pass through the TOF tube. A spectrum is generated that displays the relative intensity versus the m/z ratios of the detected peptides/proteins as shown in Figure 13.1. Although the design of this instrument does provide reasonably high sensitivities, the achievable resolution and mass accuracy are considerably less than conventional high-resolution TOF-MS instruments provide. Recently, Ciphergen introduced the

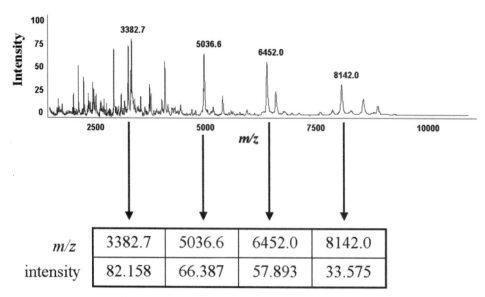

Figure 13.1. Surface enhanced laser desorption ionization time-of-flight mass spectrometry generated spectrum displaying the relative intensity versus the mass-to-charge (m/z) ratios of detected peptides/proteins.

Series 4000 Personal Edition mass spectrometer. This benchtop mass analyzer offers a number of new features designed to improve instrument sensitivity, especially in the high mass range (>100 kDa). These features include improved detector, ion source, and flight tube design.

The basic software allows the data to be displayed in three formats. The traditional chromatographic display presents the data as a series of protein peaks, referred to as a trace view (Figure 13.2A). When there are a large number of peaks that make complete visualization of the spectrum difficult, a map view (Figure 13.2B) is advantageous since this presentation reduces the complexity and background normally seen in the traditional spectrum by centroiding the peaks and showing them as thin vertical lines. One of the most useful features of the display software is the gel view (Figure 13.2C). In this view, the output is converted to a simulation of a SDS-PAGE stained gel. This "virtual gel" is often used when comparing multiple spectra. Other useful features allow spectra to be stacked, overlaid, or offset overlaid for direct data comparison.

13.3.2 Utility of Surface Enhanced Laser Desorption Ionization Time-of-Flight Mass Spectrometry

Its low resolution, low mass accuracy, and lack of MS/MS capabilities make direct protein identification by SELDI-TOF-MS challenging. These criticisms of the technique, however valid, are also somewhat biased. The mass analyzer is not designed to provide the protein identification capabilities possessed by higher end mass spectrometers. Surface enhanced laser desorption ionization TOF-MS offers a user-friendly platform for the rapid analysis of differentially abundant proteins

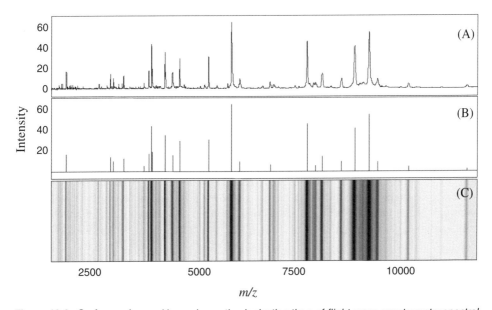

Figure 13.2. *Surface enhanced laser desorption ionization time-of-flight mass spectrometry spectral output shown as a (A) trace, (B) map, and (C) gel view.*

from complex biological mixtures. A simple protein profiling experiment consists of assaying the biofluid of choice (e.g., human serum) on each of the chemically modified proteinchip array surfaces (anionic, cationic, hydrophobic, and metal affinity capture). This experiment can be accomplished in an 8 hour workday and provides a low-resolution spectrum of the sample's protein composition. Another strength of this technique lies in its ability to acquire spectra from a large number of samples with little or no sample preparation. For example, twenty-four 8-spot proteinchip arrays (192 samples) can be prepared manually in a single workday and is easily doubled with automation. Once peaks of interest (i.e., potential biomarkers) are recognized in this high-throughput screening experiment, they may then be purified using classical chromatographic methods and identified by high-resolution MS/MS.

13.3.3   Application of Surface Enhanced Laser Desorption Ionization Time-of-Flight Mass Spectrometry in Cancer Diagnostics

Petricoin and Liotta (2004) and colleagues (Petricoin et al., 2002a) are credited with pioneering proteomic pattern analysis for the diagnosis of ovarian cancer. Ovarian cancer is the fifth most common cause of cancer-related deaths affecting 6% of the female population in the United States (American Cancer Society, 2004). In most cases, ovarian cancer is not diagnosed and treated until it has metastasized from the ovary to the abdomen (stage III) or beyond (stage IV). At this stage the five year survival rate for women is only 15–20%. Detection of ovarian cancer at stage I increases the five year survival rate to nearly 95% with surgical intervention. Unfortunately, 80% of women are not diagnosed until the disease has progressed. In their groundbreaking study, Petricoin and Liotta obtained proteomic patterns from the serum of women with ovarian cancer and a group of normal matched controls using SELDI-TOF-MS. Mass spectral data were analyzed using Proteome-Quest software from Correlogics Systems, Inc. This bioinformatics package combines elements of a genetic algorithm with cluster analysis. Input files initially generated are comprised of m/z values on the x-axis and corresponding intensities on the y-axis. Data analysis occurs in two phases: a pattern discovery phase and a pattern matching phase. In the pattern discovery phase, randomly selected spectra (a training set) from both normal and cancer-affected individuals are analyzed in an attempt to find a set of m/z values (and associated intensities) that completely segregate the two sets (i.e., a model). In this phase, the source of the sample (normal or cancer-affected) is known and this information is supplied to the algorithm. The algorithm selects small subsets (5–20) of exact m/z values along the x-axis of the mass spectra. A diagnostic pattern is formed by plotting the combined y-axis intensities of candidate subsets of key m/z values in N-dimensional space, where N is the number of m/z values found within the training set of spectra. The model that is formed is then rated for its ability to segregate normal and cancer-affected sample spectra. The highest rated models are subsequently reshuffled. This process is repeated until a model or models are generated that completely segregate normal from cancer-affected sample spectra. In the pattern matching phase, the model/models generated in phase I are used to test and validate a blinded sample set. Each of the blinded samples is classified into a cluster diagnostic of either a normal or cancer-affected individual. A new cluster will be generated only if the blinded sample does not match any of the patterns defined in the pattern discovery phase. In the Petricoin and Liotta study, the model generated in phase I was able to correctly discriminate samples obtained from healthy or ovarian cancer-affected patients with a sensitivity of 100%, specificity of 95%, and an overall PPV of 94%. Additionally, all patients with stage I ovarian cancers were correctly classified.

Since this initial study, proteomic pattern analysis using SELDI-TOF-MS has been employed in studies of breast (Pusztai et al., 2004), prostate (Wagner et al., 2004), lung (Zhukov et al., 2003), pancreatic (Koopman et al., 2004), head and neck (Wadsworth et al., 2004), and colorectal cancers (Xiao et al., 2004). The technique has even been extended beyond cancer diagnosis and has been employed for discriminating patterns of proteins obtained from the urine of interstitial cystitis and unaffected patients (Van et al., 2003, 2004) and from postmortem CSF samples obtained from schizophrenics and normal controls (Johnston-Wilson et al., 2001). Proteomic pattern analysis studies use a number of different bioinformatic algorithms, that is, decision trees, neural networks, genetic algorithms, and random forest algorithms. The goal of each of these bioinformatic tools is identical: construction of a model or models capable of segregating one group of samples from another. In most cases, the ability of this technique to diagnose disease has yielded significant sensitivities, specificities, and PPVs. While these studies are encouraging, one must remember that a minimum sensitivity and specificity of 99.6% would be required for a diagnostic test to be used as a general population screen for diseases of relatively low prevalence (Kainz, 1996). Therefore, while proteomic pattern diagnostics has the potential to reach this goal, it is currently useful only as a tool to confirm a cancer diagnosis. Its use as a screening tool is still limited.

Surface enhanced laser desorption ionization TOF-MS has been used to improve the efficacy of currently available ovarian and prostate cancer biomarkers. In a five-center case–control study, Zhang et al. (2004) analyzed serum samples obtained from healthy women, women with ovarian cancer, and women with benign pelvic masses. Surface enhanced laser desorption ionization TOF-MS analysis of serum revealed three potential biomarkers modulated at m/z 3372 (downregulated), 12,828 (upregulated), and 28,043 (upregulated). These species were purified and identified by MS/MS as follows: apolipoprotein A1 (28,043), a truncated form of transthyretin (12,828), and a cleavage fragment of inter-α-trypsin inhibitor heavy chain H4 (3372). Zhang and colleagues then combined the three markers with CA125 and applied a multivariate predictive model to the data in an attempt to discriminate early stage invasive epithelial ovarian cancer from healthy controls. They found the specificity of the model (94%) to be significantly higher than that of CA125 alone (52%). A fixed sensitivity (83%) was maintained in each group. These results illustrate the potential of SELDI TOF-MS when used to improve biomarker efficacy for the detection of early stage ovarian cancer.

Wright et al. (1999) demonstrated the utility of SELDI-TOF-MS for the direct capture and detection of PSA, prostate acid phosphatase (PAP), prostate specific peptide (PSP), and prostate specific membrane antigen (PSMA) from serum, cell lysates, and seminal plasma. These researchers developed a SELDI-TOF-MS-based immunoassay designed to capture single antigens (PSA, PSP, or PMSA) or to simultaneously capture complexed antigens from biofluids (PSA-ACT). This study was extended to include the development of a SELDI-TOF-MS-based immunoassay for the quantitation and comparison of PMSA levels in serum obtained from healthy men and men with benign or malignant prostate disease (Xiao et al., 2001). PMSA was captured from serum via anti-PMSA antibodies coupled to the proteinchip array surface. The captured PMSA was detected by SELDI-TOF-MS and quantified by comparing the integrated peak areas to a standard curve established using purified recombinant PMSA. This study demonstrated the ability of the SELDI-TOF-MS technique to detect varying levels of PMSA in sera and to discriminate benign from malignant prostate disease. Surface enhanced laser desorption ionization TOF-MS technology has also been used for the discovery of new, potential biomarkers for the detection of colon (Shiwa et al., 2003), breast (Li et al., 2002), pancreatic (Koopman et al., 2004), and endometrial (Yang et al., 2004) cancers.

13.3.4 Diagnosis of Infectious Diseases

HIV/AIDS has become a worldwide epidemic (NIH, 2004). By the end of 2004, it was estimated that 39.4 million people will be living with HIV/AIDS, 12.4% of which will be new cases (Avent.org). In 2003 alone, there were 3.1 million HIV/AIDS-related deaths, including an estimated 490,000 deaths in children younger than 15 years of age (NIAID HIV/AIDS statistics). Scientists are seeking to understand this disease and have just recently utilized SELDI TOF-MS for this purpose. One aspect of the HIV/AIDS crisis is HIV-associated dementia (HAD). Although the virus enters the brain soon after HIV infection, neurological changes manifesting in HAD are not observed until many years later, usually during the destruction of the immune system and the development of AIDS (Zheng and Gendelman, 1997). Because HIV-1 infection of the nervous system is strongly associated with the infiltration of mononuclear phagocytic cells, there is a growing body of evidence suggesting that the virus is carried to the brain by infected monocytes (Kaul et al., 2001). Luo et al. (2003) used proteomic pattern analysis to determine whether a unique protein pattern exists for HIV patients with or at risk for HAD. Monocyte populations were recovered by Percoll gradient centrifugation and cultured for a period of 7 days in the presence of macrophage colony stimulating factor. Cell lysates were then prepared and analyzed on weak cationic exchange chips using SELDI-TOF-MS. The researchers identified unique monocyte-derived macrophage (MDM) protein profiles from cognitively impaired HIV-1 positive patients. The study was extended to test whether these unique MDM proteins were affected by highly active antiretroviral therapy (HAART). Surface enhanced laser desorption ionization TOF-MS identified multiple unique protein peaks between 3 and 20 kDa present in one HAD monocyte-derived macrophage sample prior to HAART treatment. Each of these peaks disappeared after a 3 month HAART regimen. The subsequent disappearance of the peaks also corresponded with a marked improvement of this patient on neuropsychological tests, suggesting that changes in MDM expression may correlate with progression or improvement of dementia. These results not only show the utility of the SELDI-TOF-MS technique for detecting changes in MDM patient protein profiles but also suggest that the technique may be developed as a measure of the therapeutic efficiency of HAART treatment (Wojna et al., 2004).

Only a handful of papers have been published describing the application SELDI TOF-MS technology to bacterial proteomics. A study by Barzaghi and colleagues (2004) was undertaken to determine the effectiveness of SELDI-TOF-MS for profiling microorganisms as a complementary technique or as an alternative to 2D-PAGE. Preliminary experiments were performed with *Steptococcus pneumoniae* (the leading cause of bacterial pneumonia, otitis media, and bacterial meningitis) in order to develop a protocol for SELDI analysis. Optimized parameters were then used to generate expression profiles from a wide range of gram-positive and gram-negative bacteria.

The *Helicobacter* genus is associated with a wide spectrum of gastrointestinal tract infections. Hynes and colleagues (2003) have used SELDI-TOF-MS protein profiling to clearly demonstrate differences between outer membrane protein profiles of *Helicobacter* strains. These researchers feel that the technique will be useful in the area of strain characterization and as a promising tool to find host-specific bacterial virulence factors.

Microbial pathogenesis is dependent on a microbe's ability to rapidly alter protein expression in response to environmental changes within its host. Surface enhanced laser desorption ionization TOF-MS is proving to be a valuable tool to track these changes. Thulasiraman et al. (2001) demonstrated the effectiveness of this technique for monitoring the expression of temperature- and calcium-related virulence factors of *Yersinia pestis* (the causative agent of the human plague). This bacterium is carried to mammalian hosts by a flea vector and responds to the higher temperature of its new host by regulating a number of proteins. By profiling crude bacterial extracts, these researchers successfully identified five proteins modulated in response to higher temperature. Two of the proteins were subsequently identified as catalase-peroxidase and Antigen 4.

13.3.5 Diagnosis of Neurological Abnormalities

Surface enhanced laser desorption ionization TOF-MS enables investigators to discover new biomarkers for neurological diseases and to develop assays for known disease markers. To this end, Alzheimer's disease has dominated this area of research. Alzheimer's disease is a neurodegenerative dementia that accounts for 50–60% of all cases of dementia among people over the age of 65. Proteolytic processing of amyloid precursor protein (APP) results in the generation of β-amyloid peptides. It is cerebral β-amyloid peptide deposition, accumulation, and neuritic plaque formation that form the pathological signature of this disease. Ciphergen BioSystems, Inc. has developed a kit to detect multiple amyloid peptides from crude biological samples using SELDI-TOF-MS. Three purified amyloid peptides (1–16, 1–40, and 1–42) are provided, along with an anti-human β-amyloid monoclonal antibody, proteinchip arrays, a negative control, and all reagents needed for subsequent MS analysis. The experimental design for use of the kit essentially follows the principles of any SELDI proteinchip-based immunoassay, where the antigen-specific antibody is coupled to the reactive chip surface, enabling the capture of β-amyloid peptides from biologically complex mixtures. Quantification is possible based on a standard curve established using the three purified amyloid peptides. By using this approach, novel amyloid peptide variants in both cerebrospinal fluid (CSF) and postmortem brain homogenates have been discovered (Lewczuk et al., 2003, 2004).

The search for Alzheimer biomarkers has not been limited to the immunoaffinity capture of amyloid peptides. Carrette et al. (2003) employed strong anionic exchange proteinchips and detected four overexpressed and one underexpressed polypeptide in Alzheimer cerebrospinal fluid (CSF). These peptides were further identified as cystatin c, two β-2-microglobulin isoforms, a VGF polypeptide, and an unknown polypeptide of 7.69 kDa.

13.3.6   New Advances in Proteomic Pattern Diagnostics

While proteomic pattern technology has almost exclusively relied on the same technology platform since its conception, there have been some significant improvements in the process. A significant advance in proteomic pattern analysis is the use of a high-resolution hybrid quadrupole TOF (QqTOF)-MS fitted with a SELDI ion source. When fitted with a SELDI ion source, this instrument offers proteinchip array and MS/MS capabilities, ion selectivity, and high mass accuracy and resolution. The PBS-II TOF-MS and QqTOF-MS provide similar spectra; however, the resolution obtainable with the QqTOF is 60-fold greater than that obtained with the PBS-II. Conrads and co-workers hypothesized that the increased resolution provided by the QqTOF would allow a greater number of features to be detected, thereby improving the likelihood of finding discriminating proteomic patterns diagnostic for ovarian cancer (Conrads and Veenstra, 2004; Conrads et al., 2004a,b). A side-by-side study was designed to directly compare data generated from the two instruments. In this study, the same proteinchip arrays were analyzed on both instruments, thereby eliminating potential experimental variability caused by the arrays. Two hundred and forty-eight serum samples obtained from healthy and ovarian cancer-affected patients were analyzed simultaneously on both instruments. Proteomic patterns were analyzed using the ProteomeQuest software from Correlogics Systems, Inc. One hundred and eight diagnostic models were generated in the pattern discovery phase of the data analysis. These models were validated during the pattern matching phase using blinded sample spectra obtained from 37 normal and 40 ovarian cancer-affected women. Four diagnostic QqTOF models were found to be 100% sensitive and specific in the ability to correctly discriminate between serum samples obtained from healthy women and ovarian cancer patients. No false positives or false negatives occurred using these four models giving each a PPV of 100%. More importantly, 22 of 22 stage I ovarian cancer serum samples were correctly classified. No models with 100% sensitivity, specificity, and PPV were generated by the PBS-II.

13.3.7   OvaCheck™

Proteomic pattern diagnostics has been the focus of a controversial test for ovarian cancer licensed under the name of Ovacheck™ (Correlogics Systems, Inc., Bethesda, MD, USA). This technique differs from SELDI-TOF-MS in that electrospray ionization (ESI) is used for mass spectral analysis. Another key difference is that OvaCheck does not require the use of proteinchip arrays.

Serum samples are diluted .1:250 in a 50:50 mixture of acetonitrile and water, containing 0.2% formic acid. The samples are transferred to a 96 well filter plate and the samples are filtered to remove insoluble material, as shown in Figure 13.3. The 96 well plate containing filtrate is then placed into a NanoMate™ 100 (Advion BioSciences, Inc., Ithaca, NY, USA) MS interface. The NanoMate 100 is a fully automated nanoelectrospray system that is compatible with a variety of mass spectrometers, including the QqTOF. It uses nanoelectrospray ionization to create positively charged gas-phase ions that are measured by MS. A proteomic pattern is obtained representing each patient's serum sample. The raw data is analyzed and scored, resulting in a probability curve that indicates the likelihood that the patient has ovarian cancer.

The OvaCheck test for ovarian cancer can be run on a patient's sample taken from a simple fingerstick. It is designed to identify patients likely to have ovarian cancer and is initially intended for screening high-risk patients (i.e., breast cancer survivors), those with family histories of ovarian cancer, and patients with BRCA1 or BRCA2 mutations. As data accumulates, it is hoped that the test will also be useful for diagnosing ovarian cancer stages. OvaCheck is a promising test; however,

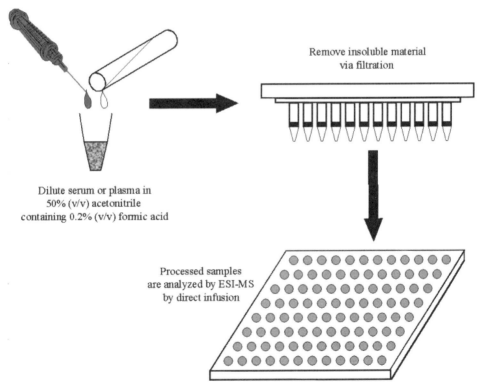

Figure 13.3. *Diagnostic analysis of serum or plasma samples using the OvaCheck diagnostic platform. In this method, raw serum or plasma is diluted using an organic solvent and filtered to remove particulates. The pattern of the molecules retained in the filtrate samples is acquired using electrospray ionization mass spectrometry (ESI-MS). Sophisticated bioinformatics is used to discover peaks that enable the source of the original sample (i.e., healthy or diseased) to be determined.*

further validation of the technology is needed before this technique can possibly be used in any type of clinical assay.

13.3.8 Magnetic Particle MALDI TOF-MS

Villanueva et al. (2004) used a novel technology platform for the simultaneous measurement of a large number of peptides from the serum of controls and brain tumor patients using reverse-phase (RP) batch processing in a magnetic particle-based format. The captured polypeptides are then analyzed by MALDI-TOF-MS. The principle of this technology is simple and can be automated. The protocol, as shown in Figure 13.4, consists of six discrete steps. A suspension of magnetic beads to which reverse-phase ligands (C-18 or C-8) are attached, is mixed with 50 μL of serum. The beads and serum are then mixed and the beads are pulled to the side of the tube by magnetic force. The supernatant is removed and discarded. The beads with the bound peptides are then washed twice with 200 μL of 0.1% TFA in

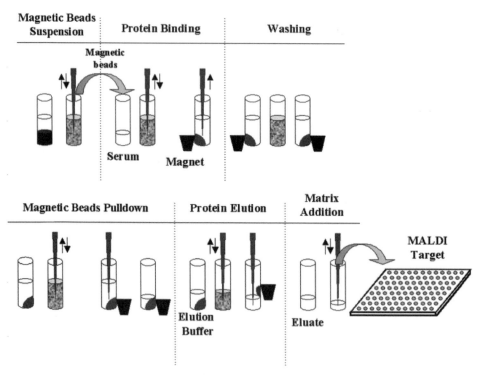

Figure 13.4. *Magnetic particle-assisted processing of serum peptides for matrix-assisted laser desorption ionization time-of-flight mass spectrometry (MALDI-TOF-MS) analysis. Magnetic beads are surface-derivatized with reverse-phase ligands. A measured volume of bead suspension is transferred to a tube containing an aliquot (typically 50 μL) of serum. After thorough mixing, the beads are separated from the solution by magnetic force, and the supernatant is removed and discarded. Washing: After several washing and magnetic bead recovery steps, a minimal volume of elution solvent is added to the bead pellet and the beads are again separated from the solution. At this stage the eluate is mixed with matrix solution and the sample is applied to a MALDI target plate and analyzed by MALDI-TOF-MS.*

water. The addition of 5 µL of 50% acetonitrile removes the peptides from the surface-modified magnetic beads. Stepwise elution of peptides from the magnetic beads can also be achieved by using various percentages of acetonitrile in the eluate. At this stage the supernatant, which contains the serum peptides of interest, is carefully transferred to a second tube and matrix solution is added to the eluate and mixed. An aliquot (1 µL) of this solution is transferred to the MALDI target and the proteomic patterns of the various samples are acquired.

This sample preparation procedure has significant advantages over the SELDI-based methodology in that it does not require the use of proteinchip arrays, and it enables serum proteome patterns to be acquired on different TOF-MS instruments, while there are only two different types of spectrometers that accept a SELDI platform. The sample preparation technique presented here enables proteomic patterns to be acquired on any MS capable of MALDI. These types of ion sources are available in almost every MS laboratory or core facility and when coupled with TOF instruments are capable of producing spectra with resolution in excess of 10,000, mass accuracy within 50 ppm, and sensitivity in the femtomole range. The sample preparation step is automatable and can easily be used to prepare hundreds of samples per day. In addition, the cost of sample preparation by SELDI is more expensive than the magnetic particle preparation.

Villanueva et al. (2004) applied this novel sample preparation technology to a pilot study with the goal of distinguishing patients with glioblastoma (GBM) from controls (i.e., no evidence of cancer). In this pilot study 34 serum samples from GBM patients and 22 from healthy volunteers were processed as described earlier and analyzed using MALDI-TOF-MS. Each MS spectrum contained more than 400 distinct peptide peaks that were unambiguously detected in each sample. After being aligned through their m/z values ("binning"), almost 1700 unique peaks were found in the 56 cumulative spectra. Two hundred and seventy-four peaks showed a statistical difference ($p < 0.05$) between GBM and control cases. These 274 peaks were then used to cross-validate the ability to discriminate the classes. Two classes were created by using 55 out of the 56 samples as a training set and using the 56th sample as a test set to verify whether it would be correctly classified. The process was repeated 56 times, that is, until all samples had been used as a test set. This analysis was able to correctly classify 53 out of 55 (96.4%) samples. Two controls were incorrectly classified (3.6%) as GBM samples, while the remaining sample was not classified. While this study was conducted using a limited number of samples, nonetheless, it is the first of its kind to employ a straight MALDI-based approach for the acquisition of serum proteomic patterns for the purpose of disease diagnostics.

13.4 BIOMARKER DISCOVERY COMBINING FRACTIONATION AND MASS SPECTROMETRY

Recently, a number of novel approaches based on use of conventional MS-based methods have been used to identify biomarkers. These approaches may provide an effective means to diagnose diseases such as cancer, without identifying all the constituents in the proteome. The methods that have been evaluated for profiling proteins for diagnostic purposes are often based on separation followed by MS detection and identification. These methods include two-dimensional polyacrylamide gel electrophoresis–MS (2D-PAGE)-MS (Bergman et al., 2000; He et al., 2004), liquid chromatography (LC)-MS (Govorukhina et al., 2003), capillary electrophoresis (CE)-MS (Neuhoff et al., 2004; Weissinger et al., 2004), and imaging of thin tissue slices by MALDI-MS (Chaurand et al., 2004a,b).

13.4.1 Two-Dimensional Polyacrylamide Gel Electrophoresis and Mass Spectrometry

Today, the method of choice for proteomic disease biomarker discovery is 2D-PAGE for the separation of complex protein mixtures and MS for their identification (Cash and Kroll, 2003). In this method, proteins from two distinct samples (diseased versus normal) are analyzed via 2D-PAGE and their protein expression patterns compared. Protein spots of interest are excised from the gel and are then proteolytically or chemically digested. The resulting peptides are identified by MALDI-TOF-MS or by ESI-MS, after which a peptide mass fingerprint is produced. This fingerprint is subsequently compared with peptide masses obtained from protein sequence databases in an attempt to identify the protein.

O'Farrell (1975) introduced 2D-PAGE for the separation of complex protein mixtures. In this approach, proteins are separated in two consecutive steps (dimensions). The first is isoelectrofocusing, whereby proteins are separated according their isoelectric point (the pH at which the net charge on that protein is zero). In the second dimension, proteins are separated by molecular mass. First-dimension gels were traditionally cast in small diameter tubes, leading to difficulties in reproducibility related to handling and stability of pH gradients. With the advent of immobilized pH gradients, separation in the first dimension has greatly improved, enabling the identification of thousands of proteins (Klose and Kobalz, 1995). As a separation technique, 2D-PAGE provides excellent resolution of complex protein mixtures. The method, however, is limited by (i) its sample loading capacity, (ii) its inability to resolve proteins at extreme isoelectric points and molecular weights, (iii) the sensitivity of conventional stains, and (iv) its labor intensiveness.

Several 2D-PAGE MS studies have been used in cancer research for comparing cancerous and unaffected protein expression profiles; we cite here a few examples. In a study by Fels et al. (2003), epithelial cell lysates from gastric cancer patients and healthy individuals were compared using 2D-PAGE MS. Several significant differences in protein expression were observed including downregulation of the CA11 protein in gastric cancer patients. CA11 is a 21,999 Da protein previously shown by Northern blot analysis and RT-PCR to be modulated at the RNA level (Yashikawa et al., 2000). Studies of B-cell chronic lymphocytic leukemia (B-CLL) by 2D-PAGE MS have shown correlations between protein-expression profiles and specific genetic aberrations. Voss and colleagues (Voss et al., 2001) have demonstrated lower levels of glutathione-S-transferase (a p53 response gene) and thioredoxin peroxidase I (an inhibitor of apoptosis) in B-CLL patients with deletions in the p53 tumor suppressor gene. In addition, increased levels of Hsp27 and reduced levels of both thioredoxin peroxidase I and protein disulfide isomerase (PDI) were associated with shorter survival times in B-CLL patients. 2D-PAGE MS also allows for the comparison of proteins in precancerous and cancerous tissue. A comparison of normal, metaplastic, and esophageal tissues revealed lower levels of Hsp27 (a protective protein against cytotoxic stress) in Barrett's metaplasia and esophageal adenocarcinomas (Soldes et al., 1999).

13.4.2 High Performance Liquid Chromatography/Mass Spectrometry

High performance (or pressure) liquid chromatography (HPLC) is a powerful analytical technique for the separation of proteins and peptides. HPLC may be operated in different modes of separation; however, the modes that have been most widely used for fractionating and separating proteins and peptides are ion exchange reverse-phase and size exclusion (Issaq et al., 2002). Complex mixtures of proteins such as cell lysates or serum proteome are best resolved by fractionation using ion exchange, isoelectric focusing (IEF), or size exclusion in the first dimension followed by reverse-phase chromatography in the second dimension. Mass spectrometry is used for detection and MS/MS for protein identification. Single-dimensional and multidimensional HPLC when interfaced on-line or off-line with the mass spectrometer is a highly sensitive, specific, accurate, and rapid diagnostic tool for the clinical laboratory that would result in the collection of large amounts of data. Advantages of 2D-HPLC/MS over 2D gel electrophoresis are automation, speed, and sensitivity. In addition, HPLC allows the analysis of acidic and basic, hydrophobic and hydrophilic proteins after enzymatic or chemical digestion. It is interesting to note that a single HPLC-MS/MS analysis would result in the separation and sequencing of hundreds of peptides and open up approaches for biomarker discovery. Advances in the ability to quantify differences between samples, healthy and diseased, and to detect for an array of post-translational modifications allow for the discovery of classes of protein biomarkers that were previously unassailable (McDonald and Yates, 2002).

In a HPLC/MS study (Govorukhina et al., 2003), serum of a cervical cancer patient was first depleted of albumin and γ-globulins, then analyzed for the presence of a specific tumor marker protein, squamous cell carcinoma antigen 1 (SCCA1). It is worth noting that human serum albumin and γ-globulins constitute approximately 65–97% of serum proteins. Removal of these proteins would allow the detection of lower abundance proteins such as SCCA1, present at 160.5 μg/L. To test the validity of LC/MS, serum from a cervical cancer patient was depleted, digested with trypsin, and analyzed for SCCA1. A control serum from a healthy donor was analyzed in comparison. A third normal serum sample was spiked with a known amount of recombinant SCCA1 to check for recovery of the tumor marker after depletion and digestion. To identify suitable tryptic peptide fragments, pure recombinant GST-SCCA1 was digested and analyzed by LC–MS, which allowed identification of seven SCCA1-specific peptides. While the presence of a peptide from SCCA1 was clearly visible in the serum of a cervical cancer patient, it was absent from the control serum of a healthy subject. This result shows that relevant tumor markers can be detected in human serum after depletion of albumin and IgG and after tryptic digestion, indicating that LC-MS is a promising technique for biomarker discovery. In the above two LC-MS studies, the biomarker was a known compound, which makes it an easy target to identify using single ion monitoring MS. The search for novel (unknown) protein disease markers by LC-MS from serum or other biological specimen requires a more complex set of experiments

TABLE 13.1. Proteome Analysis: Proteins Versus Peptides

Proteins are less soluble than peptides	Peptides are more soluble
Protein mixtures are less complex	Peptide mixtures are more complex
Proteins are harder to work with	Peptides are easier to work with
Protein loss is complete loss of information	Peptide loss is not total loss of information
Proteins are less homogeneous	Peptides are more homogeneous

involving fractionation, separation, derivatization, quantitation, MS detection, and MS/MS protein identification followed by exhaustive data analysis. The search for protein markers using MS can be achieved in two different approaches: top-down, where proteins are analyzed and identified without digestion, or bottom-up (shotgun proteomics). As mentioned earlier, the search for diagnostic protein markers from serum is not an easy task due to the large number of proteins in serum and tissue, and because a proteome is a dynamic entity. Also, for the proteome, unlike for the genome where PCR is used to increase the number of DNA copies, no such procedure is available for enriching the low abundance proteins. Most global proteomic studies to date employ the bottom-up approach developed by Yates and co-workers (Wasburn et al., 2001), where proteins are first digested chemically or proteolytically, then analyzed by 2D HPLC-MS. Each of these two approaches has its advantages and disadvantages; see Table 13.1. The top-down approach requires sophisticated MS and a loss of one protein may mean a loss of important information. In the shotgun approach, the proteins are digested, resulting in a complex mixture of peptides; however, the loss of one or more peptides does not lead to the loss of the entire protein species.

Lubman and his group (Hamler et al., 2004) at the University of Michigan elected to use a 2D liquid-phase procedure for fractionation and separation of proteins and MS for identification. In this study, 225 million cells from either normal or malignant breast epithelial cells maintained in culture were lysed in a buffer made up of 6 M urea, 2 M thiourea, 0.5% w/v n-octyl β-D-glycopyranoside, and 10 mM DTT. After vortexing for 1 hour and centrifugation at $40,000g$ at 4 °C for 20 min, the supernatant was collected and diluted to 15 mL with lysis buffer. One milliliter (6.25% v/v) of ampholyte solution was added prior to loading onto the IEF instrument, Rotofor from Bio Rad. Twenty separate fractions of approximately 550 µL each were collected. Each of these fractions was then analyzed by a nonporous silica reverse-phase HPLC column maintained at 65 °C coupled with ESI-TOF MS. Separations were performed using water/acetonitrile gradient (0.1% v/v TFA, 0.3% v/v formic acid). The gradient for acetonitrile was as follows: 5% to 15% in 1 min, 15% to 25% in 2 min, 25% to 31% in 3 min, 31% to 41% in 10 min, 41% to 47% in 10 min, 41% to 47% in 3 min, 47% to 67% in 4 min, and 67% to 100% in 1 min. A schematic of the experimental procedure is summarized in Figure 13.5. Mass spectrometry was used for protein quantitation, mass measurement, and peptide mass fingerprinting. Twenty-two proteins were more highly expressed in one or more of the malignant cell lines than in the normal cell lines. They concluded that these proteins represent potential breast cancer biomarkers that could aid in diagnosis, therapy, or drug development. This needs to be proved using cells from breast tumor tissue. In another study from the same laboratory, a similar 2D liquid-phase separation procedure was used to search for biomarker proteins from

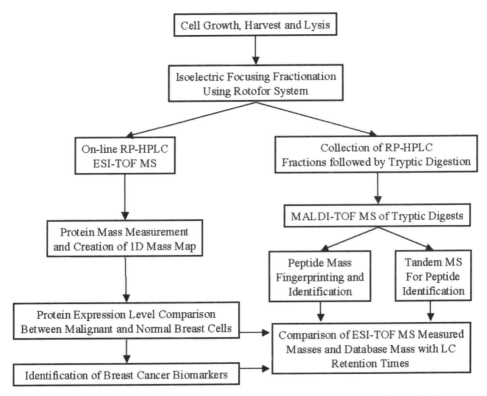

Figure 13.5. *Experimental overview of a liquid-phase two-dimensional separation technique combined with mass spectrometry for proteomic analysis of normal and fully malignant breast epithelial cell lines.*

human ovarian epithelial whole cell lysates. Two-dimensional protein expression maps were generated displaying protein isoelectric point (pI) versus intact protein molecular weight. Resulting 2D images effectively displayed quantitative differential protein expression in ovarian cancer cells versus nonneoplastic ovarian epithelial cells. Protein peak fractions were collected from the HPLC eluant, enzymatically digested, and analyzed by MS. Protein bands with significant up- or downregulation in one cell line versus another as viewed in the 2D expression maps were identified. This strategy may prove useful in identifying novel ovarian cancer marker proteins (Wang et al., 2004).

Wu et al. (2003) described a LC-MS method for the profiling of proteins in a sample prepared by laser capture microdissection (LCM) from a breast cancer cell line (SKBR-3). The captured cells (approximately 10,000) were solubilized with a denaturing buffer and digested with trypsin, and the resulting peptide mixture was fractionated by reverse-phase HPLC, and the identity of the peptides was determined by MS/MS. They were able to identify HER-2, a receptor protein kinase, and breast cancer marker and other related kinases. Unlike the previous studies, this study was performed on a limited number of cells that raises the possibility that such studies may be performed on individual patient samples obtained by needle biopsy.

Lai et al. (2001) developed a micro-HPLC-MS method to determine the presence of 17-alpha-hydroxyprogesterone (17OHP), a biomarker for congenital adrenal hyperplasia (CAH), which is caused by 21-hydroxylase deficiency. CAH is the most common inborn error of the adrenal steroid pathways. Early diagnosis and monitoring of CAH can be lifesaving. Blood samples from CAH patients or 2–5 day old infants were collected on filter paper and dried at room temperature for at least 3 h, and then stored in polypropylene bags at room temperature until analyzed. Four $\frac{1}{8}$ inch circles from each blood spot (equivalent to 11.5 μL of whole blood) were excised from a 0.5 inch (12.7 mm) diameter dried blood spot and placed into a flat-bottom 96 well block. A stock solution containing a known concentration of an internal standard was added to each vial (200 μL). The wells were capped and shaken for 50 min. Subsequently, the extracts were transferred into a clean V-bottomed 96 well plate and the solutions were evaporated to dryness under a gentle stream of dry nitrogen. The residue in each well was derivatized with 160 μL of GirardP solution, incubated at 65 °C for 50 min, and evaporated to dryness under a gentle stream of dry nitrogen. The GirP-derivatized 17OHP (GirP-17OHP) and 6α-methylprednisolone (6MP) (GirP-6MP) were reconstituted in 30 μL of 50% aqueous acetonitrile and analyzed by μHPLC/ESI-MS/MS. This high-throughput LC-MS/MS method (approximately 300 samples per day) is not only useful for both diagnosis and monitoring of treatment of CAH in different age groups but may also be used to screen babies for CAH.

13.4.3 Capillary Electrophoresis/Mass Spectrometry

Capillary electrophoresis (CE), a high-resolution microseparation technique, is suitable for the separation of proteins and peptides. In CE, proteins and peptides are driven through a suitable medium (buffer or gel) in a narrow fused silica capillary under the influence of an electric field and separated according to their size-to-charge ratio, size, or isoelectric point. The combination of capillary electrophoresis on-line with MS is a fast, sensitive, and specific technique for the separation and identification of peptides and proteins from biological specimens. In our laboratory, CE was used for the second separation dimension, after fractionation by reverse-phase HPLC, of a protein mixture digest (Issaq et al., 2001), a cancer cell line digest (He et al., 2002), and the low molecular weight serum proteome (Janini et al., 2004).

In most of the previously mentioned studies, the specimen of interest was either serum or cell lines. Urine, which is obtained in a noninvasive manner, can also be used for clinical diagnostics. The amount of proteins is less than in serum, and healthy individuals excrete less than 150 mg/day, while the urinary protein excretion of patients with renal diseases may exceed several grams per day (Shihabi et al., 1991). Thus, the evaluation of these proteins may lead to an increased understanding of renal physiology and possibly allow the differentiation of subgroups of renal diseases by the identification of these proteins via CE–MS. Such data would eventually even make renal biopsy superfluous.

Weissinger and collaborators published a series of studies (Wittke et al., 2003; Kaiser et al., 2004; Weissinger et al., 2004) dealing with the discovery of peptide/protein patterns in urine of normal and diseased subjects for clinical and diagnostic purposes using CE coupled with ESI-TOF-MS. In their earliest study (Wittke et al., 2003), urine samples of five patients with renal diseases and impaired renal function and eighteen healthy volunteers were obtained and stored at –20 °C until analysis time. Thawed urine was applied onto a C-2 column to remove urea, electrolytes, salts, and other interfering components, to decrease matrix effects that would interfere with the CE resolution and MS detection and to enrich for the polypeptides present. Polypeptides were eluted with 50% (v/v) acetonitrile in HPLC-grade water containing 0.5% (v/v) formic acid. The pretreated samples were lyophilized and resuspended in 20 μL of HPLC-grade water and analyzed by CE-ESI-TOF-MS. Analysis of the individual CE-MS spectra of the eighteen healthy volunteers, which were similar and comparable, allowed the establishment of a "normal urine polypeptide pattern," consisting of 247 polypeptides, each of which was found in more than 50% of the healthy individuals. Applying CE-MS to the analysis of urine of patients with kidney disease revealed differences in polypeptide patterns. Twenty-seven polypeptides were found exclusively in samples of patients while 13, present in controls, were missing (Wittke et al., 2003).

In a recent study (Kaiser et al., 2004), the same CE-ESI-TOF-MS method was applied for the examination of urine samples from 57 healthy individuals, 16 patients with minimal change disease, 18 patients with membranous glomerulonephritis, and 10 patients with focal segmental glomerulosclerosis. Examination of the CE-ESI-TOF-MS results indicated that 173 polypeptides were present in more than 90% of the urine samples of the healthy subjects, while 690 polypeptides were present with more than 50% probability. As in the previous study (Wittke et al., 2003), researchers were able to establish a "normal" polypeptide pattern in healthy individuals while polypeptides found in the urine of patients differed significantly from the normal controls. These differences allowed the distinction of specific protein spectra in patients with different primary renal diseases. Also, abnormal patterns of proteins were found even in urine from patients in clinical remission. It is believed that this CE-MS method provides a promising tool that permits fast and accurate identification and differentiation of protein patterns in body fluids of healthy and diseased individuals, thus enabling diagnosis (Kaiser et al., 2004).

A comparison of SELDI-TOF-MS and CE-MS methodologies for protein pattern diagnostics was undertaken (Neuhoff et al., 2004). Urine samples from patients suffering from membranous glomerulonephritis and healthy volunteers were analyzed by both techniques. A rich and complex pattern of polypeptides with high resolution and high mass accuracy was obtained by CE-MS, while the pattern obtained by SELDI-TOF-MS from the same samples showed a much lower number of peptides and lower resolution and mass accuracy. Although the SELDI-TOF-MS analysis identified three differentially expressed polypeptides, which are potential biomarkers, approximately 200 potential biomarkers could be identified by CE-MS. Thus, while SELDI-TOF-MS is easy to use and requires very little sample preparation, CE-MS generates a much richer data set. Unfortunately, in this study the number of patients and number of controls were not clearly stated. Normally, tens of samples from both groups need to be analyzed before any solid conclusions could be drawn.

13.5 IMAGING MASS SPECTROMETRY

Caprioli and his group (Chaurand et al., 2004a) developed a new concept, imaging MS (IMS), for the analysis of expressed proteins within a tissue sample. Imaging MS is a relatively new technology that takes advantage of the methodology and instrumentation of MALDI-MS, a particularly useful technique because of its potential for high throughput and ability to provide information on the localization of molecules in a sample. Protein profiles and images can be obtained directly from thin tissue sections cut on a cryostat from fresh-frozen tissue blocks. The advantage of IMS is the potential for the simultaneous analysis of different molecular species, proteins, peptides, drugs, and metabolites, present in a single tumor, opening new possibilities for the measurement of concomitant protein changes in specific tissues after systemic drug administration. Direct tissue profiling and IMS allow the visualization of 500–1000 individual protein signals in the molecular weight range from 2000 to over 200,000 from thin tissue sections. To date, profiling and IMS have been applied to multiple diseased tissues, including human non-small cell lung tumors, gliomas, and breast tumors. Interrogation of the resulting complex MS data sets using modern biocomputational tools has resulted in identification of both disease-state and patient-prognosis specific protein patterns.

The experimental setup is illustrated in Figure 13.6. Tissue biopsies or other relevant tissue samples are frozen immediately after acquisition in liquid nitrogen or isopentane to preserve the sample's morphology and minimize protein degradation through enzymatic proteolysis. For most applications, 5–20 µm thick sections are cut at −15 °C and thaw-mounted on an electrically conductive sample plate. The sections are dried in a dessicator for several minutes before MALDI matrix deposition. The MALDI matrix can be deposited either as individual droplets or as a homogeneous layer coated on the tissue section. The resulting mass spectra typically yield 300–1000 signals, of various intensities. Most of the signals detected are below m/z 50,000 because of the inability of TOF mass analyzers to resolve and efficiently detect higher molecular weight compounds. In general, the most intense signals come from the most abundant protein species. The exact sensitivity of the technology, however, is hard to estimate because exact amounts of proteins within a specific tissue are generally not well known. The profiles recovered have been found to be extremely specific to a given tissue type and, when analyzing serial sections, very reproducible.

Yanagisawa et al. (2003) used MALDI-TOF-MS to generate protein spectra directly from frozen tissue sections from 79 surgically resected lung tumors and 14 normal lung tissue for profiling of protein expression to classify lung tumors. More than 1600 protein peaks from histologically selected 1 mm diameter regions of single frozen sections from each tissue were recorded. Unsupervised and supervised hierarchical multivariant cluster analyses were employed to subclassify the samples according to their expression patterns and to look for relationships between tumor subtypes and clinical outcome. A class prediction model was created using the protein profiles from a training cohort of 34 primary lung tumors, two pulmonary metastases of previously resected non-small cell lung carcinomas (NSCLCs), one pulmonary carcinoid, five metastases to the lung from other sites, and eight normal lung samples. From among more than 1600 individual protein signals detected across all patient samples, 82 signals differentially expressed between lung

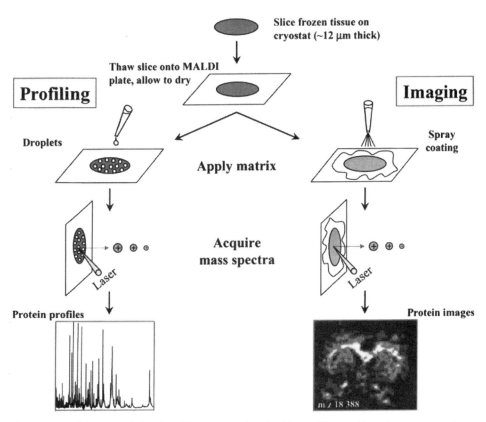

Figure 13.6. *Scheme outlining the different steps involved for profiling and imaging mass spectrometry of mammalian tissue samples. Reprinted with permission from Chaurand et al. (2004b).*

tumors and normal lung were selected as discriminators. When this model was applied to a blinded test cohort of 32 primary NSCLCs, five metastases to the lung, and six normal lung samples to estimate the rate of misclassification, their proteomic patterns correctly classified all samples as either tumor or normal. They also reported that, using similar class prediction models, primary NSCLC could be distinguished from normal lung based on 91 MS signals with 100% accuracy, and it was possible to discriminate between the major histological subtypes of NSCLCs: adenocarcinoma, squamous cell carcinoma, and large cell. These studies suggest that such proteomic information will become more and more important in assessing disease progression, prognosis, and drug efficacy.

13.6 CONCLUSION

Surface enhanced laser desorption ionization TOF-MS is not unlike any other technique; there are limitations as well as benefits. A major criticism is the inability of the mass analyzer to provide the MS/MS capabilities of higher-end mass spectrometers. As emphasized previously, SELDI-TOF-MS is not designed for this purpose. This technology's strength lies in its ease of use and its ability to rapidly screen hundreds of relatively crude samples to determine targets (key *m/z* values) for further investigation. The technique is amenable to automation, thereby doubling the number of samples processed per day.

Proteomic pattern diagnostics is an ideal application for SELDI-TOF-MS. Although this application holds great promise as a diagnostic tool, it has several critical areas that must be addressed. How the combination of MS and bioinformatics meshes to determine the pathological condition of a patient from a simple mass spectrum is still a mystery. A major criticism of proteomic pattern analysis is that the identity of the protein or peptides giving rise to the key *m/z* features is not known. It is debatable as to whether it is worthwhile to even identify these key features as they may provide little advantage in developing a diagnostic platform. For example, many of the *m/z* values that account for the diagnostic predictability are less than 10 kDa; therefore, it is possible that these species could be fragments generated from larger proteins proteolyzed either within the circulatory system or in the tumor/host microenvironment. It would be extremely difficult to generate an affinity reagent with specificity to a peptide fragment without considerable cross-reactivity to its parent protein. In addition, there are tools currently used in the medical field, such as the electrocardiogram, which rely solely on a pattern as basis for diagnosing disease. Even the identification of disease-specific biomarkers may not provide any direct insight into how a disease may arise or progress. Indeed, identifying a specific biomarker does not guarantee that this knowledge will provide any mechanistic or therapeutic insights into a certain cancer. An excellent example is PSA. This biomarker is used to indicate the possible presence of a prostate tumor, yet its role in cancer development is unknown. On the other hand, knowing the identity of the key features from a proteomic pattern analysis may provide a glimpse into the manifestation and progression of cancer. The medical community is certainly interested in identifying key diagnostic features and discerning how these proteins may or may not relate to disease physiology or progression.

Mass spectrometry-based proteomics has traditionally been focused on one major goal: to identify and characterize an increasing number of proteins from clinical samples in an attempt to find disease-specific biomarkers. Surface enhanced laser desorption ionization TOF-MS has ushered in a new era. While biomarker discovery is still an important initiative, proteomic pattern diagnostics using SELDI-TOF-MS represents a powerful tool to assist in disease diagnosis. Although there are still many critics of this technique, there are obvious benefits that cannot be ignored. The most obvious is that the technique enables the screening of large populations to detect diseases at earlier stages, thus facilitating more effective medical treatment and survival rates. It is clear that the next few years will be critical in the validation of proteomic patterns as a clinical test for disease diagnosis.

ACKNOWLEDGMENTS

This project has been funded in whole or in part with federal funds from the National Cancer Institute, National Institutes of Health, under Contract No. NO1-CO-12400.

By acceptance of this article, the publisher or recipient acknowledges the right of the U.S. Government to retain a nonexclusive, royalty-free license and to any copyright covering the article. The content of this publication does not necessarily reflect the views or policies of the Department of Health and Human Services, nor does mention of trade names, commercial products, or organizations imply endorsement by the U.S. Government.

REFERENCES

Adam BL, Qu Y, Davis JW, Ward MD, Clements MA, Cazares LH, Semmes OJ, Schellhammer PF, Yasui Y, Feng Z, Wright GL Jr. 2002. Serum protein fingerprinting coupled with a pattern-matching algorithm distinguishes prostate cancer from benign prostate hyperplasia in healthy men. *Cancer Res* 62:3609–3614.

American Cancer Society, Cancer Statistics. 2004. Available from http://www.cancer.org/docroot/STT/stt_0.asp html. Accessed December 1, 2004.

Anderson NL, Anderson NG. 2002. The human plasma proteome: history, character, and diagnostic prospects. *Mol Cell Proteomics* 1:845–867.

Banez LL, Prasanna P, Sun L, Ali A, Zou Z, Adam BL, McLeod DG, Moul JW, Srivastava S. 2003. Diagnostic potential of serum proteomic patterns in prostate cancer. *J Urol* 170:442–446.

Barzaghi D, Isbister JD, Lauer KP, Born TL. 2004. Use of surface-enhanced laser desorption/ionization–time of flight to explore bacterial proteomes. *Proteomics* 4:2624–2628.

Bergman AC, Benjamin T, Alaiya A, Waltham M, Sakaguchi K, Franzen B, Linder S, Bergman T, Auer G, Appella E, Wirth PJ, Jornvall H. 2000. Identification of gel-separated tumor marker proteins by mass spectrometry. *Electrophoresis* 21:679–686.

Bock C, Coleman M, Collins B, Davis J, Foulds G, Gold L, Greef C, Heil J, Heilig JS, Hicke B, Nelson-Hurst M, Husar GM, Miller D, Ostroff R, Petach H, Schneider D, Vant-Hull B, Waugh S, Weiss A, Wilcox SK, Zichi D. 2004. Photoaptamer arrays applied to multiplexed proteomic analysis. *Proteomics* 4:609–618.

Carrette O, Demalte I, Scherl A, Yalkinoglu O, Corthals G, Burkhard P, Hochstrasser DF, Sanchez J-C. 2003. A panel of cerebrospinal fluid potential biomarkers for the diagnosis of Alzheimer's disease. *Proteomics* 3:1486–1494.

Cash P, Kroll JS. 2003. Protein characterization by two-dimensional gel electrophoresis. *Methods Mol Med* 71:101–118.

Chaurand P, Sanders ME, Jensen RA, Caprioli RM. 2004a. Proteomics in diagnostic pathology: profiling and imaging proteins directly in tissue sections. *Am J Pathol* 165:1057–1068.

Chaurand P, Schwartz SA, Caprioli RM. 2004b. Assessing protein patterns in disease using imaging mass spectrometry. *J Proteome Res* 3:245–252.

Conrads TP, Veenstra TD. 2004. The utility of proteomic patterns for the diagnosis of cancer. *Curr Drug Targets* 4:41–50.

Conrads TP, Fusaro VA, Ross S, Johann D, Rajapakse V, Hitt BA, Steinberg SM, Kohn EC, Fishman DA, Whitely G, Barrett JC, Liotta LA, Petricoin EF III, Veenstra TD. 2004a. High resolution serum proteomic features for ovarian cancer detection. *Endocr Relat Cancer* 11:163–178.

Conrads TP, Hood BL, Issaq HJ, Veenstra TD. 2004b. Proteomic patterns as a diagnostic tool for early-stage cancer. *Mol Diagn* 8:77–85.

DiMagno EP, Reber HA, Tempero MA. 1999. Epidemiology, diagnosis and treatment of pancreatic ductal adenocarcinoma. *Gastroenterology* 117:1463–1484.

Fels LM, Buschmann T, Meuer J, Reymond MA, Lamer S, Rocken C, Ebert MPA. 2003. Proteome analysis for the identification of tumor-associated biomarkers in gastrointestinal cancer. *Dig Dis* 21:292–298.

Govorukhina NI, Keizer-Gunnink A, van der Zee AGJ, de Jong SH, de Bruijn'and WA, Bischoff WA. 2003. Sample preparation of human serum for the analysis of tumor markers: comparison of different approaches for albumin and γ-globulin depletion. *J Chromatogr A* 1009:171–178.

Hamler RL, Zhu K, Buchanan NS, Kreunin P, Kachman MT, Miller FR, Lubman D. 2004. A two-dimensional liquid-phase separation method coupled with mass spectrometry for proteomic studies of breast cancer and biomarker identification. *Proteomics* 4:562–577.

He Y, Yeung ES, Chan KC, Issaq HJ. 2002. Two dimensional mapping of cancer cell extracts by liquid chromatography–capillary electrophoresis with ultraviolet absorbance detection. *J Chromatogr A* 979(1–2):81–89.

Hutchens TW, Yip TT. 1993. New desorption strategies for the mass spectrometric analysis of macromolecules. *Rapid Commun Mass Spectrom* 7:576–580.

Hynes SO, McGuire J, Wadström T. 2003. Potential for proteomic profiling of *Helicobacter pylori* and other *Helicobacter* spp. using a ProteinChip array. *FEMS Immunol Med Microbiol* 36:151–158.

Imai K, Yamanaka H. 1998. The significance and limitation of prostate specific antigen in the mass screening for prostate cancer. *Nippon Rinsho* 56:1994–1997.

Issaq HJ, Chan KC, Liu CS, Li Q. 2001. Multidimensional high performance liquid chromatography—capillary electrophoresis separation of a protein digest: an update. *Electrophoresis* 22:1133–1135.

Issaq HJ, Conrads TP, Janini GM, Veenstra TD. 2002. Methods for fractionation, separation and profiling of proteins and peptides. *Electrophoresis* 23:3048–3061.

Issaq HJ, Conrads TP, Veenstra TD. 2003. Proteomic patterns: A potential test for early diagnosis of ovarian cancer. *The Female Patient* 28:42–52.

Jacobs IJ, Skates SJ, MacDonald N, Menon U, Rosenthal AN, Davies AP, Woolas R, Jeyarajah AR, Sibley K, Lowe DG, Oram DH. 1999. Screening for ovarian cancer: a pilot randomised controlled trial. *Lancet* 353:1207–1210.

Janini GM, Chan KC, Conrads TP, Issaq HJ, Veenstra TD. 2004. Two-dimensional liquid chromatography–capillary zone electrophoresis–sheathless electrospray ionization–mass spectrometry: evaluation for peptide analysis and protein identification. *Electrophoresis* 25:1973–1980.

Johnston-Wilson NL, Bouton CMLS, Pevsner J, Breen JJ, Torrey EF, Yolken RH, The Stanley Neurovirology Working Group. 2001. Emerging technologies for large-scale screening of human tissues and fluids in the study of severe psychiatric disease. *Int J Neuropsychopharm* 4:83–92.

Jones MB, Krutzsch H, Shu H, Zhao Y, Liotta LA, Kohn EC, Petricoin EF III. 2002. Proteomic analysis and identification of new biomarkers and therapeutic targets for invasive ovarian cancer. *Proteomics* 2:76–84.

Kainz C. 1996. Early detection and preoperative diagnosis of ovarian carcinoma. *Wien Med Wochenschr* 146:2–7.

Kaiser T, Wittke S, Just I, Krebs R, Bartel S, Fliser D, Mischak H, Weissinger EM. 2004. Capillary electrophoresis coupled to mass spectrometer for automated and robust polypeptide determination in body fluids for clinical use. *Electrophoresis* 25:2044–2055.

Kaul M, Garden GA, Lipton SA. 2001. Pathways to neuronal injury and apoptosis in HIV-associated dementia. *Nature* 410:988–994.

Klose J, Kobalz U. 1995. Two dimensional electrophoresis of proteins. An updated protocol and implications for a functional analysis of the genome. *Electrophoresis* 16:1034–1059.

Koopman J, Zhang Z, White N, Rosenzweig J, Fedarko N, Jagannath S, Canto MI, Yeo CT, Chan DW, Goggins M. 2004. Serum diagnosis of pancreatic adenocarcinoma using surface-enhanced laser desorption and ionization mass spectrometry. *Clin Cancer Res* 10:860–868.

Lagendijk JH, Mullink H, VanDeist PJ, Meijer GA, Meijer CJLM. 1998. Tracing the origin of adenocarcinomas with unknown primary using immunochemistry. *Human Pathol* 29:491–497.

Lai CC, Tsai CH, Tsai FJ, Lee CC, Lin WD. 2001. Rapid monitoring of congenital adrenal hyperplasia with microbore high-performance liquid chromatography/electrospray ionization tandem mass spectrometry from dried blood spots. *Rapid Commun Mass Spectrom* 15:2145–2151.

Lawrie LC, Curran S, McLeod HL, Fothergill JE, Murray GI. 2001. Application of laser capture microdissection and proteomics in colon cancer. *Mol Pathol* 54:253–258.

Lewczuk P, Esselmann H, Markus M, Wollscheid V, Neumann M, Otto M, Maler JM, Ruther E, Kornhuber J, Wiltfang J. 2003. The amyloid-β (Aβ) peptide pattern in cerebrospinal fluid in Alzheimer's disease: evidence of a novel carboxyterminally elongated Aβ peptide. *Rapid Commun Mass Spectrom* 17:1291–1296.

Lewczuk P, Esselmann H, Groemer TW, Bibl M, Maler JM, Steinacker P, Otto M, Kornhuber J, Wiltfang J. 2004. Amyloid β peptides in cerebrospinal fluid as profiled with surface enhanced laser desorption/ionization time-of-flight mass spectrometry: evidence of novel biomarkers in Alzheimer's disease. *Biol Psychiatry* 55:524–530.

Li J, Zhang Z, Rosenzweig JM, Wang YY, Chan DW. 2002. Proteomics and bioinformatics to detect breast cancer. *Clin Chem* 48:1296–1304.

Li J, White N, Zhang Z, Rosenzweig J, Mangold LA, Partin AW, Chan DW. 2004. Detection of prostate cancer using serum proteomics pattern in a histologically confirmed population. *J Urol* 171:1782–1787.

Luo X, Carlson KA, Wojna V, Mayo R, Biskup TM, Stoner J, Anderson J, Gendelman HE, Melendez LM. 2003. Macrophage proteomic fingerprinting predicts HIV-1-associated cognitive impairment. *Neurology* 60:1931–1937.

MacNeil JS. 2004. Better biomarkers for the diagnostics labyrinth. *Genome Technol* 24–33.

Martin SA, Rosenthal RS, Biemann K. 1987. Fast atom bombardment mass spectrometry and tandem mass spectrometry of biologically active peptidoglycan monomers from *Neisseria gonorrhoeae. J Biol Chem* 262:7514–7522.

McDonald WH, Yates JR III. 2002. Shotgun proteomics and biomarker discovery. *Dis Markers* 18:99–105.

National Cancer Institute News Center. 2004. Proteomics and Cancer. http://www.cancer.gov/newscenter/pressreleases/proteomicsQandA.

Neuhoff NV, Kaiser T, Wittke S, Krebs R, Pitt A, Burchard A, Sundmacher A, Schlegelberger B, Kolch W, Mischak H. 2004. Mass spectrometry for the detection of differentially expressed proteins: a comparison of surface-enhanced laser desorption/ionization and capillary electrophoresis/mass spectrometry. *Rapid Commun Mass Spectrom* 18:149–156.

NIH. 2004. US Department of Health and Human Services, NIAID HIV/AIDS Statistics, 2004. http://www.niaid.nih.gov/factsheets/aidsstat.html.

Nishizuka S, Chen S-T, Gwadry FG, Alexander J, Major SM, Scherf U, Reinhold WC, Waltham M, Charboneau L, Young L, Bussey KJ, Kim S, Lababidi S, Lee JK, Pittaluga S, Scudiero DA, Sausville EA, Munson PJ, Petricoin EF III, Liotta LA, Hewitt SM, Raffeld M, Weinstein JN. 2003. Diagnostic markers that distinguish colon and ovarian adenocarcinomas: identification by genomic, proteomic and tissue array profiling. *Cancer Res* 63:5243–5250.

O'Farrell PH. 1975. High resolution two-dimensional electrophoresis of proteins. *J Biol Chem* 250:4007–4021.

Petricoin EF III, Ardekani AM, Hitt BA, Levine PJ, Fusaro VA, Steinberg SM, Mills GB, Simone C, Fishman DA, Kohn EC, Liotta LA. 2002a. Use of proteomic patterns in serum to identify ovarian cancer. *Lancet* 359:572–577.

Petricoin EF III, Ornstein DK, Paweletz CP, Ardekani A, Hackett PS, Hitt BA, Velassco A, Trucco C, Wiegand L, Wood K, Simone CB, Levine PJ, Linehan WM, Emmert-Buck MR, Steinberg SM, Kohn EC, Liotta LA. 2002b. Serum proteomic patterns for detection of prostate cancer. *J Natl Cancer Inst* 94:1576–1578.

Petricoin EF, Liotta LA. 2004. SELDI TOF-based serum proteomic pattern diagnostics for early detection of cancer. *Curr Opin Biotechnol* 15:24–30.

Pieper R, Su Q, Gatlin CL, Huang ST, Anderson NL, Steiner S. 2003. Multi-component immunoaffinity subtraction chromatography: an innovative step towards a comprehensive survey of the human plasma proteome. *Proteomics* 3:422–432.

Pusztai L, Gregory BW, Baggerly KA, Pang B, Koopman J, Kuerer HM, Esteva FJ, Symmons WF, Wagner P, Hortobagyi GN, Laronga C, Semmes OJ, Wright GL Jr, Drake RR, Vlahou A. 2004. Pharmacoproteomic analysis of prechemotherapy and postchemotherapy plasma samples from patients receiving neoadjuvant chemotherapy for breast carcinoma. *Cancer* 100:1814–1822.

Rai AJ, Zhang Z, Rosenzweig J, Shih IM, Pham T, Fung ET, Sokoll LJ, Chan DW. 2002. Proteomic approaches to tumor marker discovery. *Arch Pathol Lab Med* 126:1518–1526.

Shihabi ZK, Konen JC, O'Connor ML. 1991. Albuminuria versus urinary total protein for detecting chronic renal disorders. *Clin Chem* 37:621–624.

Shiwa M, Nishimura Y, Wakatabe R, Fukawa A, Arikuni H, Ota H, Kato Y, Yamori T. 2003. Rapid discovery and identification of a tissue-specific tumor biomarker from 39 human cancer cell lines using the SELDI ProteinChip platform. *Biochem Biophys Res Commun* 309:18–25.

Soldes OS, Kuick RD, Thompson IA II, Hughes SJ, Orringer MB, Iannettini MD, Hanash SM, Beer DG. 1999. Differential expression of Hsp27 in normal oesophageal, Barrett's metaplasia, and oesophageal adenocarcinomas. *Br J Cancer* 79:595–603.

Thulasiraman V, McCutchen-Maloney SL, Motin VL, Garcia E. 2001. Detection and identification of virulence factors in *Yersinia pestis* using SELDI ProteinChip System. *Biotechniques* 30:428–432.

Uchida T, Fukawa A, Uchida M, Fujita K, Saito K. 2002. Application of a novel protein biochip technology for detection and identification of rheumatoid arthritis biomarkers in synovial fluid. *J Proteome Res* 1:495–499.

Van QN, Klose JR, Lucas DA, Prieto DA, Luke BT, Collin J, Burt SK, Chmurny GN, Issaq HJ, Conrads TP, Veenstra TD. 2003, 2004. The use of urine proteomic and metabonomic patterns for the diagnosis of interstitial cystitis and bacterial cystitis. *Dis Markers* 19:169–183.

Villanueva J, Philip J, Entenberg D, Chaparro CA, Tanwar MK, Holland EC. Tempst P. 2004. Serum peptide profiling by magnetic particle-assisted, automated sample processing and MALDI-TOF mass spectrometry. *Anal Chem* 76:1560–1570.

Vlahou A, Schellhammer PF, Mendrinos S, Patel K, Kondylis FI, Gong L, Nasim S, Wright GL Jr. 2001. Development of a novel proteomic approach for the detection of transitional cell carcinoma of the bladder in urine. *Am J Pathol* 158:1491–1502.

Voss T, Ahorn H, Haberl P, Dohner H, Wilgenbus K. 2001. Correlation of clinical data with proteomics profiles in 24 patients with B-cell chronic lymphocytic leukemia. *Int J Cancer* 91:180–186.

Wadsworth JT, Somers KD, Cazares LH, Malik G, Adam BL, Stack BC Jr., Semmes OJ. 2004. Serum protein profiles to identify head and neck cancer. *Clin Cancer Res* 10:1625–1632.

Wagner M, Naik DN, Pothan A, Kasukurti S, Devineni RR, Adam BL, Semmes OJ, Wright GL Jr. 2004. Computational protein biomarker prediction: a case study for prostate cancer. *BMC Bioinformatics* 5:26.

Wang H, Kachman MT, Schwartz DR, Cho KR, Lubman DM. 2004. Comprehensive proteome analysis of ovarian cancers using liquid phase separation, mass mapping and tandem mass spectrometry: a strategy for identification of candidate cancer biomarkers. *Proteomics* 4:2476–2495.

Washburn MP, Wolters D, Yates JR. 2001. Large-scale analysis of the yeast proteome by multidimensional protein identification technology. *Nat Biotechnol* 19:242–247.

Wauters CCAP, Smedts F, Gerrits LGM, Bosman FT, Ramaekers FCS. 1995. Keratins 7 and 20 as diagnostic markers of carcinomas metastatic to the ovary. *Human Pathol* 26:852–855.

Weissinger EM, Wittke S, Kaiser T, Haller H, Bartel S, Krebs R, Golovko I, Rupprecht HD, Haubitz M, Hecker H, Mischak H, Fliser D. 2004. Proteomic patterns established with capillary electrophoresis and mass spectrometry for diagnostic purposes. *Kidney Int* 265:2426–2434.

Wittke SH, Fliser D, Haubitz M, Bartel S, Krebs H, Hausadel F, Hillmann M, Golovko I, Koester P, Haller H, Kaiser T, Mischak H, Weissinger EM. 2003. Determination of peptides and proteins in human urine with capillary electrophoresis–mass spectrometry, a suitable tool for the establishment of new diagnostic markers. *J Chromatogr A* 1013:173–181.

Wojna V, Carlson KA, Luo X, Mayo R, Melendez LM, Kraiselburd E, Gendelman HE. 2004. Proteomic fingerprinting of human immunodeficiency virus type 1-associated dementia from patient monocyte-derived macrophages: a case study. *J NeuroVir* 10:74–81.

Wong YF, Cheung TH, Lo KW, Wang VW, Chan CS, Ng TB, Chung TK, Mok SC. 2004. Protein profiling of cervical cancer by protein-biochips: proteomic scoring to discriminate cervical cancer from normal cervix. *Cancer Lett* 211:227–234.

Wright GL, Cazarres LH, Leung S-M, Nasim S, Adam BL, Yip TT, Schellhammer PF, Gong L, Vlahou A. 1999. ProteinChip surface enhanced laser desorption/ionization (SELDI) mass spectrometry: a novel protein biochip technology for detection of prostate cancer biomarkers in complex protein mixtures. *Prostate Cancer Prostatic Dis* 2:264–276.

Wu S-L, Hancock WS, Goodrich GG, Kunitake ST. 2003. An approach to the proteomic analysis of a breast cancer cell line (SKBR-3). *Proteomics* 3:1037–1046.

Wulfkuhle JD, McLean KC, Paweletz CP, Sgroi DC, Trock BJ, Steeg PS, Petricoin EF III. 2001. New approaches to proteomic analysis of breast cancer. *Proteomics* 1:1205–1215.

Xiao Z, Adam BL, Cazares LH, Clements MA, Davis JW, Schellhammer PF, Dalmasso EA, Wright GL Jr. 2001. Quantitation of serum prostate-specific antigen by a novel protein biochip immunoassay discriminates benign from malignant prostate disease. *Cancer Res* 61:6029–6033.

Xiao X, Liu D, Tang Y, Guo F, Xia L, Liu J, He D. 2003. Development of proteomic patterns for detecting lung cancer. *Dis Markers* 19:33–39.

Xiao Z, Luke BT, Izmirlian G, Umar A, Lynch PM, Phillips RKS, Patterson S, Conrads TP, Veenstra TD, Greenwald P, Hawk ET, Ali IU. 2004. Serum proteomic profiles suggest celecoxib-modulated targets and response predictors. *Cancer Res* 64:1–6.

Yanagisawa K, Shyr Y, Xu BJ, Massion PP, Larsen PH, White BC, Roberts JR, Edgerton M, Gonzalez A, Nadaf S, Moore JH, Caprioli RM, Carbone DP. 2003. Proteomic patterns of tumor subsets in non-small-cell lung cancer. *Lancet* 362:433–439.

Yang ECC, Guo J, Diehl G, DeSouza L, Rodrigues MJ, Romaschin AD, Colgan TJ, Siu KWM. 2004. Protein expression profiling of endometrial malignancies reveals a new tumor marker: chaperonin 10. *J Proteome Res* 3:636–643.

Yashikawa Y, Mukai H, Hino F, Asada K, Kato I. 2000. Isolation of two novel genes downregulated in gastric cancer. *Jpn J Cancer Res* 91:459–463.

Yeo CT, Cameron JL, Lillemoe KD, Sitzman JV. 1995. Pancreatic duodenectomy for cancer of the head of the pancreas. *Ann Surg* 221:721–731.

Zhang L, Yu W, He T, Yu J, Caffrey RE, Dalmasso EA, Fu S, Pham T, Mei J, Ho JJ, Zhang W, Lopez P, Ho DD. 2002. Contribution of human alpha-defensin 1, 2, and 3 to the anti-HIV-1 activity of CD8 antiviral factor. *Science* 298:995–1000.

Zhang Z, Bast RC Jr, Yu Y, Li J, Sokoll LJ, Rai AJ, Rosenzweig JM, Cameron B, Wanf YY, Meng X-Y, Berchuck A, Van Haaften-Day C, Hacker NF, deBruijn HWA, van der Zee AGJ, Jacobs JI, Fung ET. 2004. Three biomarkers identified from serum proteomic analysis for the detection of early stage ovarian cancer. *Cancer Res* 64:5882–5890.

Zheng J, Gendelman HE. 1997. The HIV-1 associated dementia complex: a metabolic encephalopathy fueled by viral replication in mononuclear phagocytes. *Curr Opin Neurol* 10:319–325.

Zhukov TA, Johnson RA, Cantor AB, Clark RA, Jockman MS. 2003. Discovery of distinct protein profiles specific for lung tumors and pre-malignant lung lesions by SELDI mass spectrometry. *Lung Cancer* 40:267–279.

14

Automation in Proteomics

Timothy D. Veenstra

*Laboratory of Proteomics and Analytical Technologies, SAIC-Frederick, Inc.,
National Cancer Institute at Frederick, Frederick, Maryland*

14.1   INTRODUCTION

Automation will come to anything that people find necessary, enjoyable, or fasci-
nating. This rule applies to everything from automobile production to making ice
cream. It was inevitable that automation would find proteomics since many of the
necessary ingredients were already in place before this field of science became
popularized. The proteome is a vast frontier whose complexity is yet to be fully
understood and will require almost an infinite number of experiments before a true
grasp of its nature is attained. Combine this with the impatience of most scientists
and the equation equals automation. The term proteomics was originally coined to
refer to the identification of all of the proteins expressed within a cell, tissue, or
organism at a particular time and under a given set of conditions (Wilkens et al.,
1996). This term is now used to refer to almost everything that previously fell under
the guise of protein science. For example, in the past structural biologists have used
such tools as X-ray crystallography and nuclear magnetic resonance (NMR) spectroscopy to determine the three-dimensional structure of a protein. Now they do
structural proteomics (Jung and Lee, 2004). What used to be the determination of
the phosphorylation state of a protein is now referred to as phosphoproteomics
(Stern, 2001). The one thing that separates protein science from proteomics in the
minds of many, however, is the scale at which each is conducted. While protein
science evokes the connotation of a single protein being studied at great length,
proteomics suggests the analysis of many (possibly thousands of) proteins almost
simultaneously or within a short time frame. The scale at which proteomics is conducted may be the one characteristic that distinguishes it from protein science.

Proteomics for Biological Discovery, edited by Timothy D. Veenstra and John R. Yates.
Copyright © 2006 John Wiley & Sons, Inc.

Let's consider that one of the goals of proteomics is to completely sequence all of the proteins expressed within a cell at any given time under a specific set of conditions. A recent report has placed the number of known or predicted human proteins at 25,931 (http://www.proteomesci.com/content/1/1/5). The most efficient method to obtain protein sequences is through the analysis of their peptides produced via tryptic digestion and analyzing these fragments using tandem mass spectrometry (MS/MS) (Lin et al., 2003). Therefore, the maximum number of total tryptic peptides that would need to be sequenced is 1,175,015. Even assuming that only 30% of the genome is expressed at any given time, this would still require the analysis of approximately 7800 proteins and 350,000 peptides. Currently, the best available MS-based technologies are able to routinely sequence on the order of 10,000 peptides corresponding to about 2000–3000 unique proteins from a sample of human cells (Yu et al., 2004). Obviously, today's technologies are capable of providing significant coverage on the protein level, but poor coverage on the peptide level. This result means we often obtain a minimal description of most of the proteins within a complex proteome sample.

Studies to identify proteins within proteome samples through the use of peptide surrogates are important and can reveal not only the presence but also the relative abundance of a protein. The character of a protein, however, cannot be fully described without detailing other critical features about it. As shown in Figure 14.1, direct translation of a gene sequence to a putative protein sequence is inadequate as many different post-transcriptional and post-translational processes can occur

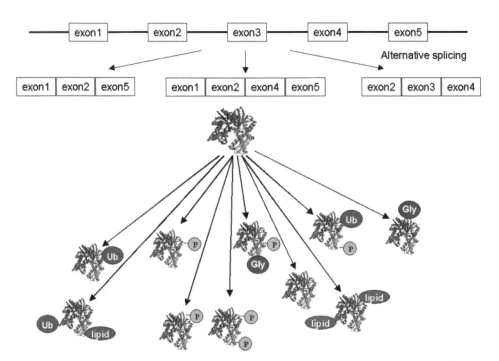

Figure 14.1. *Figure showing the ultimate protein complexity that can arise from the transcription and translation of a single gene. (See color insert.)*

during a gene's travels to becoming a mature protein. Indeed, a single gene can give rise to multiple forms of a specific protein through splicing of the transcript or differential modification of the translated protein. The complete primary structure of a protein can be used to calculate its isoelectric point (pI), search for possible structural domains, and indicate if any post-translational or post-transcriptional processing has occurred to create the mature form of the protein. Complete sequence mapping can also identify any amino acids that have been covalently modified. How do we increase our overall peptide coverage so that the knowledge obtained for every protein is more complete? While many different strategies can be employed that are directed at gathering individual bits of information across all of the proteins within a proteome, one fact cannot be denied: the amount of information acquired concerning a particular proteome sample is roughly proportional to the number of analyses conducted on that sample.

14.2   THE HUMAN GENOME PROJECT

When considering the approximate numbers that provide a conservative reflection of the complexity of the human proteome, it is obvious that an extremely challenging analytical problem lies ahead. In many ways the challenge is not unlike that probably felt by the pioneers of the human genome project. The sequencing of the human genome was first proposed at a meeting in May of 1985 in Santa Cruz, California. The meeting was headed by the distinguished molecular biologist Robert Sinsheimer and was attended by a number of biologists active in genetics and gene mapping, including Nobelist Walter Gilbert (Sinsheimer, 1989). They proposed that the entire human genome should not only be mapped but also sequenced, meaning determining the precise order of each A, G, C, and T. A single copy of DNA contains approximately 3 billion bases (A, G, C, or T). In 1985, the analysis of a single base cost over $10 and a good scientist could sequence only 50–100 bases per day. Therefore, the challenge to sequence the entire genome was not only imaginative and ambitious, but also prohibitively expensive. By 1991, gene sequencing technology had improved to the level that 10,000 bases could be sequenced per day at a cost of about $1 a base (Hunkapiller et al., 1991). Within a couple of years, sequencing was being done by several laboratories using automated robotic analyzers at a rate of 500,000 bases daily, at a cost of 10–15 cents/base. The primary factors that increased the sequencing rate, while lowering the cost, were the development of automated sequencers and bioinformatic tools that could "piece" together the available sequences into annotated genes.

While the number of bases (3 billion) that required sequencing to complete the human genome is far greater than the number of amino acids present in the number of presently known human proteins (approximately 12 million, assuming each of the 1,175,015 tryptic peptides contain on average 10 residues), there are many attributes of proteomics that make it far more challenging than genomics, as listed in Table 14.1. First, the genome is made up of four different bases while the proteome is comprised of twenty different types of amino acid residues. Second, while bases can be covalently modified, the extent and number of different types of modifications seen within proteins produces much more complexity. Third, the chemical behavior of genes occupies a narrow range since the fundamental archi-

TABLE 14.1. Comparison of Properties of Genes and Proteins

Property	Genome	Proteome
Subunits	Four nucleotide bases	At least 20 amino acid residues
Modifications	Few	~300
Biophysical property	Homogeneous	Heterogeneous
Form	Static	Dynamic
Function	One main function	Unknown number of functions

tecture (i.e., four different bases) of the genome is quite simple; however, the chemical behavior of the proteins span a wide, divergent range, reflective of the chemical diversity of their constituent building blocks. Fourth, a single gene product can result in multiple protein products through such events as gene splicing and frameshift mutations. Lastly, and probably the most underappreciated, the sequencing of the genome has a defined endpoint, while the dynamic nature and the uncertainty of the species present within the proteome means that it is impossible to ascertain with any certainty when it has been completely characterized for any particular species. While certainly not doing it justice, Figure 14.1 captures a brief glimpse into the complexity of the proteome compared to the genome or transcriptome.

14.3 THE SHIFT TO PROTEOMICS

With the success of the various genome sequencing projects and the development of techniques to measure differences in gene expression at the transcription level (Chittur, 2004), considerable focus has shifted toward proteomics—the characterization of gene expression at the protein level (Wilkins et al., 1996). Proteomic analysis represents a much more complex characterization than either genome sequencing or mRNA profiling. In genomics, the goal is to sequence the vast string of deoxynucleotide bases that comprise the genome of a particular species. In mRNA array analysis, the goal is to measure the relative abundance of the gene transcripts from two or more different cell types. While each of these areas represents tremendous technical achievements, they both are called upon to provide one basic characteristic, either the genome sequence or the abundance of the gene products. Proteomics, on the other hand, requires the analysis of a group of biomolecules with an almost undefined number of characteristics. Compared to DNA and RNA, which are comprised of four different monomeric subunits, proteins are comprised of at least twenty different amino acids. At present, no high-throughput instrument has been developed that sequences proteins in a comparable fashion to DNA. Proteins come in a wide variety of sizes and structures and are localized throughout the cell. Proteins can be acidic or basic, soluble or insoluble in an aqueous environment, monomeric or oligomeric, and so on. The chemical heterogeneity of proteins represents one of the major factors that make proteomics a greater analytical challenge than genomics or transcriptomics.

Almost any technology that is geared to the analysis of proteins now falls under the umbrella of proteomics. This being the case, there are a plethora of automation technologies that could be described in this chapter. For the sake of brevity, this chapter focuses on automation as it pertains to the characterization of differences in protein abundances on a global level.

14.3.1 High-Throughput Protein Identification

The whole proteomics revolution was in many respects spurred on by the protein identification capabilities of mass spectrometry (MS). While Edman degradation, which has been around for several decades, has a similar capability for protein identification, the uniqueness of MS approaches is that it allows proteins within crude mixtures to be successfully identified. It is this unique character that separates MS from other protein identification techniques that rely on either preknowledge of the protein's identity (i.e., Western blotting) or a highly purified protein preparation. In addition, Edman sequencing typically requires an hour for the sequencing of a single amino acid residue, while an entire peptide can be identified by MS/MS in a matter of 1–2 seconds. As described in some of the earlier chapters, identification is typically done at the peptide level, requiring the intact proteins to be digested either chemically (i.e., cyanogen bromide) or enzymatically (i.e., trypsin). While this digestion results in an approximately two orders of magnitude increase in the complexity of the mixture, it is still possible to obtain sequence information on many of the peptide species in this incredibly complex mixture.

Automation impacts almost every step of MS-based protein identification of proteomic samples. The key to characterization of a significant fraction of any given proteome is to "divide and conquer." This statement means that its complexity is such that any analytical tool will only provide a small glimpse of the total proteome if it is presented with it all at once. Therefore, methods have been devised to sub-fractionate proteome samples prior to analysis using such tools as MS (Issaq et al., 2002). Unfortunately, fractionation results in a multiplicity of samples that need to be analyzed; however, automation techniques have been developed that allow for the unattended analysis of said samples in most cases.

How is MS able to identify proteins and peptides in such complex mixtures? Whether the peptides are identified via peptide mapping or tandem MS, fractionation of the sample into analyzable pieces is critical. One of the great developments that have allowed components within complex mixtures to be characterized in a high-throughput manner was the direct coupling of liquid chromatography (LC) with MS (Covey et al., 1991). As with many empowering methods in science, the impact of LC-MS is reflective of the ever-increasing number of published references to this technique. The number of document articles in the U.S. National Library of Medicine that discuss LC-MS was 738 in 1991 and 2285 in 2001, with a total of 13,147 for the whole ten year period, an increase of 310%. While MS characterization of simple mixtures could be conducted by direct infusion, the preseparation of compounds directly prior to MS analysis revolutionized this area of science. Without this coupling, proteomics as it exists today would be radically different and would exclusively be accomplished through the use of 2D-PAGE fractionation (O'Farrell et al., 1977). LC-MS has become such an essential tool in MS-based proteomics that it is rare to find a proteomics laboratory that does not have an operational LC-MS system. Even MALDI, which is not strictly compatible with on-line LC fractionation prior to MS analysis, is embracing the attributes of LC for the separation of complex proteome mixtures prior to individually spotting on MALDI target plates (Zhang et al., 2004).

The combination of LC-MS is so seamless today that many LC systems can be controlled directly through the software used to control the MS itself. The LC method and the number of samples to be analyzed are often entered directly into the MS software so that the operation of both instruments is kept in proper synchronization. Samples to be analyzed are most often automatically injected by the LC system, allowing unattended analysis of hundreds of samples. This type of automation has afforded the opportunity to severely fractionate a proteome sample into many different components prior to LC-MS analysis, since the MS is now able to acquire data around the clock and the necessary personnel can be freed up to focus on other tasks.

Fractionation is also a key component in the identification of proteins via peptide mapping (Lescuyer et al., 2004). Peptide maps are typically recorded using a MALDI-based MS system and can be acquired in a very high-throughput manner. While there is no direct on-line separation available for MALDI analysis, there are commonly used fractionation methods that are applied prior to MALDI peptide map spectral acquisition. The most common of these is 2D-PAGE. It could be argued that 2D-PAGE is one of the cornerstones of proteomics, since it has the capability of resolving thousands of proteins as well as providing quantitative measurements of protein abundances from two distinct systems. While some groups contend that 2D-PAGE is laborious and time consuming, fortunately many of the steps necessary to fractionate and visualize proteins via this method have been automated (Hille et al., 2001). In industrial settings, robotics is used to load, run, and stain the gels in essentially a manual-free method. A simplistic representation of the entire 2D-PAGE procedure for differential proteomic analysis is shown in Figure 14.2. Once the gels are visualized, they are then automatically scanned using visualization software that also aligns gels and measures the intensity of respective protein spots so that multiple gels can be compared to one another. Proteins of interest are then automatically excised from the gel, using a spot-picking robotic system, and placed within the wells of microtiter plates. The microtiter plates are then transferred to a robotic liquid handling system that supplies the reagents necessary to digest the protein, extract the resulting peptides from the gel piece, and spot them onto a MALDI target plate with the proper amount of matrix mixed in. While the entire process is still relatively time consuming, a combination of automation and time management makes this scheme high-throughput much in the same manner that assembly line production makes automobile manufacturing high-throughput. As mentioned briefly before, LC has also been recently used in combination with MALDI-MS analysis. In this method, a proteome sample is digested into peptides and this mixture is separated by reverse-phase LC and the eluant is directly spotted, along with added matrix, onto a MALDI target plate. This spotting can be automated, allowing for the collection of approximately 100 individuals fractions per hour. Unfortunately, this LC-MALDI combination is still in its infancy and the impact it will have in the area of high-throughput proteomics is still not clear.

While fractionation methods and other necessary sample preparation methods are automated, peptide identification at the MS instrumental level is still key to make the entire process high-throughput. Fortunately, significant advancements in spectrometer development have occurred over the past few years to make this pos-

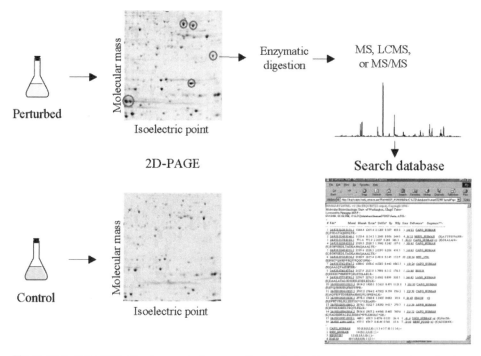

Figure 14.2. *Schematic showing general comparative analysis of two proteome samples using two-dimensional polyacrylamide gel electrophoresis (2D-PAGE). Samples are initially separated by 2D-PAGE and protein spots that are of greater intensity on one of the gels are excised, digested into peptides, and analyzed using mass spectrometry (MS). The MS results are searched against an appropriate database to identify the differentially abundant protein. (See color insert.)*

sible. For systems that couple LC separations directly on-line with MS analysis, the spectrometer can be programmed to operate in a data-dependent MS/MS mode in which peptides are automatically select for MS/MS, enabling sequence information to be obtained for peptides in complex mixtures (Blonder et al., 2004). In this mode the instrument selects the peaks that will be subjected to MS/MS independent of the operator. Prior to the experiment, the operator inputs the number of peaks to be selected prior to a subsequent MS scan, as well as the intensity characteristics of the peaks to be selected. For example, in Figure 14.3, the instrument is set up to select the three peptides that give the highest signal in the MS scan for subsequent MS/MS analysis. This selection parameter can be set to select for any number of high-intensity peptide signals (3–5 being the normal range) or to select for peptide signals that are not the most intense. In a typical LC-MS/MS experiment, anywhere from 2000 to 10,000 automated MS/MS experiments are typically conducted, in which anywhere from 10% to 20% may provide spectra of sufficient quality that enable the confident identification of a peptide.

14.3.2 Data Analysis

Another potential bottleneck in proteomics research is data analysis. Considering that a single MS/MS experiment will be made up of anywhere from 2000 to 10,000

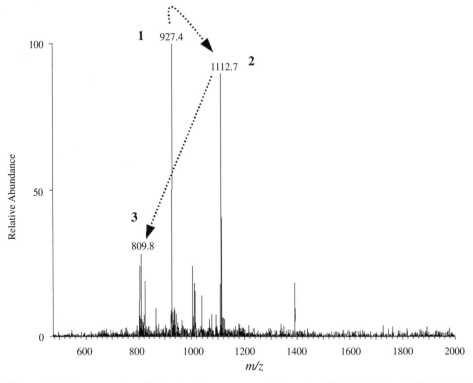

Figure 14.3. *Mass spectrum illustrating the sequence of three ions (labeled 1, 2, and 3) selected for collision induced dissociation in a routine data-dependent tandem mass spectrometry (MS/MS) operation of a conventional ion-trap mass spectrometer. In this mode of operation, the mass spectrometer is directed to select the three most abundant peptide molecular ions for MS/MS measurement.*

sequencing attempts, and complete studies may encompass over 2,000,000 such experiments, the need for automated data analysis is clear. As described in earlier chapters, proteins are identified primarily using MS either by peptide mapping or MS/MS (Loo, 2003). In peptide mapping, proteins are identified by matching a series of experimentally measured peptide masses with calculated masses generated from each protein within a database. In MS/MS, the correlation between the CID spectrum of a peptide and an in silico fragmentation of all of the peptides within a database is used for identification purposes. A typical LC-MS/MS raw data file takes approximately 4 hours to search against the appropriate database using a desktop personal computer. This time may vary depending on the processing speed of the computer and the database that the raw MS data is being searched against. More complex databases require more time. In our laboratory, we search such files using an 18-node computer cluster that cuts the data analysis time down from 4 hours to 20 minutes. As with sample analysis by MS, however, the raw data files that need to be searched against a database can be queued so that they are analyzed sequentially with little operator intervention. This unattended operation allows 24/7 data analysis.

Obviously, automated data analysis is necessary for protein identification. Results obtained from other proteomic types of experiments also require automated means of analysis. A good example of how investigators have developed pipelines that receive data, analyze it, and provide a specified output is provided by work performed in the laboratory of Coral del Val (del Val et al., 2004). This group designed an automatic pipeline to analyze annotated open reading frames (ORFs) stemming from full-length cDNAs (Figure 14.4). The annotated ORFs are then cloned into vectors and expressed for use in large-scale assays to determine such characteristics of the protein as its subcellular localization, secondary structure, and kinase reaction specificity. The pipeline's main function is to subject the identified ORFs to an exhaustive bioinformatic analysis such as similarity searches, protein domain architecture determination, and prediction of physicochemical characteristics and secondary structure. These analyses are conducted using a wide variety of bioinformatic methods that are linked to the most up-to-date public databases such as PRINTS, BLOCKS, INTERPRO, PROSITE, and SWISSPROT. A MS SQL-Server relational database is then used to store all of the data and experimental results obtained from this integrated bioinformatic analysis. By automating this type of comprehensive data analysis, investigators are able to gather a greater

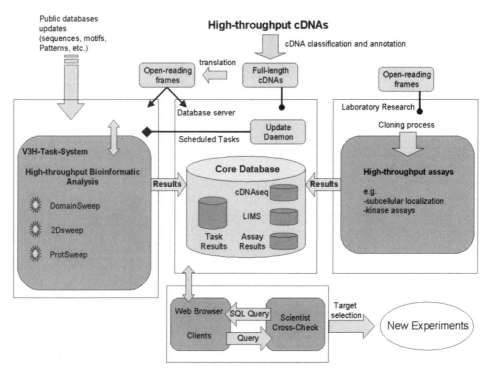

Figure 14.4. *Schematic of a data analysis pipeline for an integrated and automated strategy to obtain and administer data from high-throughput investigations of cDNAs. The data input is full-length cDNAs. Verified open reading frames (ORFs) are cloned and expressed for their use in experimental assays to measure such parameters as subcellular location and structure. All identified ORFs move through the pipeline, where they are exhaustively analyzed by running automated tasks such as DomainSweep, 2Dsweep, and ProtSweep. Computational and experimental results are integrated into a relational database (Core Database). The core database contains several single databases allowing researchers to cross-check various protein features in silico through the execution of SQL queries via web browsers or clients. (See color insert.)*

wealth of information, allowing them to find answers to biological questions easily. This ease of data gathering speeds up the selection of, and confidence in, targets for further analysis. While this example is specific for this particular study, it provides a template by which other computational architectures are designed to move large amounts of data through a pipeline in which questions are tailored to meet the studies' needs.

14.3.3 Automation in Other Areas of Proteomics

While a majority of this chapter has been devoted to describing automation as it pertains to the identification of proteins in a global proteomic analysis, many of the same principles apply to other areas of proteomics. For instance, quantitative proteomics follows many of the same automation steps as far as fractionation, MS analysis, and database searching as proteome-wide identification. Robotic systems have been developed for determining the optimum conditions for crystallization of proteins in structural proteomics. Automated dispensers are available to spot minuscule amounts of protein lysates or other solutions onto membrane or solid surfaces for protein array analyses. Robotic workstations have been programmed to prepare protein chips for acquiring proteomic patterns for the diagnosis of cancer. Automation has touched every conceivable area of proteomics and the versatility of robotic workstations allows them to be programmed for almost every task that a researcher with a pipette is capable of doing and more. While described in another chapter of this book, automation in the development and application of protein arrays is critical to the ability to interrogate hundreds or thousands of different proteins species in an affinity-based platform.

14.4 CONCLUSION

Reflecting on what has enabled such projects as the human genome project to be completed and what enables us to consider proteomics as a viable venture, it is in many ways due to the advent of automation in the scientific workplace. In many ways, proteomics projects in which the goal is to provide extensive coverage of the proteins present within a cellular system is analogous to digging 100 fence post holes. The task is difficult and laborious, but the tools are available. The task can then be looked at in two ways: pick up the shovel and just start digging or invent a better way to dig. Proteomics has encompassed both approaches. In cases where a global identification of as many proteins as possible is the aim, it is necessary to simply fractionate the sample into many different aliquots and analyze each of these. Fortunately, as described earlier, these laborious approaches have become highly automated. While the tasks are still time consuming, the ability to automate allows effective organization and time management permits a steady flow of results to be continually produced. Meanwhile, proteomic laboratories continue to develop and refine approaches to better interrogate the proteome and focus more specifically on particular segments, such as the phosphoproteome.

While data analysis can be automated, ideas on what to do with the plethora of data cannot. Probably the biggest challenge facing proteomics today is how to make biological sense out of the results of global cell analyses. Probably the biggest anomaly between genomics and proteomics is the dynamic nature of the proteome. While providing job security for the proteome scientist, this lack of an endpoint also makes it difficult to ascertain when the final conclusion of a global analysis has been achieved. It may be that, in the future, proteomic analyses follow an expansion and contraction pathway, in which a great expanse of data is acquired and it is analyzed with a contracting mentality in which only a fraction of the data is utilized to make some biological inference about the system under study.

ACKNOWLEDGMENTS

This project has been funded in whole or in part with federal funds from the National Cancer Institute, National Institutes of Health, under Contract No. NO1-CO-12400.

By acceptance of this article, the publisher or recipient acknowledges the right of the U.S. Government to retain a nonexclusive, royalty-free license and to any copyright covering the article. The content of this publication does not necessarily reflect the views or policies of the Department of Health and Human Services, nor does mention of trade names, commercial products, or organizations imply endorsement by the U.S. Government.

REFERENCES

Blonder J, Rodriguez-Galan MC, Lucas DA, Young HA, Issaq HJ, Veenstra TD, Conrads TP. 2004. Proteomic investigation of natural killer cell microsomes using gas-phase fractionation by mass spectrometry. *Biochim Biophys Acta* 1698:87–95.

Chittur SV. 2004. DNA microarrays: tools for the 21st century. *Comb Chem High Throughput Screen* 7:531–537.

Covey TR, Huang EC, Henion JD. 1991. Structural characterization of protein tryptic peptides via liquid chromatography/mass spectrometry and collision-induced dissociation of their doubly charged molecular ions. *Anal Chem* 63:1193–2000.

del Val C, Mehrle A, Falkenhahn M, Seiler M, Glatting KH, Poustka A, Suhai S, Wiemann S. 2004. High-throughput protein analysis integrating bioinformatics and experimental assays. *Nucleic Acids Res* 32:742–748.

Hille JM, Freed AL, Watzig H. 2001. Possibilities to improve automation, speed and precision of proteome analysis: a comparison of two-dimensional electrophoresis and alternatives. *Electrophoresis* 22:4035–4052.

Hunkapiller T, Kaiser RJ, Koop BF, Hood L. 1991. Large-scale and automated DNA sequence determination. *Science* 254:59–67.

Issaq HJ, Conrads TP, Janini GM, Veenstra TD. 2002. Methods for fractionation, separation and profiling of proteins and peptides. *Electrophoresis* 23:3048–3061.

Jung JW, Lee W. 2004. Structure-based functional discovery of proteins: structural proteomics. *J Biochem Mol Biol* 37:28–34.

Lescuyer P, Hochstrasser DF, Sanchez JC. 2004. Comprehensive proteome analysis by chromatographic protein prefractionation. *Electrophoresis* 25:1125–1135.

Lin D, Tabb DL, Yates JR III. 2003. Large-scale protein identification using mass spectrometry. *Biochim Biophys Acta* 1646:1–10.

Loo JA. 2003. The tools of proteomics. *Adv Protein Chem* 65:25–56.

O'Farrell PZ, Goodman HM, O'Farrell PH. 1977. High resolution two-dimensional electrophoresis of basic as well as acidic proteins. *Cell* 12:1133–1142.

Sinsheimer RL. 1989. The Santa Cruz Workshop, May 1985. *Genomics* 5:954–956.

Stern DF. 2001. Phosphoproteomics. *Exp Mol Pathol* 70:327–331.

Wilkins MR, Sanchez JC, Williams KL, Hochstrasser DF. 1996. Current challenges and future applications for protein maps and post-translational vector maps in proteome projects. *Electrophoresis* 17:830–838.

Yu L-R, Conrads TP, Uo T, Kinoshita Y, Morrison RS, Lucas DA, Chan KC, Blonder J, Issaq HJ, Veenstra TD. 2004. Global analysis of the cortical neuron proteome. *Mol Cell Proteomics* 3:896–907.

Zhang N, Li N, Li L. 2004. Liquid chromatography MALDI MS/MS for membrane proteome analysis. *J Proteome Res* 3:719–727.

15

Bioinformatics Tools for Proteomics

Daniel C. Liebler

Proteomics Laboratory, Mass Spectrometry Research Center, Vanderbilt University School of Medicine, Nashville, Tennessee

15.1 INTRODUCTION

This chapter describes bioinformatics tools for identification, quantitation, and mapping modifications on proteins with MS data. They include (i) algorithms and software to identify protein sequences from MS data, (ii) software that enables quantitative comparisons based on MS data, (iii) algorithms and software that identify modified and variant protein and peptide forms from MS data, and (iv) software and database tools to evaluate and organize the results of protein identifications. Together, these bioinformatics tools enable the analysis of complex samples and systems and the extraction and organization of the data in a useful biological context.

The bioinformatics tools of proteomics may appear as complex as the systems they help to analyze. However, most of these tools perform operations that could be done manually. The reason these tools are needed is that the volume of data in proteomics analyses is so large that manual analysis is practically impossible. For example, a typical liquid chromatography–tandem MS (LC-MS/MS) analysis of a peptide digest may generate 2000–3000 MS/MS spectra within 60 minutes. Manual interpretation of a peptide sequence from a *single* spectrum may take a trained analyst an hour or more. However, cross-correlation analysis of the entire data set against an indexed protein database with the Sequest program can be done in minutes on a multiprocessor computer system. The resulting list of peptide

Proteomics for Biological Discovery, edited by Timothy D. Veenstra and John R. Yates.
Copyright © 2006 John Wiley & Sons, Inc.

sequences assigned to the MS/MS spectra usually requires further evaluation and organization to enable interpretation in the context of the original experiment. This could be done by cutting and pasting results from several programs onto spreadsheets and sorting or filtering the data, but this also is time consuming to the point of being impractical. Thus, other software and database systems have emerged to automate these higher-level organization tasks.

A great challenge in bioinformatics is to streamline the processes of data analysis, interpretation, and organization while preserving the opportunity for critical evaluation of the data. There is an inevitable trade-off between quality and throughput in the interpretation of proteomics data. This reflects the variability in data quality and the fact that all data analysis tools are subject to error. Completely automated evaluation of MS data will always incorrectly identify some proteins while failing to detect others.

A major problem in the proteomics field is that there are no universally accepted standards for the identification of proteins or peptide sequences from MS data. This reflects intense marketplace competition between bioanalytical instrumentation and software providers and the resulting diversity of tools employed by researchers. Although some conventions are beginning to emerge among users of the most popular analytical platforms, this situation places a burden of uncertainty on those who will evaluate studies based on proteomics analyses.

15.2 BIOINFORMATICS TOOLS AND PROTEOMICS WORKFLOW

Perhaps the easiest way to understand how bioinformatics tools are integrated into proteome analyses is to consider their use in the workflow of "typical" proteomics analyses (Figure 15.1). Proteomics analyses begin with one or more protein samples of varying complexity. These can be subject to either of two general analytical approaches. In the first, intact proteins are first separated by 2D SDS-PAGE and the gels are then imaged and analyzed. Analyses may entail evaluation of features on a single gel and comparisons of sets of features on multiple gels (Figure 15.1). (Tools for 2D gel image analyses have been reviewed elsewhere (Hunter et al., 2002; Raman et al., 2002; Rogers et al., 2003) and will not be discussed further here.) The major alternative approach is referred to as "shotgun" analysis, in which a complex mixture is directly digested without prior fractionation into individual protein components. (There are variations of this approach that include prior fractionation of protein samples, but shotgun analyses all begin with mixtures of proteins.) The resulting highly complex peptide mixtures are subjected to multidimensional chromatography (e.g., ion exchange, reverse phase) and MS/MS analysis. Both shotgun-based and 2D gel-based analyses ultimately yield MS data, which can be analyzed to identify the peptides and proteins represented in the sample (Figure 15.1).

MS data produced in proteomics laboratories usually include measurements of peptide masses and MS/MS spectra, which encode peptide sequences, as well as information on modifications. Several data analysis tools can extract information from these primary MS data. Most commonly, peptide masses or MS/MS spectra are compared to database sequences to identify the corresponding peptides and proteins (Figure 15.1). Where database information is insufficient or lacking, other

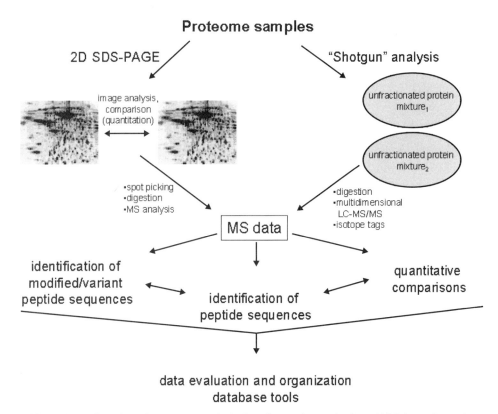

Figure 15.1. *Overview of proteome analysis data flow and organization of bioinformatics tasks.*

programs permit de novo sequence interpretation from MS/MS spectra. Other tools identify MS/MS spectra with characteristics of modified or sequence variant peptides (Figure 15.1). In analyses where proteins or peptide levels are to be compared between samples, isotopic labels can be employed to differentially label the samples. MS data from these analyses can be analyzed with other software to yield quantitative comparisons (Figure 15.1). Finally, other software can evaluate the output of the software tools employed in Figure 15.1 and provide integration of this "derived" data in a useful biological context.

Another useful way to understand that bioinformatics tools underlying protein identification from MS data is to consider a hierarchy of data analysis and integration (Figure 15.2). The MS and MS/MS analysis of peptides yields spectra, which are the *primary* data on which identifications are based. Software to identify peptides from mass measurements and MS/MS spectra yields a second tier data set, in which scores or statistical parameters define the association of spectra and sequences. Other secondary data sets include outputs from programs that identify spectra corresponding to modified or variant peptides. These "identifications" may or may not be correct and require further evaluation or filtering to remove poor quality matches. In analyses of complex samples, the proteins may be "spread out" by fractionation (as in multidimensional chromatography), such that different peptides from a single protein may appear in multiple analysis files.

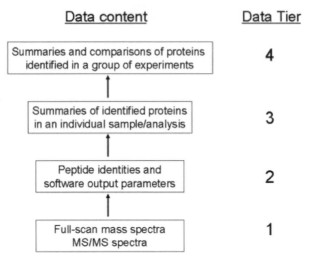

Figure 15.2. Hierarchy of data analysis in proteomics studies.

The task of evaluating the outputs of the secondary data sets and collecting together the results of the multiple analyses that comprise one sample represents the third tier of data analysis tools, which operate on the outputs of these programs. Finally, a fourth tier of analysis integrates proteome analyses replicated over multiple samples and data analysis conditions. In new bioinformatics data systems for proteomics analysis, these third- and fourth-tier data analysis tools often are linked to databases and provide the end-user interface for users of proteomics laboratories.

A fundamental challenge facing the field of proteomics is the lack of a clearly agreed upon set of standards for confirming protein identifications, defining the conditions of proteomics analyses, and integrating the work of different laboratories into common databases. This problem is similar to that faced by investigators using microarray technology for gene expression analyses. The microarray community eventually developed the MIAME (Minimum Information About a Microarray Experiment) standards for reporting and sharing data from microarray experiments (Brazma et al., 2001). Data sets that meet these standards now are required for publication and an increasing number of major journals. An important advance in the proteomics field is the recent publication of a proposed model, called Proteomics Experiment Data Repository (PEDRo) (Taylor et al., 2003). The PEDRo model contains information on the source of a sample, experimental details, sample processing, analytical instrumentation, the primary or raw data, and the secondary or derived data. This model implements a unified modeling language (UML) approach to describing proteomics data and uses extensible markup language (XML) and structured query language (SQL) to capture, store, and disseminate information. The work by Taylor et al. (2003) is particularly significant as it represents a joint effort by a number of leaders in proteomics and may form the basis of a unified model for dealing with proteomics data.

15.3 TOOLS TO IDENTIFY PEPTIDE SEQUENCES WITH MS DATA

15.3.1 Protein Identification by Peptide Mass Fingerprinting

In this approach, proteins are obtained from simple samples containing no more than 2–3 proteins. The proteins are digested, typically with trypsin, and then the digest is analyzed by MALDI–TOF MS. The resulting spectra contain peaks that represent the singly charged masses ($[M + H]^+$ ions) of the peptides. Peptide mass fingerprinting software then compares these masses to mass lists of tryptic peptides generated in silico from a database of protein or translated nucleic acid sequences (Pappin et al., 1993; James et al., 1994). Accurate mass measurements (within 10 ppm) of the peptide $[M + H]^+$ ions permit successful matching to database entries (Clauser et al., 1999). Thus, spectra are obtained in reflectron mode rather than linear mode. Typically, a good MALDI–TOF spectrum of a protein digest contains multiple peaks that can be matched to database peptides, thus increasing the confidence of the association.

Peptide mass fingerprinting by MALDI–TOF MS is most commonly done in the identification of protein spots on 2D gels. Samples containing more than 2–3 proteins provide complex spectra that are less successfully handled by available algorithms and software. Spectral peaks corresponding to peptides bearing known modifications (e.g., serine/threonine/tyrosine phosphorylation or lysine acetylation) can be successfully matched to database peptide sequences if the user specifies the possibility of modification. However, unanticipated modifications will lead to incorrect assignments. This is not normally a major barrier to successful protein identification if other unmodified peptides give peaks that can be matched to database peptides.

Several programs provide peptide mass fingerprinting data analysis (Table 15.1). All of these programs require input of lists of peptide ion peak lists acquired by MS analysis. They also may use information on the apparent pI, MW, source organism, and related information on the protein sample. The earliest approaches ranked protein "hits" based on the number of matches between spectrum peaks and database proteins. This system is biased toward identifications of higher MW proteins, which have a greater number of potential matches. The MOWSE algorithm takes into account the number of tryptic peptides from database protein entries in the MW range of the protein being analyzed (Perkins et al., 1999). The newest peptide

TABLE 15.1. Software for Protein Identification by Peptide Mass Fingerprinting

Program	Scoring	Source
Mascot	Probability	http://www.matrixscience.com
ProFound	Probability	http://prowl.rockefeller.edu/cgi-bin/ProFound; http://www.genomicsolutions.com
Protein Lynx	Probability	http://www.waters.com
MS-Fit	MOWSE	http://prospector.ucsf.edu
IonIQ	MOWSE	http://www.proteomesystems.com
PeptIdent	# Matches	http://us.expasy.org/tools/peptident.html
Peptide Search	# Matches	http://www.mann.embl-heidelberg.de/GroupPages/PageLink/peptidesearchpage.html

mass fingerprinting software tools (e.g., Mascot, ProFound) employ probability-based scoring algorithms to better distinguish correct assignments from random matches (Fenyo, 2000).

Most of the tools in Table 15.1 are accessible via the World Wide Web at no cost. Users can simply enter data into a browser window and perform searches. Higher-throughput analyses are done by batched searches with either local servers or subscription access web servers. In many cases, the choice of peptide mass fingerprinting software is dictated by packages bundled with MS instruments by vendors. However, all of the available programs can be used with data from essentially any instrument, as long as search data are appropriately formatted. Aside from the differences in algorithm types described earlier, there is little objective basis for selecting one software tool over others except for ease of integration with MS datasystems or downstream informatics resources.

15.3.2   Protein Identification by Analysis of Uninterpreted MS/MS Data

 MS/MS spectra result from the fragmentation of peptide ions in a MS instrument. Fragmentation along the peptide backbone yields product ions that can be predictably related to sequences (Roepstorff and Fohlman, 1984). In most commonly used tandem mass analyzers, low-energy collision induced dissociation predominantly yields b- and y-ions, which correspond to peptide fragments that retain charge at their N and C termini, respectively (Figure 15.3). Other fragment ions, including immonium ions, a-, c-, x- and z-ions, as well as associated fragments due to loss of water and ammonia are less commonly seen, although these are more prominently formed by high-energy collision induced dissociation (Baldwin et al., 2001). In addition, many amino acid modifications produce characteristic features in MS/MS

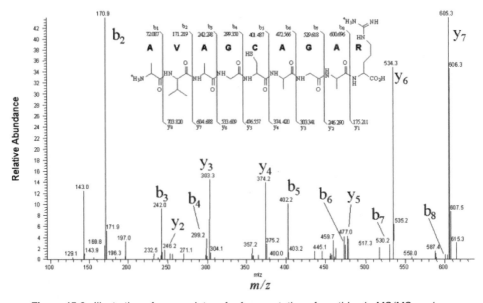

Figure 15.3. Illustration of nomenclature for fragmentation of peptides in MS/MS analyses.

spectra (Liebler, 2002). All of these signals provide information about the sequence of a peptide and the sequence specificity of modifications.

De novo interpretation of spectra would seem to be the most straightforward route to sequence information. However, problems with de novo sequencing algorithms and data quality led to the adoption of an alternate approach, in which uninterpreted MS/MS spectra were compared to database protein sequences to identify possible matches. An early approach was the peptide sequence tag method of Mann and Wilm (1994), in which partial de novo interpretation of the spectrum yields a short sequence "tag," which is searched against a database together with the mass of the precursor peptide. The method accurately identifies corresponding database sequences, but extraction of the peptide sequence tags required manual de novo interpretation, which greatly limited throughput. Recently, the Yates group at Scripps developed a new sequence tag analysis tool called GutenTag that automates the extraction of sequence tags from MS/MS spectra and searches multiple tag sets against databases. The software is available upon request (http://fields. scripps.edu/GutenTag/index.html).

Several software tools identify database peptide sequences corresponding to uninterpreted MS/MS spectra (Table 15.2). The first program to successfully analyze MS/MS data in a high-throughput manner was Sequest, which was developed by Yates and Eng in 1994 (Eng et al., 1994). Sequest first matches the precursor ion mass in a MS/MS spectrum to possible database peptide sequences of the same nominal mass. These candidate sequences are then used to generate theoretical MS/MS spectra, which are then compared to the actual spectrum with a cross-correlation algorithm. The best matches are scored by the cross-correlation algorithm and ranked. Good quality spectra of peptides whose sequences exist in databases yield correct assignments. Spectra of modified peptides can be identified correctly if possible modifications (e.g., phosphorylation of serine/threonine/tyrosine) are specified by the user at the outset of data analysis (Yates et al., 1995; MacCoss et al., 2002). However, poor quality spectra, unanticipated modifications, and database errors lead to erroneous assignments. Thus, Sequest search outputs often require additional review and inspection (see below). Sequest analyses require that the raw MS/MS data be in a specific file format (.dta files), although programs and scripts for conversion of various MS instruments' raw datafile formats to .dta files are available.

TABLE 15.2. Software for Protein Identification by Comparison of MS/MS Data with Database Sequences

Program	Search Method and Scoring	Source
SEQUEST	Cross-correlation	http://www.thermo.com
Mascot	MOWSE; probability	http://www.matrixscience.com
Sonar MS/MS	Vector algebra; probability	http://www.genomicsolutions.com
MS-Tag	Fragment ion tags	http://prospector.ucsf.edu
Pep-Frag	Fragment ion tags	http://prowl.rockefeller.edu/PROWL/pepfragch.html
Pep-Sea	Peptide sequence tag	http://www.narrador.embl-heidelberg.de/GroupPages/Homepage.html

Another widely used tool for analysis of MS/MS data is Mascot, which provides probability-based searches of peptide database sequences by an extension of the MOWSE algorithm for peptide mass fingerprinting (Perkins et al., 1999). As with Sequest, specification of variable modifications can enable the detection of modified peptides. In contrast to Sequest, Mascot provides a probability-based assignment of spectra to database sequences, which provides the user guidance in distinguishing correct from incorrect assignments. Mascot operates on several different raw data file formats generated by major MS instrument data systems, including .dta files.

The recently introduced Sonar MS/MS program uses multidimensional vector algebra to represent MS/MS data and to assess the correspondence of spectra to theoretical vector constructs for protein sequences in databases (Field et al., 2002). Statistical evaluation then provides a probability-based scoring of sequence-to-spectrum comparisons. The MS-Tag (Clauser et al., 1999) and Pep-Frag (Fenyo, 2000) programs use various fragment ions together with the precursor ion mass to identify possible matches in a database but do not calculate scores or probabilities of false-positive matches.

Although Sequest and Mascot are the most widely used tools for database searches of MS/MS data, this reflects their history and the fact that they are bundled with bioanalytical software sold with the most widely used MS instruments in proteomics laboratories. Widespread use of Mascot, MS-Tag, and Pep-Frag also reflects their free availability via web interfaces. Anecdotal information indicates that there are some differences in how all of the above tools perform with equivalent data sets. However, there is not sufficient information to enable any of them to be characterized as "best" for MS/MS data evaluation.

An increasingly important factor in the selection of software for sequence-to-spectrum comparisons is computational speed and throughput. Use of multidimensional LC-MS/MS approaches (Link et al., 1999; Washburn et al., 2001; Wolters et al., 2001) and automated instrumentation makes it possible for some proteomics laboratories to generate upward of a million MS/MS spectra per week. This volume of data production is not compatible with simple web-based versions of search tools (e.g., Mascot, MS-Fit, Pep-Frag), which are designed for manual data upload and offer limited search speeds. Similarly, Sequest searches on single processor workstations may take substantially longer to perform than did the actual LC-MS/MS analyses. The solution to these problems is the use of parallel computing versions of these programs on multiprocessor (computer cluster) systems (Sadygov et al., 2002). Multiprocessor versions of both Sequest and Mascot are available commercially and are bundled with computer cluster hardware for use in proteomics laboratories.

15.3.3   Software for De Novo Sequence Interpretation from MS/MS Data

De novo sequence interpretation differs from the tools described previously in that a sequence is determined directly from a spectrum without any type of database matching. The development and application of de novo sequence analysis tools has lagged behind other approaches for three reasons. First, de novo identification of sequence requires very high quality MS/MS data, including accurate mass measurements of both precursor ions and product ions. Thus, most de novo sequence software can only operate effectively with data from a limited number of high-end MS instruments. Second, the algorithms developed for de novo sequencing are more computationally intensive than database search tools. Finally, few MS/MS spectra display complete *b*- or *y*-ion series needed for unambiguous sequence assignment. In contrast, good quality matches of incomplete spectra to database sequences are more easily obtained with database search software. For these reasons, de novo sequence analysis has been done mainly by investigators who are working with proteins from organisms with unsequenced genomes or poorly documented collections of proteins (Liska and Shevchenko, 2003).

Producers of quadrupole time-of-flight MS instruments provide de novo sequencing programs as part of the bioinformatics software packages for these instruments. The proprietary MassSeq program, part of the ProteinLynx software package provided with Micromass Q-Tof MS instruments (www.waters.com), uses a Bayesian strategy to score spectra against random peptide sequences, whereas the BioAnalyst software provided with ABI QSTAR (www.appliedbiosystems.com) contains a different de novo sequencing tool. Both incorporate BLAST searching to perform database searches of possible sequences against databases. In both systems, the mass accuracy and resolution of the quadrupole time-of-flight mass analyzer are required to provide accurate measurements on which the algorithms depend. Data from ion trap LC-MS/MS instruments have generally been regarded as unsuitable for automated de novo sequence interpretation because of limitations in mass accuracy and resolution. Nevertheless, ThermoFinnigan recently released DeNovoX, a probability-based de novo sequence interpretation utility as part of the Bioworks software package (www.thermo.com). This program can generate sequences from lower resolution MS/MS spectra produced by ion trap MS instruments.

15.4 TOOLS TO EVALUATE THE RESULTS OF SEQUENCE-TO-SPECTRUM COMPARISONS

The software tools described earlier generate lists of spectra, associated peptide sequences, and output scores. All of these programs also provide individual spectrum annotations or tabular data summaries to indicate which signals matched those predicted for the matched peptide. The outputs of Mascot and Sonar MS/MS are probability-based scores, which allow users to assess the probability that the assigned sequence-to-spectrum associations are stronger than would occur by chance alone. However, even these probability-based assignments can be in error and some amount of manual curation of the data is necessary.

Outputs from Sequest, MS-Fit, and Pep-Frag do not provide probability-based assessment for sequence-to-spectrum matches but instead consist of algorithm scores and/or lists of matching sequences. The user is then left to evaluate the quality of the matches and determine which results are correct. At best, this is a time-consuming task and with large data sets, the challenge can be overwhelming. This has led to development of secondary data evaluation tools, particularly for the widely used Sequest program, which enable automated or semiautomated evaluation of the outputs from Sequest searches.

15.4.1   Tools for Analysis of Sequest Search Outputs

Because of the widespread use of Sequest in many proteomics laboratories, considerable effort has gone into tools to facilitate the evaluation of Sequest outputs. Sequest searches generate a summary page that lists the spectra, associated sequences, and scoring parameters (Figure 15.4). The cross-correlation score (Xcorr) is most useful in assessing the quality of a sequence-to-spectrum comparison. Other parameters include the delta Cn (dCn), which indicates the difference in normalized correlation score between the top and second ranked matches and the preliminary score (Sp) from initial match of the spectrum to database sequences. Several laboratories have reported the use of filter approaches, in which these parameters are used in combinations to filter out poor quality matches (Link et al., 1999; Washburn et al., 2001; Han et al., 2001). The choice of thresholds and parameters (e.g., minimum Xcorr and dCn scores) was largely empirical and reflected an arbitrary trade-off between sensitivity (likelihood of missing true matches) and error (likelihood of accepting false-positive matches). Peng et al. (2003) systematically evaluated the use of different Xcorr thresholds in filtering a carefully validated data set from a multidimensional LC-MS/MS analysis of a tryptic digest of the yeast proteome. For doubly charged ions of fully tryptic peptides, Xcorr minimum thresholds of 1.5, 2.0, and 3.0 gave estimated false-positive identification rates of 0.80%, 0.46%, and 0.00%, respectively. Further estimates of the depen-

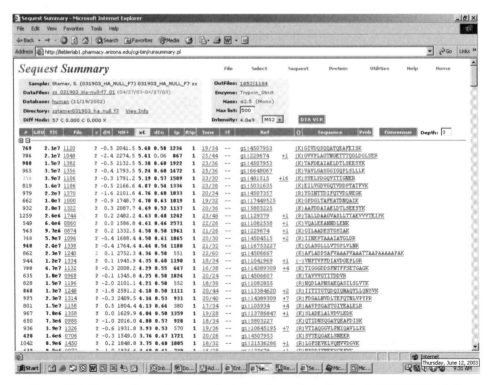

Figure 15.4. Sample Sequest search output page for analysis of a single LC-MS/MS datafile.

dence of false-positive rate on dCn indicated that values of approximately 0.1 or greater led to a false-positive identification rate of 0.5% or less regardless of Xcorr. Thus, the use of Xcorr in combination with dCn provided a relatively simple approach to filter Sequest outputs.

An alternative to use of parameter thresholds is to develop statistical models to evaluate the distributions of scores for correct and incorrect assignments with Sequest. The statistical model then can be used to estimate the probabilities of correct versus incorrect assignments. Keller et al. (2002) used a tryptic digest of a mixture of 18 proteins to produce a data set of MS/MS spectra from LC-MS/MS analyses. Sequest searches against a database containing those protein sequences and other sequences generated a set of both correct and incorrect matches. Discriminant function analysis was applied to Sequest output scores to model the distributions of scores for correct and incorrect assignments. This enabled calculation of probabilities of correct versus incorrect assignments based on Sequest output parameters. A mixture model was then developed using the expectation maximum algorithm to enable estimation of score distributions from additional data sets. Comparison of the performance of the model with simple threshold settings for combinations of Sequest output parameters (see above) indicated that the model offered substantially greater sensitivity and lower error rates. The software (Peptide Prophet™) is available as an open-source application (www.systemsbiology.org). Another approach to analysis was employed by Anderson et al. (2003), who used support vector machine classification to classify Sequest search matches as correct or incorrect. They used a training set of validated correct and incorrect assignments to train the algorithm based on Sequest output parameters. Support vector machine classification proved superior to simple thresholds for distinguishing correct from incorrect matches.

15.5 TOOLS TO ORGANIZE, INTEGRATE, AND COMPARE THE RESULTS OF MULTIPLE PROTEOMICS ANALYSES

The increasing use of fractionation and multidimensional separations in proteome analyses generates collections of data sets that require reassembly to have any biological meaning. For example, in a multidimensional (ion exchange/reverse phase) LC-MS/MS analysis, peptides from a complex mixture may be resolved into perhaps 10–20 or more ion exchange fractions, each of which yields a separate LC-MS/MS datafile (Peng et al., 2003). Each protein may be represented by peptides in several fractions. Thus, reassembly of all the protein information requires collection of the associated sequence–spectra pairs, their probability scores, and/or other output parameters. This operation provides lists of putative protein identifications and links to their supporting information. (The problem of what constitutes a protein identification is discussed later.) This collected information can be stored in a relational database for further use. These functions represent the third tier of data analysis depicted in Figure 15.2.

In addition to collecting the data for individual samples from multiple analyses, collections of proteins identified in multiple samples also must be compared. For example, identification of the members of a multiprotein complex often entails isolation of the complex by affinity capture methods, followed by multidimensional

LC-MS/MS analysis of peptides from a digest of the captured protein mixture. Because nonspecifically associated proteins are incidentally captured in such analyses, a control sample (e.g., immunoprecipitation with a nonspecific antibody) is needed for comparison and this sample generates a second data set. Identification of the proteins that specifically associate with the complex requires subtraction of the proteins from the control data set. In other types of experiments, comparisons of multiple analyses of the same or similar samples are required to determine which proteins reproducibly are detected under a given experimental condition. In either case, the comparison of multiple data sets involves the application of fourth-tier data analysis tools (Figure 15.2). Such tools have recently emerged as (i) integrated bioinformatics software packages for proteomics and (ii) stand-alone programs for the third- and fourth-tier analysis of Sequest search results.

15.5.1   Integrated Bioinformatics Software Packages for Proteomics

Major manufacturers of MS instrumentation used for proteomics and bioinformatics software companies have recognized the need to provide users with software tools to manage the prodigious output of data generated by new instruments. A summary of proteomics bioinformatics software packages is presented in Table 15.3. A noteworthy feature of these systems is that most utilize the widely available peptide mass fingerprinting programs and MS/MS search engines described earlier. Two exceptions are ProteinLynx and Spectrum Mill, which use propriety search engines for both peptide mass fingerprinting and MS/MS data searches. Although the listed search engines are the primary tools available to users of each package, most search engines can be used with data from most major MS instruments. For example, Mascot accepts data from several data system formats. Although Sequest requires input of data as .dta files, conversion of data to this format is possible through simple scripts or conversion programs available from some instrument manufacturers or produced by individual proteomics laboratories. The ability of search engines and related tools to use data from different instruments is important, given that most proteomics laboratories combine MS instrumentation from different manufacturers.

Another useful development is the bundling of applications that span tiers 1–4 of data analysis, as depicted in Figure 15.2. The RADARS and GPS Explorer software incorporate tools for filtering raw MS data (see below) together with MS data search engines, tools for organizing and reviewing search results, and a back-end relational database system. The integration of all of these elements is a helpful trend in proteomics that ultimately will help biologists make more effective use of proteomics laboratories and instrumentation.

15.5.2   Stand-alone Tools for Analysis of Sequest Search Results

As noted earlier, Sequest outputs from large sets of MS/MS data require additional review to distinguish good sequence-to-spectrum assignments from poor ones. With small data sets (e.g., individual LC-MS/MS runs), one can manually review the assignments, but larger data sets (e.g., from multiple multidimensional LC-MS/MS analyses) make this impractical. Moreover, Sequest outputs are presented on individual webpages, which makes it difficult to collect and organize the results

TABLE 15.3. Integrated Bioinformatics Systems for Proteomics

Name	Vendor	Peptide Mass Fingerprinting[a]	MS/MS Search Engine	De novo Sequencing	Isotope Tag Quantitation	Modification Detection[b]	Data Integration/ Database Utility
Bioworks	ThermoFinnigan www.thermo.com	None[c]	Sequest	DeNovoX	XPRESS	SALSA	None
GPS Explorer	Applied Biosystems www.appliedbiosystems.com	Mascot	Mascot	Proprietary	ProICAT (proprietary)	None	Yes
ProteinLynx	Waters/Micromass www.waters.com	Proprietary	Proprietary	MassSeq (proprietary)	None	AutoMod (proprietary)	None
RADARS	Genomic Solutions/Proteometrics www.genomicsolutions.com	ProFound	Sonar MS/MS	None	None	None	Yes
Spectrum Mill	Agilent www.agilent.com	Proprietary	Proprietary	Sherenga	Proprietary	None	Yes

[a] Peptide mass fingerprinting analysis from MALDI-TOF data.
[b] Homology-based or error tolerant application of MS/MS database searching can detect modified forms in some cases.
[c] A peptide mapping utility "Pepmap" uses LC-MS full scan spectra to identify peptides from protein sequences entered into the program.

derived from multiply fractionated samples. Several laboratories have developed programs to facilitate the filtering, review, and organization of Sequest outputs.

The first such programs, DTA Select and Contrast, were developed by Tabb and Yates (Tabb et al., 2002). DTA Select operates on Sequest output directories resulting from multiple analyses (e.g., representing multiple LC-MS/MS runs in a multidimensional analysis). The program first collects the identified peptide sequences determined by Sequest and produces a list of proteins (or gene loci) represented by the data set. Then user-selectable filters are applied to retain or reject assignments on the basis of Sequest output parameters, such as Xcorr and dCn values, to purge duplicate spectra, and to include or reject protein identifications based on numbers of peptides identified or other user-selectable criteria. Although DTA Select was developed for use with Sequest outputs, support for analysis of Mascot search outputs was recently implemented. DTA Select also includes other tools useful for review of Sequest outputs, including a graphical user interface with an improved spectrum viewer, and tools for viewing protein coverage and sequence alignments. Contrast provides comparisons of data sets to identify proteins differentially present in replicate analyses (e.g., controls versus specific experimental conditions). The program also enables comparisons of data sets following DTA Select under different sets of evaluation criteria. Although a user-friendly graphical user interface is not currently available, Contrast is quite powerful and can compare up to 63 data sets simultaneously. DTA Select and Contrast are available upon request (http://fields.scripps.edu/DTASelect/).

Another tool for third- and fourth-tier analysis of Sequest search outputs is INTERACT, which was developed by Eng, Aebersold, and colleagues at the Institute for Systems Biology (Han et al., 2001). INTERACT provides a user-friendly graphical user interface to collect and evaluate outputs from Sequest directories representing multiple searches. A comparison utility allows comparisons of up to three data sets simultaneously. This permits assessment of differences between control and experimental protein sets. INTERACT is available from the Institute for Systems Biology (http://www.systemsbiology.org).

Gururaja et al. (2002) reported another interesting approach to comparative analysis of complex data sets. They developed an automated data analysis system called Medusa, which is a web-based, proprietary Oracle database system that evaluates both Sequest outputs and MS/MS spectra. Comparison of data sets was done with an algorithm called MS2filter that identified and removed MS/MS spectra that were apparently identical in the two sets. Elimination of the "redundant" spectra reduced the data set to spectra representing species that were differentially present. These spectra then were analyzed with Sequest and the results were filtered based on Xcorr scores and other Sequest output parameters. Medusa also reassembled the protein identification data from the results to provide a list of differentially expressed proteins. Although Medusa is a proprietary product, the approach illustrated by this application should be applicable to other data sets with some of the other tools described earlier.

15.6   WHAT CONSTITUTES PROTEIN IDENTIFICATION?

Virtually all proteomics problems require identification of proteins from MS data and use the tools described in earlier sections. Nevertheless, there are no universally accepted criteria for just what constitutes protein identification in proteomics studies. Operationally, *peptide* identifications are made when data mining algorithms and software deliver sequence-to-spectrum comparisons with acceptable scores or probability rankings. However, there is no clear standard that defines "acceptable" in this context. As noted earlier, the application of a data filter approach to accept Sequest search results that meet minimum score thresholds is subject to a trade-off between sensitivity and false-positive errors. Likewise, more sophisticated probability-based scoring and machine learning approaches are themselves subject to error. Visual inspection has long been considered a "gold standard" for confirming both spectral quality and assignments of spectral features to peptide fragmentations, but this inevitably introduces an element of bias into the analysis. The emergence of increasingly accurate algorithms and sophisticated, probability-based scoring for sequence-to-spectrum comparisons should continue to improve the quality of peptide identification.

Protein identification derives in large part from the sum of peptide identifications. Multiple peptide identifications increase the probability that a correct protein identification is made. However, many peptide sequences map to more than one protein. Thus, even high-quality peptide identifications made from high-quality data do not always unambiguously identify a protein. Frequency of identification of peptides in replicate analyses of the same sample provides another factor by which protein identification can be assessed.

As with peptide identifications, the problem of identifying proteins from peptide identifications is a probability problem. Nesvizhskii and colleagues (2003) recently reported the development of a statistical model for validation of protein identity assignments from peptide identifications. The model, called Protein Prophet™, builds on the Peptide Prophet™ program and statistical model described earlier and combines the probabilities of identifications of individual peptides to calculate the probability of a false-positive protein identification. This open source application is available from the Institute for Systems Biology (www.systemsbiology.org).

Two final points about protein identification are worth considering. First, protein identifications done as described earlier are most often generic, in that they identify a protein polypeptide that was present in the sample. The actual "protein" represented in a biological system may be a collection of several unmodified, modified, and variant (e.g., amino acid substituted or truncated) forms. Identification only distinguishes individual forms when data are obtained on characteristic modified and variant peptides (see below). Second, the level of confidence required for protein identifications is inherently variable because the purposes of proteome analyses are variable. Obviously, no one wants missed or false-positive protein identifications in a data set. Nevertheless, some identifications made in proteomics analyses will generate follow-up experiments, such as analysis with antibodies or measurements of activities that will provide corroboration of the presence of the protein(s). In this case, false positives arising from proteomics analyses eventually can be excluded. Some investigators may wish to err on the side of greater sensitivity for detection of proteins with a concomitantly higher false-positive rate in order to detect novel components. In other cases, the goals may entail fairly high-throughput analyses that will not have much or any follow-up with other techniques. In this case, a more conservative approach to protein identification may attempt to minimize false-positive identification with a concomitant loss of sensitivity.

15.7   BIOINFORMATICS TOOLS FOR SPECIALIZED ANALYSES

Essentially all of the tools described earlier are directed at protein identification from MS data. Proteomics analyses also address other questions, most notably (i) how much of particular proteins or protein forms are present in one system versus another and (ii) what modifications are present on proteins. The analytical platforms used to address these questions are essentially the same as those used for protein identification, but with some modifications.

15.7.1   Quantitative Analyses from LC-MS/MS Data with Stable Isotope Tags

Both relative and absolute quantitation of proteins in proteomics analyses are done with the aid of stable isotope tags, which are used to differentially label proteins or peptides derived from the samples to be compared (Gygi et al., 1999; Conrads et al., 2001; Regnier et al., 2002; Mason and Liebler, 2003). One sample is tagged with an unlabeled reagent (light isotope form) and the other is tagged with a stable isotope (heavy isotope) form. The samples are combined and analyzed by LC-MS/MS (or by MALDI-MS/MS). In the isotope-coded affinity tag (ICAT) approach originally reported by Gygi and colleagues (1999), data-dependent acquisitions of full scan and MS/MS spectra are used to generate isotope ratios for the light:heavy tagged forms. These ratios indicate the ratio of the protein between the two samples. Eng and colleagues developed the XPRESS software to perform the analysis of isotopomer distributions from LC-MS/MS data (Han et al., 2001). The program recognizes light:heavy isotope pairs in full scan data and performs integration and quantitative comparison through an interactive graphical user interface. The XPRESS program is licensed to Thermo Electron and distributed as part of the Bioworks software suite for proteomics analysis (www.thermo.com). A newer program called ASAPRatio that performs these functions together with more sophisticated data smoothing and statistical analyses is in development at the Institute for Systems Biology (www.systemsbiology.org). Although these tools were first applied with the deuterated, thiol-reactive ICAT reagents, the approach can employ several other types of tagging chemistries directed against other chemical moieties in peptides.

15.7.2   Detection of Post-translational Modifications and Variant Forms of Peptides

MS-based analyses are particularly well suited to the analysis of post-translational protein modifications, xenobiotic adducts, and variant sequences. The single greatest barrier to success in such analyses is the problem of obtaining MS data on the modified peptide in a digest, particularly when the modification is nonstoichiometric, unstable, and/or part of a complex mixture containing many other sequences. The general approach to mapping modifications is to digest the proteins in a sample, perform MS or MS/MS analyses, and then mine the data for evidence of modified peptide forms. Peptide mass fingerprinting can detect modified peptides either with manual data inspection to identify mass shifted peptide ions or with freely available software tools such as FindMod (http://us.expasy.org/tools/

findmod). Other peptide mass fingerprinting tools discussed earlier have provisions for including variable or fixed modifications in searches.

MS/MS data are much more valuable in the analysis of peptide modifications because the MS/MS spectrum encodes not only the sequence of the peptide but also the sequence location of the modification. The MS/MS data search programs discussed earlier, such as Sequest, Mascot, and Sonar, can incorporate fixed and variable modifications of specific amino acids into searches. For example, specification of variable phosphorylation at Ser, Thr, and Tyr residues allows both phosphorylated and unmodified peptide sequences to be correctly associated with their MS/MS spectra.

One problem with protein modifications is that they may occur in unanticipated sites and, in the case of modifications by reactive intermediates, may lead to unexpected mass modifications. The SALSA (Scoring ALgorithm for Spectral Analysis) program provides a means of identifying such unanticipated modifications (Hansen et al., 2001). SALSA searches datafiles for spectra that display user-specified characteristics. For example, SALSA can identify spectra that display characteristic product ions, neutral or charged losses of groups of signals that are indicative of a particular modification. The latter type of analysis is particularly useful, as groups of signals (e.g., corresponding to *b*- or *y*-ion series) are characteristic of peptide sequence motifs. SALSA searches for peptide sequence motifs can identify MS/MS spectra that correspond not only to the unmodified peptide but also to spectra of variant and modified forms, which nevertheless share groupings of *b*- and *y*-ion signals in common (Liebler et al., 2002).

REFERENCES

Anderson DC, Li W, Payan DG, Noble WS. 2003. A new algorithm for the evaluation of shotgun peptide sequencing in proteomics: support vector machine classification of peptide MS/MS spectra and SEQUEST scores. *J Proteome Res* 2:137–146.

Baldwin MA, Medzihradsky KF, Lock CM, Fisher B, Settineri TA, Burlingame AL. 2001. Matrix-assisted laser desorption/ionization coupled with quadrupole/orthogonal acceleration time-of-flight mass spectrometry for protein discovery, identification and structural analysis. *Anal Chem* 73:1707–1720.

Brazma A, Hingamp P, Quackenbush J, Sherlock G, Spellman P, Stoeckert C, Aach J, Ansorge W, Ball CA, Causton HC, Gaasterland T, Glenisson P, Holstege FC, Kim IF, Markowitz V, Matese JC, Parkinson H, Robinson A, Sarkans U, Schulze-Kremer S, Stewart J, Taylor R, Vilo J, Vingron M. 2001. Minimum information about a microarray experiment (MIAME)—toward standards for microarray data. *Nat Genet* 29:365–371.

Clauser KR, Baker P, Burlingame AL. 1999. Role of accurate mass measurement (±10 ppm) in protein identification strategies employing MS or MS/MS and database searching. *Anal Chem* 71:2871–2882.

Conrads TP, Alving K, Veenstra TD, Belov ME, Anderson GA, Anderson DJ, Lipton MS, Pasa-Tolic L, Udseth HR, Chrisler WB, Thrall BD, Smith RD. 2001. Quantitative proteome analysis of bacterial and mammalian proteomes using a combination of cysteine affinity tags and ^{15}N-metabolic labeling. *Anal Chem* 73:2132–2139.

Eng JK, McCormack AL, Yates JR III. 1994. An approach to correlate tandem mass spectral data of peptides with amino acid sequences in a protein database. *J Am Soc Mass Spectrom* 5:976–989.

Fenyo D. 2000. Identifying the proteome: software tools. *Curr Opin Biotechnol* 11:391–395.

Field HI, Fenyo D, Beavis RC. 2002. RADARS, a bioinformatics solution that automates proteome mass spectral analysis, optimizes protein identification, and archives data in a relational database. *Proteomics* 2:36–47.

Gururaja T, Li W, Bernstein J, Payan DG, Anderson DC. 2002. Use of MEDUSA-based data analysis and capillary HPLC-ion-trap mass spectrometry to examine complex immunoaffinity extracts of RBAp48. *J Proteome Res* 1:253–261.

Gygi SP, Rist B, Gerber SA, Turecek F, Gelb MH, Aebersold R. 1999. Quantitative analysis of complex protein mixtures using isotope-coded affinity tags. *Nat Biotechnol* 17:994–999.

Han DK, Eng J, Zhou H, Aebersold R. 2001. Quantitative profiling of differentiation-induced microsomal proteins using isotope-coded affinity tags and mass spectrometry. *Nat Biotechnol* 19:946–951.

Hansen BT, Jones JA, Mason DE, Liebler DC. 2001. SALSA: a pattern recognition algorithm to detect electrophile-adducted peptides by automated evaluation of CID spectra in LC-MS-MS analyses. *Anal Chem* 73:1676–1683.

Hunter TC, Andon NL, Koller A, Yates JR, Haynes PA. 2002. The functional proteomics toolbox: methods and applications. *J Chromatogr B Analyt Technol Biomed Life Sci* 782:165–181.

James P, Quadroni M, Carafoli E, Gonnet G. 1994. Protein identification in DNA databases by peptide mass fingerprinting. *Protein Sci* 3:1347–1350.

Keller A, Nesvizhskii AI, Kolker E, Aebersold R. 2002. Empirical statistical model to estimate the accuracy of peptide identifications made by MS/MS and database search. *Anal Chem* 74:5383–5392.

Liebler DC. 2002. Proteomic approaches to characterize protein modifications: new tools to study the effects of environmental exposures. *Environ Health Perspect* 110(Suppl1):3–9.

Liebler DC, Hansen BT, Davey SW, Tiscareno L, Mason DE. 2002. Peptide sequence motif analysis of tandem MS data with the SALSA algorithm. *Anal Chem* 74:203–210.

Link AJ, Eng J, Schieltz DM, Carmack E, Mize GJ, Morris DR, Garvik BM, Yates JR III. 1999. Direct analysis of protein complexes using mass spectrometry. *Nat Biotechnol* 17:676–682.

Liska AJ, Shevchenko A. 2003. Expanding the organismal scope of proteomics: cross-species protein identification by mass spectrometry and its implications. *Proteomics* 3:19–28.

MacCoss MJ, McDonald WH, Saraf A, Sadygov R, Clark JM, Tasto JJ, Gould KL, Wolters D, Washburn M, Weiss A, Clark JI, Yates JR III. 2002. Shotgun identification of protein modifications from protein complexes and lens tissue. *Proc Natl Acad Sci USA* 99: 7900–7905.

Mann M, Wilm M. 1994. Error-tolerant identification of peptides in sequence databases by peptide sequence tags. *Anal Chem* 66:4390–4399.

Mason DE, Liebler DC. 2003. Quantitative analysis of modified proteins by LC-MS-MS of peptides labeled with phenyl isocyanate. *J Proteome Res* 2:265–272.

Nesvizhskii AI, Keller A, Kolker E, Aebersold R. 2003. A statistical model for identifying proteins by tandem mass spectrometry. *Anal Chem* 75:4646–4658.

Pappin DJC, Hojrup P, Bleasby AJ. 1993. Rapid identification of proteins by peptide mass fingerprinting. *Curr Biol* 3:327–332.

Peng J, Elias JE, Thoreen CC, Licklider LJ, Gygi SP. 2003. Evaluation of multidimensional chromatography coupled with tandem mass spectrometry (LC/LC-MS/MS) for large scale protein analysis: the yeast proteome. *J Proteome Res* 2:43–50.

Perkins DN, Pappin DJ, Creasy DM, Cottrell JS. 1999. Probability-based protein identification by searching sequence databases using mass spectrometry data. *Electrophoresis* 20:3551–3567.

Raman B, Cheung A, Marten MR. 2002. Quantitative comparison and evaluation of two commercially available, two-dimensional electrophoresis image analysis software packages, Z3 and Melanie. *Electrophoresis* 23:2194–2202.

Regnier FE, Riggs L, Zhang R, Xiong L, Liu P, Chakraborty A, Seeley E, Sioma C, Thompson RA. 2002. Comparative proteomics based on stable isotope labeling and affinity selection. *J Mass Spectrom* 37:133–145.

Roepstorff P, Fohlman J. 1984. Proposal for a common nomenclature for sequence ions in mass spectra of peptides. *Biomed Mass Spectrom* 11:601.

Rogers M, Graham J, Tonge RP. 2003. Using statistical image models for objective evaluation of spot detection in two-dimensional gels. *Proteomics* 3:879–886.

Sadygov RG, Eng J, Durr E, Saraf A, McDonald H, MacCoss MJ, Yates JR III. 2002. Code developments to improve the efficiency of automated MS/MS spectra interpretation. *J Proteome Res* 1:211–215.

Tabb DL, McDonald WH, Yates JR III. 2002. DTASelect and Contrast: tools for assembling and comparing protein identifications from shotgun proteomics. *J Proteome Res* 1:21–26.

Taylor CF, Paton NW, Garwood KL, Kirby PD, Stead DA, Yin Z, Deutsch EW, Selway L, Walker J, Riba-Garcia I, Mohammed S, Deery MJ, Howard JA, Dunkley T, Aebersold R, Kell DB, Lilley KS, Roepstorff P, Yates JR, Brass A, Brown AJ, Cash P, Gaskell SJ, Hubbard SJ, Oliver SG. 2003. A systematic approach to modeling, capturing, and disseminating proteomics experimental data. *Nat Biotechnol* 21:247–254.

Washburn MP, Wolters D, Yates JR III. 2001. Large-scale analysis of the yeast proteome by multidimensional protein identification technology. *Nat Biotechnol* 19:242–247.

Wolters DA, Washburn MP, Yates JR III. 2001. An automated multidimensional protein identification technology for shotgun proteomics. *Anal Chem* 73:5683–5690.

Yates JR, Eng JK, McCormack AL, Schieltz D. 1995. Method to correlate tandem mass spectra of modified peptides to amino acid sequences in the protein database. *Anal Chem* 67:1426–1436.

Index

Proteomics for Biological Discovery, edited by Timothy D. Veenstra and John R. Yates.
Copyright © 2006 John Wiley & Sons, Inc.